Polymers/Their Hybrid Materials for Optoelectronic Applications

Polymers/Their Hybrid Materials for Optoelectronic Applications

Guest Editors

Tengling Ye
Bixin Li

Basel • Beijing • Wuhan • Barcelona • Belgrade • Novi Sad • Cluj • Manchester

Guest Editors

Tengling Ye
School of Chemistry and
Chemical Engineering
Harbin Institute
of Technology
Harbin
China

Bixin Li
School of Physics
and Chemistry
Hunan First Normal
University
Changsha
China

Editorial Office
MDPI AG
Grosspeteranlage 5
4052 Basel, Switzerland

This is a reprint of the Special Issue, published open access by the journal *Polymers* (ISSN 2073-4360), freely accessible at: www.mdpi.com/journal/polymers/special_issues/Conducting_Organic_Polymer_Materials_Photovoltaic_Application.

For citation purposes, cite each article independently as indicated on the article page online and using the guide below:

Lastname, A.A.; Lastname, B.B. Article Title. *Journal Name* **Year**, *Volume Number*, Page Range.

ISBN 978-3-7258-3380-1 (Hbk)
ISBN 978-3-7258-3379-5 (PDF)
https://doi.org/10.3390/books978-3-7258-3379-5

© 2025 by the authors. Articles in this book are Open Access and distributed under the Creative Commons Attribution (CC BY) license. The book as a whole is distributed by MDPI under the terms and conditions of the Creative Commons Attribution-NonCommercial-NoDerivs (CC BY-NC-ND) license (https://creativecommons.org/licenses/by-nc-nd/4.0/).

Contents

Preface vii

Xin Wang, Zongtao Wang, Mingwei Li, Lijun Tu, Ke Wang and Dengping Xiao et al.
A New Dibenzoquinoxalineimide-Based Wide-Bandgap Polymer Donor for Polymer Solar Cells
Reprinted from: *Polymers* 2022, 14, 3590, https://doi.org/10.3390/polym14173590 1

Guoping Zhang, Lihong Wang, Chaoyue Zhao, Yajie Wang, Ruiyu Hu and Jiaxu Che et al.
Efficient All-Polymer Solar Cells Enabled by Interface Engineering
Reprinted from: *Polymers* 2022, 14, 3835, https://doi.org/10.3390/polym14183835 10

Bin Du, Kun He, Xiaoliang Zhao and Bixin Li
Defect Passivation Scheme toward High-Performance Halide Perovskite Solar Cells
Reprinted from: *Polymers* 2023, 15, 2010, https://doi.org/10.3390/polym15092010 21

Shiwei Ren, Yubing Ding, Wenqing Zhang, Zhuoer Wang, Sichun Wang and Zhengran Yi
Rational Design of Novel Conjugated Terpolymers Based on Diketopyrrolopyrrole and Their Applications to Organic Thin-Film Transistors
Reprinted from: *Polymers* 2023, 15, 3803, https://doi.org/10.3390/polym15183803 52

Shiwei Ren, Wenqing Zhang, Zhuoer Wang, Abderrahim Yassar, Zhiting Liao and Zhengran Yi
Synergistic Use of All-Acceptor Strategies for the Preparation of an Organic Semiconductor and the Realization of High Electron Transport Properties in Organic Field-Effect Transistors
Reprinted from: *Polymers* 2023, 15, 3392, https://doi.org/10.3390/polym15163392 65

A. Saad, N. Hamad, Rasul Al Foysal Redoy, Suling Zhao and S. Wageh
Enhancing Blue Polymer Light-Emitting Diode Performance by Optimizing the Layer Thickness and the Insertion of a Hole-Transporting Layer
Reprinted from: *Polymers* 2024, 16, 2347, https://doi.org/10.3390/polym16162347 77

Nurul Amira Shazwani Zainuddin, Yusuke Suizu, Takahiro Uno and Masataka Kubo
The Synthesis and Optical Property of a Ternary Hybrid Composed of Aggregation-Induced Luminescent Polyfluorene, Polydimethylsiloxane, and Silica
Reprinted from: *Polymers* 2024, 16, 3331, https://doi.org/10.3390/polym16233331 92

Bin Xu, Jiankang Zhou, Chengran Zhang, Yunfu Chang and Zhengtao Deng
Research Progress on Quantum Dot-Embedded Polymer Films and Plates for LCD Backlight Display
Reprinted from: *Polymers* 2025, 17, 233, https://doi.org/10.3390/polym17020233 103

Wei Zhao, Jianguo Zhang, Fanjun Kong and Tengling Ye
Application of Perovskite Nanocrystals as Fluorescent Probes in the Detection of Agriculture- and Food-Related Hazardous Substances
Reprinted from: *Polymers* 2023, 15, 2873, https://doi.org/10.3390/polym15132873 127

Wenhao Li, Jingyu Jia, Xiaochen Sun, Sue Hao and Tengling Ye
A Light/Pressure Bifunctional Electronic Skin Based on a Bilayer Structure of PEDOT:PSS-Coated Cellulose Paper/$CsPbBr_3$ QDs Film
Reprinted from: *Polymers* 2023, 15, 2136, https://doi.org/10.3390/polym15092136 167

Carmen R. Tubio, Laura Garea, Bárbara D. D. Cruz, Daniela M. Correia, Verónica de Zea Bermudez and Senentxu Lanceros-Mendez
Environmentally Friendly Photoluminescent Coatings for Corrosion Sensing
Reprinted from: *Polymers* 2025, 17, 389, https://doi.org/10.3390/polym17030389 179

Oxana Gribkova, Varvara Kabanova, Ildar Sayarov, Alexander Nekrasov and Alexey Tameev
Near-Infrared Responsive Composites of Poly-3,4-Ethylenedioxythiophene with Fullerene Derivatives
Reprinted from: *Polymers* **2024**, *17*, 14, https://doi.org/10.3390/polym17010014 . . . . **190**

Bixin Li, Shiyang Zhang, Lan Xu, Qiong Su and Bin Du
Emerging Robust Polymer Materials for High-Performance Two-Terminal Resistive Switching Memory
Reprinted from: *Polymers* **2023**, *15*, 4374, https://doi.org/10.3390/polym15224374 . . . . **207**

Preface

In the rapidly evolving field of materials science, polymers and their hybrid materials have emerged as pivotal components for optoelectronic applications. This reprint, *Polymers/Their Hybrid Materials for Optoelectronic Applications*, edited by Tengling Ye and Bixin Li, serves as a comprehensive resource that highlights recent advancements in this dynamic area. The contributions within these pages reflect cutting-edge research from leading experts across the globe, providing readers with an in-depth understanding of how polymers/their hybrid materials are being tailored to meet the demands of modern optoelectronics.

The chapters contained herein explore diverse aspects of polymer-based materials, ranging from novel conjugated polymers designed for polymer solar cells/organic thin-film transistors to defect passivation schemes aimed at enhancing perovskite solar cells/perovskite fluorescent probes. Each contribution not only presents innovative solutions but also delves into the underlying principles that govern material performance, offering valuable insights for both academic researchers and industry professionals alike.

Moreover, this volume addresses practical applications such as blue polymer light-emitting diodes, quantum dot-embedded polymer films for LCD backlight displays, e-skin sensors, and environmentally friendly photoluminescent coatings for corrosion sensing. These examples underscore the versatility and potential impact of polymer-based technologies in addressing real-world challenges. Additionally, the inclusion of topics like near-infrared responsive composites and two-terminal resistive switching memory showcases emerging trends that promise to shape future developments in the field.

As we stand on the brink of exciting breakthroughs in materials science, we hope that this compilation will inspire further exploration and innovation among its readers. By bridging fundamental research with applied technology, *Polymers/Their Hybrid Materials for Optoelectronic Applications* aims to serve as a catalyst for advancing knowledge and fostering collaboration within the scientific community.

We extend our gratitude to all contributing authors whose dedication and expertise have made this publication possible. Their collective efforts exemplify the spirit of inquiry and discovery that drives progress in our discipline.

Tengling Ye and Bixin Li
Guest Editors

Article

A New Dibenzoquinoxalineimide-Based Wide-Bandgap Polymer Donor for Polymer Solar Cells

Xin Wang [1,†], Zongtao Wang [2,3,†], Mingwei Li [1], Lijun Tu [1], Ke Wang [1], Dengping Xiao [1], Qiang Guo [2,*], Ming Zhou [4], Xianwen Wei [1,*], Yongqiang Shi [1,4,*] and Erjun Zhou [3,*]

1. Key Laboratory of Functional Molecular Solids, Ministry of Education, and School of Chemistry and Materials Science, Anhui Normal University, Wuhu 241002, China
2. School of Materials Science and Engineering, Henan Institute of Advanced Technology, Zhengzhou University, Zhengzhou 450001, China
3. CAS Center for Excellence in Nanoscience, National Center for Nanoscience and Technology, Beijing 100190, China
4. State Key Laboratory of Oil and Gas Reservoir Geology and Exploitation, School of New Energy and Materials, Southwest Petroleum University, Chengdu 610500, China
* Correspondence: guoqiang@zzu.edu.cn (Q.G.); xwwei@mail.ahnu.edu.cn (X.W.); shiyq@ahnu.edu.cn (Y.S.); zhouej@nanoctr.cn (E.Z.)
† These authors contributed equally to this work.

Citation: Wang, X.; Wang, Z.; Li, M.; Tu, L.; Wang, K.; Xiao, D.; Guo, Q.; Zhou, M.; Wei, X.; Shi, Y.; et al. A New Dibenzoquinoxalineimide-Based Wide-Bandgap Polymer Donor for Polymer Solar Cells. *Polymers* **2022**, *14*, 3590. https://doi.org/10.3390/polym14173590

Academic Editor: Tengling Ye

Received: 26 July 2022
Accepted: 26 August 2022
Published: 30 August 2022

Publisher's Note: MDPI stays neutral with regard to jurisdictional claims in published maps and institutional affiliations.

Copyright: © 2022 by the authors. Licensee MDPI, Basel, Switzerland. This article is an open access article distributed under the terms and conditions of the Creative Commons Attribution (CC BY) license (https:// creativecommons.org/licenses/by/ 4.0/).

Abstract: The molecular design of a wide-bandgap polymer donor is critical to achieve high-performance organic photovoltaic devices. Herein, a new dibenzo-fused quinoxalineimide (BPQI) is successfully synthesized as an electron-deficient building block to construct donor–acceptor (D–A)-type polymers, namely P(BPQI-BDT) and P(BPQI-BDTT), using benzodithiophene and its derivative, which bears different side chains, as the copolymerization units. These two polymers are used as a donor, and the narrow bandgap (2,20-((2Z,20Z))-((12,13-bis(2-ethylhexyl)-3,9-diundecyl-12,13-dihydro-[1,2,5]thiadiazolo [3,4-e]thieno[2,″30′:4′,50]thieno[20,30:4,5]pyrrolo[3,2g]thieno[20,30:4,5]thieno[3,2-b]indole-2,10 diyl)bis(methanylylidene))bis(5,6-difluoro-3-oxo-2,3-dihydro-1H-indene-2,1-diylidene)) dimalononitrile) Y6 is used as an acceptor to fabricate bulk heterojunction polymer solar cell devices. Y6, as a non-fullerene receptor (NFA), has excellent electrochemical and optical properties, as well as a high efficiency of over 18%. The device, based on P(BPQI-BDTT):Y6, showed power conversion efficiencies (PCEs) of 6.31% with a J_{SC} of 17.09 mA cm^{-2}, an open-circuit voltage (V_{OC}) of 0.82 V, and an FF of 44.78%. This study demonstrates that dibenzo-fused quinoxalineimide is a promising building block for developing wide-bandgap polymer donors.

Keywords: wide bandgap; donor–acceptor; imide; polymer solar cells

1. Introduction

Organic solar cells (OSCs) have attracted a lot of attention from both academia and industry due to their versatile advantages, such as their light weight, stretchability, and solution processability [1–6]. Following great efforts in recent years, power conversion efficiencies (PCEs) were achieved in the range of 18–19% [7–9]. The remarkable progress in organic photovoltaics (OPVs) mainly depends on the development of new polymer donors and non-fullerene small molecule electron acceptors (NFSMEAs) [10–15]. Recently, various narrow-bandgap NFSMEAs have been designed and have achieved excellent performances in OPVs when paired with wide-bandgap polymer donors. Most NFSMEAs show intense absorption in the visible and near-infrared (NIR) region (650–950 nm) for enhancing the light-harvesting ability [16–18]. Compared to NFSMEAs, the development of wide-bandgap (WBG) polymer donors is still limited; only a few donor–acceptor (D–A)-type WBG polymer donors have been reported, such as PM6 and D18 [19–21]. Therefore, it is imperative to design novel WBG polymer donors, which absorb the solar energy in the wavelength range of 300–650 nm to further improve the OPV's performances [22–25]. Simultaneously, polymers can also be a carrier transport layer in perovskite solar cells [26,27]. Specially, the polymer donors of organic solar cells are often used as the hole transport layers in the perovskite solar cells due to their extraordinary p-type characteristics.

The ideal WBG polymer donors can be synthesized following the donor–acceptor strategy, where benzodithiophene (BDT) [28–30] and its derivatives are usually used as the donor units, and some electron-deficient moieties, such as quinoxaline [31–33], benzothiadiazole, benzotriazole, and imide, are used as the acceptor co-units [34–36]. In addition to these acceptor units, quinoxaline-based polymers have some characteristics that are useful for constructing WBG polymer donors: (1) quinoxaline derivatives possess strong electron-withdrawing properties that can adjust the energy levels; (2) the quinoxaline moiety can be easily modified by introducing sidechains and other functional groups to modulate the physical, chemical, and electronic properties; and (3) quinoxaline has a rigid and large planar structure that favors close π-π stacking. Li et al. [12], reported a low cost quinoxaline-based D–A-type polymer donor, PTQ10, that contains an alkoxy-substituted difluoro quinoxaline acceptor unit and thiophene donor unit on the backbone; when blended with IDIC, a PCE value of 12.7% was achieved for PSCs. Recently, Hou and coworkers reported a WBG polymer donor, PBQx-TCl, based on the thiophene-fused quinoxaline acceptor unit DTQx; the device based on PBQx-TCl:BTP-eC9 exhibited a PCE of 16.0% [14]. Therefore, quinoxaline is a promising building block for constructing high-performance WBG polymer donors in OPVs.

Herein, we design and synthesize a new dibenzo-fused quinoxalineimide (BPQI) building block, which has a rigid and planar structure. By using BDT as the copolymerization unit with BPQI, two new polymer donors, P(BPQI-BDT) and P(BPQI-BDTT), were synthesized. Both polymer donors possess wide bandgaps and deep highest occupied molecular orbital (HOMO) levels, due to the strong electron-withdrawing ability of the BPQI moiety, and have a good compatibility with the Y6 acceptor. As a result, the optimized P(BPQI-BDTT):Y6 device delivers a higher PCE of 6.31% with a J_{SC} of 17.09 mA cm^{-2}, a V_{OC} of 0.82 V, and an FF of 44.78%, compared with the P(BPQI-BDT):Y6 system (PCE of 1.48% with a J_{SC} of 6.23 mA cm^{-2}, a V_{OC} of 0.68 V, and an FF of 34.61%).

2. Results and Discussion

2.1. Materials Synthesis and Characterization

The synthetic routes to the monomer BPQI and polymers are displayed in Scheme 1. The detailed synthetic procedures are shown in the Supplementary Materials (Figures S1 and S2). Both polymers P(BPQI-BDT) and P(BPQI-BDTT) are obtained by polymerization of BPQI and the co-monomer BDT derivative. The resulting polymers were purified by Soxhlet extraction with methanol, acetone, hexane, and dichloromethane to remove the catalyst and low molecular weight fractions. The final chloroform fractions were collected and used to test the device's performance. The numerical average molecular weights (M_ns) and polydispersity indices (PDIs) of P(BPQI-BDT) and P(BPQI-BDTT) were determined by using high-temperature gel permeation chromatography (HT-GPC) at 150 °C with 1,2,4-trichlorobenzene as the eluent. The GPC results are summarized in Table 1: P(BPQI-BDT) has an M_n of 24 kDa with a PDI of 2.2, and P(BPQI-BDTT) shows an M_n of 37 kDa with a PDI of 2.0. The thermal properties of P(BPQI-BDT) and P(BPQI-BDTT) were characterized using thermogravimetric analyses (TGA) and differential scanning calorimetry (DSC). Both polymers exhibit excellent thermal stability (5% weight loss) over 300 °C, which indicates good thermal stability for PSC fabrication. Based on the DSC thermograms, no obvious exotherm or endotherm peaks were observed for both polymers P(BPQI-BDT) and P(BPQI-BDTT) in the temperature range from 50 to 300 °C, which likely indicates a low degree of crystallinity in both polymers.

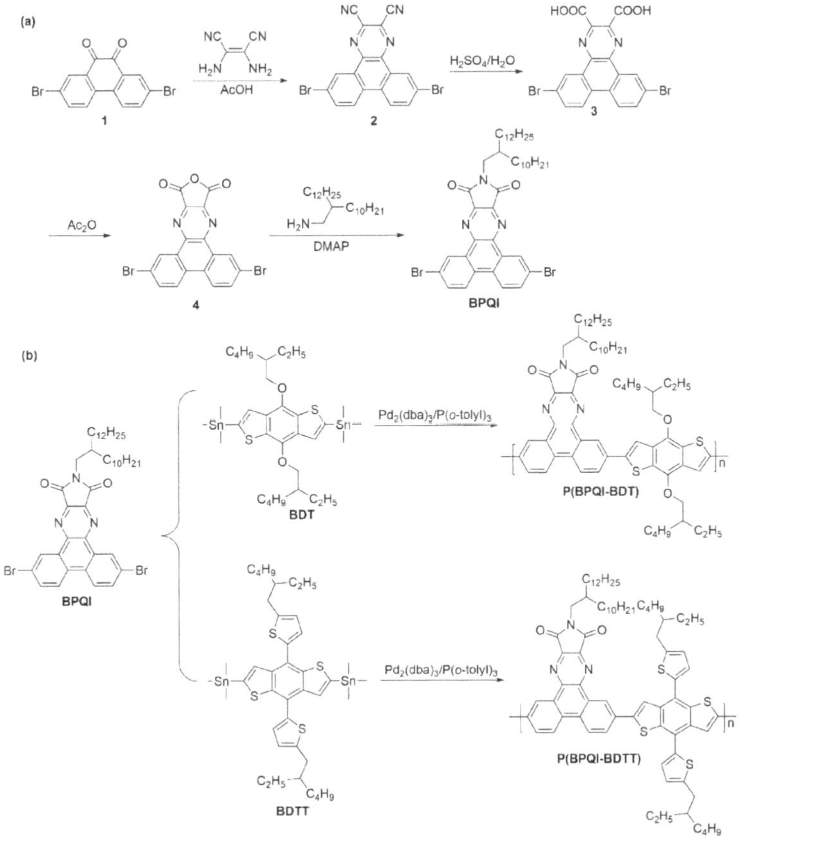

Scheme 1. Synthetic route to monomer BPQI (**a**,**b**) and corresponding polymers P(BPQI-BDT) and P(BPQI-BDTT).

Table 1. Molecular Weights and Optical and Electrochemical Properties of Polymers.

Polymer	M_n (kDa) [a]	PDI [a]	λ_{onset}^{film} (nm) [b]	E_g^{opt} (eV) [c]	HOMO (eV) [d]	LUMO (eV) [e]
P(BPQI-BDT)	24	2.2	584	2.12	−5.41	−3.29
P(BPQI-BDTT)	37	2.0	588	2.10	−5.51	−3.41

[a] Measured from GPC versus polystyrene standard; trichlorobenzene as the eluent at 150 °C. [b] Absorption spectra of the pristine film from chloroform solution. [c] Optical bandgap estimated from the absorption onset of the as-cast polymer film using the equation: $E_g^{opt} = 1240/\lambda_{onset}$ (eV). [d] $E_{HOMO} = -e(E_{ox}^{onset} + 4.80)$ eV. [e] $E_{LUMO} = E_{HOMO} + E_g^{opt}$.

2.2. Optical and Electrochemical Properties

The UV-vis absorption spectra of P(BPQI-BDT) and P(BPQI-BDTT) in a chloroform solution and as thin films were shown in Figure 1a, and the corresponding data are summarized in Table 1. In solution, both polymers show two absorption shoulder peaks in the range of 350–500 nm, and the absorption maxima (λ_{max}) of P(BPQI-BDT) and P(BPQI-BDTT) in chloroform solution are located at 393 and 432 nm, respectively. In the thin film state, the similar absorption with a slightly red-shifted absorption peak was observed for both polymers, which should be due to close packing in the solid state. The optical bandgaps of P(BPQI-BDT) and P(BPQI-BDTT) are 2.12 and 2.10 eV, respectively, according to the absorption onsets of the polymer thin films.

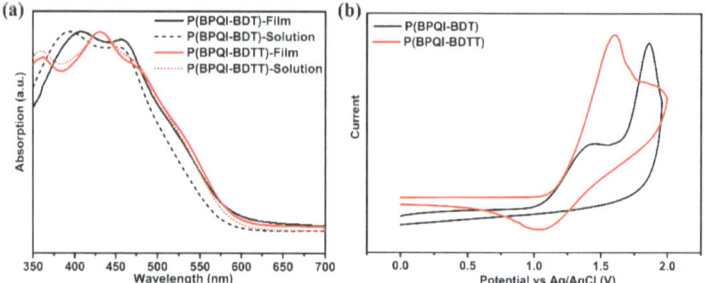

Figure 1. (a) Normalized UV-vis absorption spectra of polymers P(BPQI-BDT) and P(BPQI-BDTT) in chloroform solution and as thin films. (b) Cyclic voltammogram curves of P(BPQI-BDT) and P(BPQI-BDTT) thin films measured in 0.1 M (n-Bu)$_4$NPF$_6$ acetonitrile solution at a scan rate of 50 mV s^{-1}.

The electrochemical properties of P(BPQI-BDT) and P(BPQI-BDTT) were characterized by cyclic voltammetry. As shown in Figure 1b, the HOMO energy levels of P(BPQI-BDT) and P(BPQI-BDTT) were calculated from the onset potential of the oxidation peak, which are −5.41 and −5.51 eV (Table 1), respectively. By using the equation of $E_{LUMO} = (E_{HOMO} + E_g)$ eV, the LUMO energy levels of P(BPQI-BDT) and P(BPQI-BDTT) were found to be −3.29 and −3.41 eV, respectively. The E_{HOMO}/E_{LUMO} of both polymers match well with acceptor Y6, indicating that these two polymers can be used as polymer donor in PSCs.

2.3. Theoretical Calculations

Theoretical computation on the backbone geometry of P(BPQI-BDT) and P(BPQI-BDTT) were performed by density functional theory (DFT) using the B3LYP/6-31G level method (Figure 2). The long side chains on the BPQI and BDT were replaced by a methyl group for simplifying the computations, respectively. The DFT-calculated HOMO/LUMO are −4.76/−2.50 and −5.13/−2.56 eV for P(BPQI-BDT) and P(BPQI-BDTT), respectively. The trend of the HOMO/LUMO energy level from the theoretical energy levels are in good agreement with the values obtained by CV measurements.

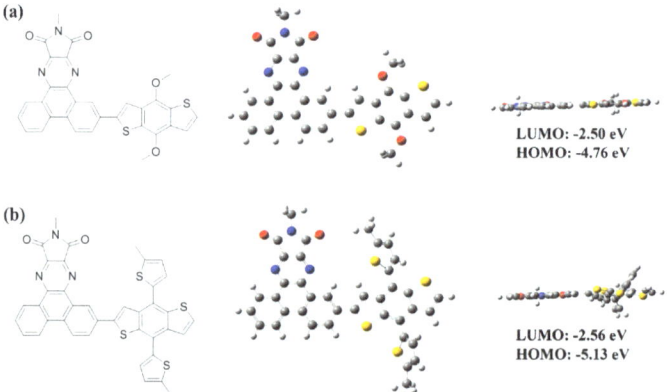

Figure 2. Chemical structures and FMO energy levels of (a) P(BPQI-BDT) and (b) P(BPQI-BDTT).

2.4. Photovoltaic Properties

To investigate the photovoltaic properties of P(BPQI-BDT) and P(BPQI-BDTT) in PSCs, the conventional device structure indium tin oxide (ITO)/poly(3,4-ethyl-enedioxythiophene):

poly(styrene sulfonic acid) (PEDOT:PSS)/active layer/sliver(Ag)/gold(Au) was fabricated. Y6 was chosen as the acceptor due to its excellent performance in the field of OSCs, and the device architecture is shown in Figure 3a. The molecular energy levels of P(BPQI-BDT) and P(BPQI-BDTT) are shown in Figure 3b. The active layer was deposited by spin-coating the P(BPQI-BDT)/P(BPQI-BDTT):Y6 mixed solution in chloroform at a D/A weight ratio of 1/1, and the solution concentration is 10 mg mL^{-1}. The current density–applied voltage (J–V) curves are shown in Figure 3c, and the corresponding parameters are summarized in Table 2. As shown in Figure 3c, the optimized P(BPQI-BDTT):Y6-based device delivers a PCE of 6.31% with a J_{SC} of 17.09 mA cm^{-2}, a V_{OC} of 0.82 V, and an FF of 44.78%, whereas the optimized P(BPQI-BDT):Y6-based device displays a low PCE of 1.48% with a J_{SC} of 6.23 mA cm^{-2}, a V_{OC} of 0.68 V, and an FF of 34.61%. The increased V_{OC} for the P(BPQI-BDTT)-based PSC is due to the deep-positioned HOMO energy level, since the V_{OC} is determined by the HOMO of a donor's and the LUMO of an acceptor's energy offset. The external quantum efficiency (EQE) spectra are shown in Figure 3d: the larger J_{SC} of the P(BPQI-BDTT):Y6 devices is ascribed to the higher EQE response in the wavelength range from 400 to 850 nm. In contrast, the EQE value of the P(BPQI-BDT):Y6 cells is lower than 25% in this range, resulting in a smaller J_{SC}. The relationship between J_{SC} and the light intensity is revealed in Figure S3a: the α values of the P(BPQI-BDT):Y6 and P(BPQI-BDTT):Y6 devices are 0.92 and 0.974, respectively, which reflects that the P(BPQI-BDTT):Y6 based devices have a smaller bimolecular recombination effect. Figure S3b shows the light-intensity dependence of V_{OC}. The n values of P(BPQI-BDT):Y6 and P(BPQI-BDTT):Y6 are 2.37 and 1.78, respectively. It is shown that the device structure of P(BPQI-BDTT):Y6 produces fewer trap-assisted recombination centers, which are responsible for facilitating charge extraction and improving the PV performance. In order to investigate the device properties of the polymers in more depth, we constructed P(BPQI-BDTT):Y6 devices with a surface area of 1 cm^2, shown in Figure 3e, and the corresponding parameters are shown in Table 3. The PCE of large device is 5%, with a V_{OC} of 0.8 V, a J_{SC} of 13.92 mA cm^{-2}, and an FF of 44.77%. It demonstrated that polymers have the prospect of large-area applications in devices. The commercially polymer PTB7-Th was selected to compare the new polymer P(BPQI-BDTT). The photovoltaic performances of two devices are shown in Figure S4 and Table S1. In Figure S5, we demonstrate the performance of the device in the dark; the variations in the J_{SC} and V_{OC} are slight, thus showing the stability of the device's performance in a dark environment.

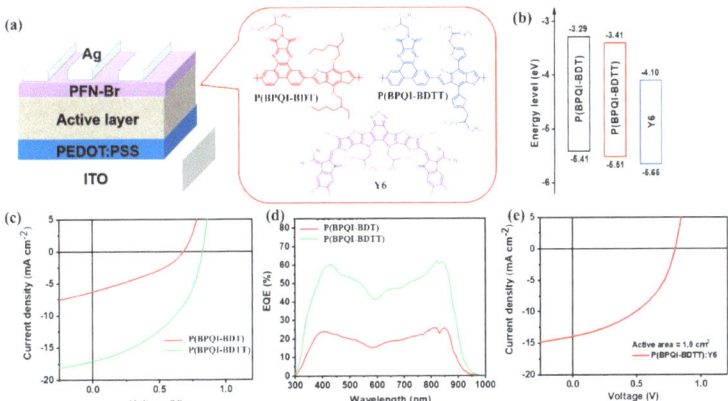

Figure 3. (a) Structure of P(BPQI-BDT) or P(BPQI-BDTT) as the polymer donor and Y6 as the acceptor; (b) the molecular energy levels of BPQI-BDT, BPQI-BDTT, and Y6; (c) vurrent density–applied voltage (J–V); (d) external quantum efficiency (EQE) curves of P(BPQI-BDT):Y6 or P(BPQI-BDTT):Y6 devices; (e) P(BPQI-BDTT):Y6 (J–V) curve when active area is 1 cm^2.

Table 2. Photovoltaic parameters of OSCs based on P(BPQI-BDT):Y6's and P(BPQI-BDTT):Y6's active layer. [a] Average data from 10 independent devices.

Active Layer	V_{OC} (V)	J_{SC} (mA cm^{-2})	J_{cal} (mA cm^{-2})	FF (%)	PCE_{max}/PCE_{ave} [a] (%)
P(BPQI-BDT):Y6	0.68	6.23	6.71	34.61	1.48/1.35 ± 0.10
P(BPQI-BDTT):Y6	0.82	17.09	17.00	44.78	6.31/5.97 ± 0.29

Table 3. Photovoltaic parameters of OSCs based on P(BPQI-BDTT):Y6's active layer with its surface area of 1 cm^2.

Active Layer	V_{OC} (V)	J_{SC} (mA cm^{-2})	FF (%)	PCE (%)
P(BPQI-BDTT):Y6	0.80	13.92	44.77	5.00

2.5. Polymer Film Morphology

Atomic force microscopy (AFM) measurements were taken to study the surface and phase separation morphology of the blend films. As shown in Figure 4, both P(BPQI-BDT):Y6 and P(BPQI-BDTT):Y6 have a relatively smooth surface. The root mean square (RMS) values of roughness for both the P(BPQI-BDT):Y6 and P(BPQI-BDTT):Y6 blend films were 1.28 and 0.9 nm, respectively. It is clear that the P(BPQI-BDT):Y6 blend film has a higher roughness than the P(BPQI-BDTT):Y6 blend, indicating that the P(BPQI-BDT):Y6 blend film enhances the aggregation and phase separation. The P(BPQI-BDTT):Y6 blend has minimal roughness, which is beneficial for charge carrier transportation, charge collection, and less charge recombination, leading to the enhancement of the FF and PCE in OPVs. X-ray diffraction (XRD) was applied to investigate the crystalline properties of these polymers. As shown in Figure S6, the two polymers show wide diffraction and high peaks located at 2θ = 37.32° and 2θ = 43.62°, which should derive from the two polymers exhibiting similar crystalline surfaces and analogous crystal structures. To further investigate the film morphologies of the polymers, SEM was used to study the surface of the P(BPQI-BDT):Y6 and P(BPQI-BDTT):Y6 films. As shown in Figure S7, it can be seen that these two polymers are irregularly flat films, while P(BPQI-BDTT):Y6 has a more uniform surface morphology, which is favorable for its charge transfer and suppresses charge recombination, effectively improving its device performance.

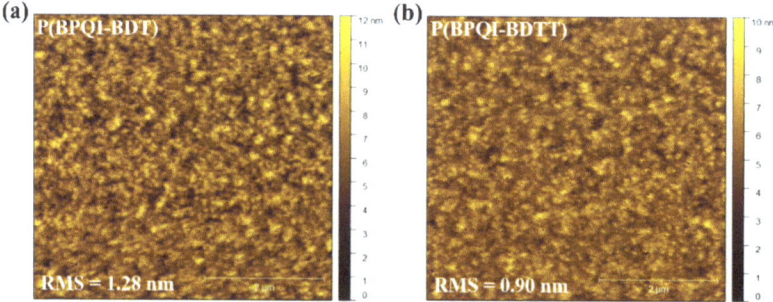

Figure 4. AFM phase images of (**a**) P(BPQI-BDT) and (**b**) P(BPQI-BDTT) blend films, respectively.

3. Conclusions

In summary, a new dibenzo-fused quinoxalineimide (BPQI) is successfully synthesized as an electron-deficient building block to construct donor–acceptor (D–A)-type polymers, namely P(BPQI-BDT) and P(BPQI-BDTT). Both polymers possess deep HOMO energy levels and wide bandgaps. These two polymers were used as donor materials for polymer solar cells. The PSC based on the P(BPQI-BDTT):Y6 blend showed a higher PCE (6.31%) with a V_{OC} of 0.82 V, a J_{SC} of 17.09 mA cm^{-2}, and an FF of 44.78% as compared to P(BPQI-BDT):Y6 (1.48%) with a V_{OC} of 0.68 V, a J_{SC} of 6.23 mA cm^{-2}, and an FF of 34.61%. The higher PCE is mainly attributed to improvements in the J_{SC} and FF. These results demonstrate that BPQI is a promising building block for developing a wide-bandgap polymer donor in OPVs. In addition, the acceptor BPQI can also be applied to other organic opto-electronic devices, such as organic field effect transistors (OFET) and perovskite solar cells (PSCs), due to its large conjugated backbone and electron-deficient characteristics. In future work, a series of quinoxalineimide electron-deficient units will be synthesized, and we believe that quinoxalineimides will become a promising building block for constructing organic semiconductors.

Supplementary Materials: The following supporting information can be downloaded at: https://www.mdpi.com/article/10.3390/polym14173590/s1, Figure S1: ^{1}H NMR spectrum of compound 5; Figure S2: ^{13}C NMR spectrum of compound 5; Figure S3: (a) light intensity dependence of J_{SC}, (b) light intensity dependence of V_{OC}; Figure S4: (a) The $J-V$ curves and (b) EQE spectra; Figure S5: The $J-V$ curves in dark; Figure S6: X-ray diffraction pattern of P(BPQI-BDT) and P(BPQI-BDTT); Figure S7: SEM of film state of P(BPQI-BDT):Y6 and P(BPQI-BDTT):Y6. Table S1: Photovoltaic parameters of OSCs based on P(BPQI-BDTT):Y6 and PTB7-Th:Y6.

Author Contributions: Data, X.W. (Xin Wang); data curation, Z.W., M.L., L.T., K.W., D.X. and Q.G.; methodology, M.Z.; resources, E.Z. and X.W. (Xianwen Wei); supervision, Y.S.; writing—original draft, Y.S. All authors have read and agreed to the published version of the manuscript.

Funding: Y.S. thanks the National Natural Science Foundation of China (22105004). This work was also supported by Open Fund (PLN2021-08) of the State Key Laboratory of Oil and Gas Reservoir Geology and Exploitation (Southwest Petroleum University) and the Innovation Entrepreneurship Training Program for Undergraduates of Anhui Normal University (202110370036, s202110370037).

Institutional Review Board Statement: Not applicable.

Informed Consent Statement: Informed consent was obtained from all subjects involved in the study.

Conflicts of Interest: The authors declare no conflict of interest.

References

1. Guo, X.; Zhou, N.; Lou, S.J.; Smith, J.; Tice, D.B.; Hennek, J.W.; Ortiz, R.P.; Navarrete, J.T.L.; Li, S.; Strzalka, J.; et al. Polymer solar cells with enhanced fill factors. *Nat. Photon.* **2013**, *7*, 825–833. [CrossRef]
2. Meng, L.; Zhang, Y.; Wan, X.; Li, C.; Zhang, X.; Wang, Y.; Ke, X.; Xiao, Z.; Ding, L.; Xia, R. Organic and solution-processed tandem solar cells with 17.3% efficiency. *Science* **2018**, *361*, 1094–1098. [CrossRef] [PubMed]
3. Liu, Q.; Jiang, Y.; Jin, K.; Qin, J.; Xu, J.; Li, W.; Xiong, J.; Liu, J.; Xiao, Z.; Sun, K. 18% Efficiency organic solar cells. *Sci. Bull.* **2020**, *65*, 272–275. [CrossRef]
4. Shi, Y.; Li, W.; Wang, X.; Tu, L.; Li, M.; Zhao, Y.; Wang, Y.; Liu, Y. Isomeric Acceptor–Acceptor Polymers: Enabling Electron Transport with Strikingly Different Semiconducting Properties in n-Channel Organic Thin-Film Transistors. *Chem. Mater.* **2022**, *34*, 1403–1413. [CrossRef]
5. Shi, Y.; Ma, R.; Wang, X.; Liu, T.; Li, Y.; Fu, S.; Yang, K.; Wang, Y.; Yu, C.; Jiao, L.; et al. Influence of Fluorine Substitution on the Photovoltaic Performance of Wide Band Gap Polymer Donors for Polymer Solar Cells. *ACS Appl. Mater. Interfaces* **2022**, *14*, 5740–5749. [CrossRef]
6. Shi, Y.; Guo, H.; Huang, J.; Zhang, X.; Wu, Z.; Yang, K.; Zhang, Y.; Feng, K.; Woo, H.Y.; Ortiz, R.P.; et al. Distannylated Bithiophene Imide: Enabling High-Performance n-Type Polymer Semiconductors with an Acceptor-Acceptor Backbone. *Angew. Chem. Int. Ed. Engl.* **2020**, *59*, 14449–14457. [CrossRef]
7. Zhu, L.; Zhang, M.; Xu, J.; Li, C.; Yan, J.; Zhou, G.; Zhong, W.; Hao, T.; Song, J.; Xue, X.; et al. Single-junction organic solar cells with over 19% efficiency enabled by a refined double-fibril network morphology. *Nat. Mater.* **2022**, *21*, 656–663. [CrossRef]
8. Zou, Y.; Chen, H.; Bi, X.; Xu, X.; Wang, H.; Lin, M.; Ma, Z.; Zhang, M.; Li, C.; Wan, X.; et al. Peripheral halogenation engineering controls molecular stacking to enable highly efficient organic solar cells. *Energy Environ. Sci.* **2022**, *15*, 3519–3533. [CrossRef]
9. Wu, J.; Fan, Q.; Xiong, M.; Wang, Q.; Chen, K.; Liu, H.; Gao, M.; Ye, L.; Guo, X.; Fang, J.; et al. Carboxylate substituted pyrazine: A simple and low-cost building block for novel wide bandgap polymer donor enables 15.3% efficiency in organic solar cells. *Nano Energy* **2020**, *82*, 105679. [CrossRef]

10. Zhao, J.; Li, Q.; Liu, S.; Cao, Z.; Jiao, X.; Cai, Y.-P.; Huang, F. Bithieno[3,4-c]pyrrole-4,6-dione-Mediated Crystallinity in Large-Bandgap Polymer Donors Directs Charge Transportation and Recombination in Efficient Nonfullerene Polymer Solar Cells. *ACS Energy Lett.* **2020**, *5*, 367–375. [CrossRef]
11. Shen, Q.; He, C.; Li, S.; Zuo, L.; Shi, M.; Chen, H. Design of Non-fused Ring Acceptors toward High-Performance, Stable, and Low-Cost Organic Photovoltaics. *Acc. Mater. Res.* **2022**, *3*, 644–657. [CrossRef]
12. Sun, C.; Pan, F.; Bin, H.; Zhang, J.; Xue, L.; Qiu, B.; Wei, Z.; Zhang, Z.-G.; Li, Y. A low cost and high performance polymer donor material for polymer solar cells. *Nat. Commun.* **2018**, *9*, 743. [CrossRef]
13. Zhang, G.; Ning, H.; Chen, H.; Jiang, Q.; Jiang, J.; Han, P.; Dang, L.; Xu, M.; Shao, M.; He, F.; et al. Naphthalenothiophene imide-based polymer exhibiting over 17% efficiency. *Joule* **2021**, *5*, 931–944. [CrossRef]
14. Xu, Y.; Cui, Y.; Yao, H.; Zhang, T.; Zhang, J.; Ma, L.; Wang, J.; Wei, Z.; Hou, J. A New Conjugated Polymer that Enables the Integration of Photovoltaic and Light-Emitting Functions in One Device. *Adv. Mater.* **2021**, *33*, 2101090. [CrossRef]
15. Zhu, C.; Meng, L.; Zhang, J.; Qin, S.; Lai, W.; Qiu, B.; Yuan, J.; Wan, Y.; Huang, W.; Li, Y. A Quinoxaline-Based D-A Copolymer Donor Achieving 17.62% Efficiency of Organic Solar Cells. *Adv. Mater.* **2021**, *33*, e2100474. [CrossRef]
16. Wang, Y.; Guo, H.; Ling, S.; Arrechea-Marcos, I.; Wang, Y.; Navarrete, J.T.L.; Ortiz, R.P.; Guo, X. Ladder-type Heteroarenes: Up to 15 Rings with Five Imide Groups. *Angew. Chem. Int. Ed. Engl.* **2017**, *56*, 9924–9929. [CrossRef]
17. Luo, Z.; Ma, R.; Yu, J.; Liu, H.; Liu, T.; Ni, F.; Hu, J.; Zou, Y.; Zeng, A.; Su, C.-J.; et al. Heteroheptacene-based acceptors with thieno[3,2-b]pyrrole yield high-performance polymer solar cells. *Natl. Sci. Rev.* **2022**, *9*, nwac076. [CrossRef]
18. Li, T.; Wu, Y.; Zhou, J.; Li, M.; Wu, J.; Hu, Q.; Jia, R.; Pan, X.; Zhang, M.; Tang, Z.; et al. Butterfly Effects Arising from Starting Materials in Fused-Ring Electron Acceptors. *J. Am. Chem. Soc.* **2020**, *142*, 20124–20133. [CrossRef]
19. Shi, Y.; Tang, Y.; Yang, K.; Qin, M.; Wang, Y.; Sun, H.; Su, M.; Lu, X.; Zhou, M.; Guo, X. Thiazolothienyl imide-based wide bandgap copolymers for efficient polymer solar cells. *J. Mater. Chem. C* **2019**, *7*, 11142–11151. [CrossRef]
20. An, N.; Cai, Y.; Wu, H.; Tang, A.; Zhang, K.; Hao, X.; Ma, Z.; Guo, Q.; Ryu, H.S.; Woo, H.Y.; et al. Solution-Processed Organic Solar Cells with High Open-Circuit Voltage of 1.3 V and Low Non-Radiative Voltage Loss of 0.16 V. *Adv. Mater.* **2020**, *32*, e2002122. [CrossRef]
21. An, Q.; Wang, J.; Ma, X.; Gao, J.; Hu, Z.; Liu, B.; Sun, H.; Guo, X.; Zhang, X.; Zhang, F. Two compatible polymer donors contribute synergistically for ternary organic solar cells with 17.53% efficiency. *Energy Environ. Sci.* **2020**, *13*, 5039–5047. [CrossRef]
22. Guo, X.; Fan, Q.; Wu, J.; Li, G.; Peng, Z.; Su, W.; Lin, J.; Hou, L.; Qin, Y.; Ade, H.; et al. Optimized Active Layer Morphologies via Ternary Copolymerization of Polymer Donors for 17.6% Efficiency Organic Solar Cells with Enhanced Fill Factor. *Angew. Chem. Int. Ed. Engl.* **2021**, *60*, 2322–2329. [CrossRef]
23. Ha, J.-W.; Kim, H.S.; Song, C.E.; Park, H.J.; Hwang, D.-H. Thienoquinolinone as a new building block for wide bandgap semiconducting polymer donors for organic solar cells. *J. Mater. Chem. C* **2020**, *8*, 12265–12271. [CrossRef]
24. Luo, Y.; Luo, Y.; Huang, X.; Liu, S.; Cao, Z.; Guo, L.; Li, Q.; Cai, Y.; Wang, Y. A New Ester-Substituted Quinoxaline-Based Narrow Bandgap Polymer Donor for Organic Solar Cells. *Macromol. Rapid Commun.* **2021**, *42*, e2000683. [CrossRef]
25. Xie, Q.; Liu, Y.; Liao, X.; Cui, Y.; Huang, S.; Hu, L.; He, Q.; Chen, L.; Chen, Y. Isomeric Effect of Wide Bandgap Polymer Donors with High Crystallinity to Achieve Efficient Polymer Solar Cells. *Macromol. Rapid Commun.* **2020**, *41*, 2000454. [CrossRef]
26. Khadka, D.B.; Shirai, Y.; Yanagida, M.; Noda, T.; Miyano, K. Tailoring the Open-Circuit Voltage Deficit of Wide-Band-Gap Perovskite Solar Cells Using Alkyl Chain-Substituted Fullerene Derivatives. *ACS Appl. Mater. Interfaces* **2018**, *10*, 22074–22082. [CrossRef]
27. Nakanishi, R.; Nogimura, A.; Eguchi, R.; Kanai, K. Electronic structure of fullerene derivatives in organic photovoltaics. *Org. Electron.* **2014**, *15*, 2912–2921. [CrossRef]
28. Zhang, Y.; Liu, D.; Lau, T.; Zhan, L.; Shen, D.; Fong, P.W.K.; Yan, C.; Zhang, S.; Lu, X.; Lee, C.-S.; et al. A Novel Wide-Bandgap Polymer with Deep Ionization Potential Enables Exceeding 16% Efficiency in Ternary Nonfullerene Polymer Solar Cells. *Adv. Funct. Mater.* **2020**, *30*, 1910466. [CrossRef]
29. Zhang, T.; Zeng, G.; Ye, F.; Zhao, X.; Yang, X. Efficient Non-Fullerene Organic Photovoltaic Modules Incorporating As-Cast and Thickness-Insensitive Photoactive Layers. *Adv. Energy Mater.* **2018**, *8*, 1801387. [CrossRef]
30. Zhang, W.; Huang, J.; Xu, J.; Han, M.; Su, D.; Wu, N.; Zhang, C.; Xu, A.; Zhan, C. Phthalimide Polymer Donor Guests Enable over 17% Efficient Organic Solar Cells via Parallel-Like Ternary and Quaternary Strategies. *Adv. Energy Mater.* **2020**, *10*, 2001436. [CrossRef]
31. Keshtov, M.L.; Khokhlov, A.R.; Godovsky, D.Y.; Ostapov, I.E.; Alekseev, V.G.; Xie, Z.; Chayal, G.; Sharma, G.D. Novel Pyrrolo [3,4-b] Dithieno [3, 2-f:2″,3″-h] Quinoxaline-8,10 (9H)-Dione Based Wide Bandgap Conjugated Copolymers for Bulk Heterojunction Polymer Solar Cells. *Macromol. Rapid Commun.* **2022**, *43*, e2200060. [CrossRef] [PubMed]
32. Busireddy, M.R.; Chen, T.-W.; Huang, S.-C.; Nie, H.; Su, Y.-J.; Chuang, C.-T.; Kuo, P.-J.; Chen, J.-T.; Hsu, C.-S. Fine Tuning Alkyl Substituents on Dithienoquinoxaline-Based Wide-Bandgap Polymer Donors for Organic Photovoltaics. *ACS Appl. Mater. Interfaces* **2022**, *14*, 22353–22362. [CrossRef] [PubMed]
33. Sun, C.; Zhu, C.; Meng, L.; Li, Y. Quinoxaline-Based D–A Copolymers for the Applications as Polymer Donor and Hole Transport Material in Polymer/Perovskite Solar Cells. *Adv. Mater.* **2022**, *34*, 2104161. [CrossRef] [PubMed]
34. Yong-qiang, S.; Wang, Y.; Guo, X. Recent Progress of Imide-Functionalized N-Type Polymer Semiconductors. *Acta. Polym. Sin.* **2019**, *50*, 873–889.

35. Shi, Y.; Guo, H.; Qin, M.; Wang, Y.; Zhao, J.; Sun, H.; Wang, H.; Wang, Y.; Zhou, X.; Facchetti, A.; et al. Imide-Functionalized Thiazole-Based Polymer Semiconductors: Synthesis, Structure–Property Correlations, Charge Carrier Polarity, and Thin-Film Transistor Performance. *Chem. Mater.* **2018**, *30*, 7988–8001. [CrossRef]
36. Shi, Y.; Guo, H.; Qin, M.; Zhao, J.; Wang, Y.; Wang, H.; Wang, Y.; Facchetti, A.; Lu, X.; Guo, X. Thiazole Imide-Based All-Acceptor Homopolymer: Achieving High-Performance Unipolar Electron Transport in Organic Thin-Film Transistors. *Adv. Mater.* **2018**, *30*, 1705745. [CrossRef]

Article

Efficient All-Polymer Solar Cells Enabled by Interface Engineering

Guoping Zhang [1,†], Lihong Wang [1,†], Chaoyue Zhao [1,†], Yajie Wang [1], Ruiyu Hu [1], Jiaxu Che [1], Siying He [1], Wei Chen [2,*], Leifeng Cao [2], Zhenghui Luo [3,*], Mingxia Qiu [1,*], Shunpu Li [1,*] and Guangye Zhang [1,*]

[1] College of New Materials and New Energies, Shenzhen Technology University, Shenzhen 518118, China
[2] College of Engineering Physics, Shenzhen Technology University, Shenzhen 518118, China
[3] College of Materials Science and Engineering, Shenzhen University, Shenzhen 518060, China
* Correspondence: chenwei@sztu.edu.cn (W.C.); zhhuiluo@szu.edu.cn (Z.L.); qiumingxia@sztu.edu.cn (M.Q.); lishunpu@sztu.edu.cn (S.L.); zhangguangye@sztu.edu.cn (G.Z.)
† These authors contributed equally to this work.

Abstract: All-polymer solar cells (all-PSCs) are organic solar cells in which both the electron donor and the acceptor are polymers and are considered more promising in large-scale production. Thanks to the polymerizing small molecule acceptor strategy, the power conversion efficiency of all-PSCs has ushered in a leap in recent years. However, due to the electrical properties of polymerized small-molecule acceptors (PSMAs), the FF of the devices is generally not high. The typical electron transport material widely used in these devices is PNDIT-F3N, and it is a common strategy to improve the device fill factor (FF) through interface engineering. This work improves the efficiency of all-polymer solar cells through interfacial layer engineering. Using PDINN as the electron transport layer, we boost the FF of the devices from 69.21% to 72.05% and the power conversion efficiency (PCE) from 15.47% to 16.41%. This is the highest efficiency for a PY-IT-based binary all-polymer solar cell. This improvement is demonstrated in different all-polymer material systems.

Keywords: organic photovoltaics; all-polymer solar cells; power conversion efficiency; electron transport layer

Citation: Zhang, G.; Wang, L.; Zhao, C.; Wang, Y.; Hu, R.; Che, J.; He, S.; Chen, W.; Cao, L.; Luo, Z.; et al. Efficient All-Polymer Solar Cells Enabled by Interface Engineering. *Polymers* 2022, *14*, 3835. https://doi.org/10.3390/polym14183835

Academic Editor: Tengling Ye

Received: 22 August 2022
Accepted: 13 September 2022
Published: 14 September 2022

Publisher's Note: MDPI stays neutral with regard to jurisdictional claims in published maps and institutional affiliations.

Copyright: © 2022 by the authors. Licensee MDPI, Basel, Switzerland. This article is an open access article distributed under the terms and conditions of the Creative Commons Attribution (CC BY) license (https://creativecommons.org/licenses/by/4.0/).

1. Introduction

Organic solar cells (OSCs) are advantageous for distributed photovoltaic applications due to their flexibility, semitransparency, high indoor light matching, high power generation per unit weight, patternable design, etc. [1–7]. While the classical OSCs active layer consists of a polymer donor and a fullerene derivative acceptor [8], with the development of non-fullerene acceptor materials, the mainstream of research has now shifted to a material system based on a polymer donor and a non-fullerene acceptor [9–17]. Compared with fullerene acceptors, non-fullerene acceptors have flexible chemical structure designs and thus easily tunable optoelectronic properties, endowing them with more compatible absorbance spectra and electrical properties with the donor. Additionally, the resulting devices usually exhibit lower voltage losses [18], which improve device efficiency [19]. In terms of materials, non-fullerene acceptors are divided into two main categories: small-molecule acceptors and polymer acceptors. In contrast to the small molecule acceptor, when the acceptor is polymer, all-polymer solar cells (All-PSCs) can be prepared by combining with a polymer donor. In addition to similar advantages to small-molecule acceptor-based OSCs, All-PSCs possess better mechanical properties [20,21], such as higher tensile and flexural toughness, and potentially higher thermal stability, which provide them with better prospects for large-scale production [22]. The all-PSCs field has evolved over 20 years, starting with the earliest polymer donor materials based on poly(p-phenylene vinylidene) (PPV) units. The widespread use of aromatic imide repeating units, e.g., Naphthalimide (NDI) and Perylenediimide (PDI), has led to significant advances in the power conversion efficiency (PCE) of all-PSC devices. Nevertheless, the fill factor (FF) of all-polymer solar cell devices is relatively low (Figure 1) compared to their fullerene or small molecule acceptor counterparts, mainly due to complicated morphology control. Even the highly efficient devices developed in the past two years have difficulty in surpassing 70% FF.

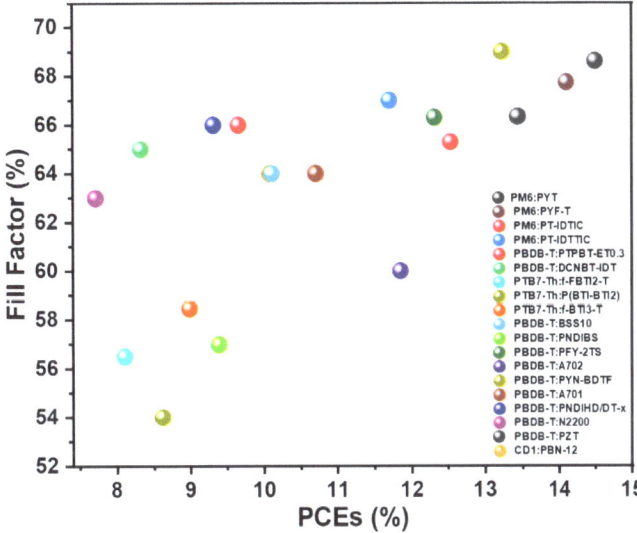

Figure 1. Fill Factors and PCEs for several high-performance all-polymer solar cells in recent years.

The current molecular design strategy, namely, polymerizing the high-performance small molecular acceptor, led to a surge of polymer acceptors with excellent photovoltaic performance, which significantly improved device efficiency in this field. However, despite the fact that the state-of-the-art devices based on the polymerized small molecule acceptors (PSMAs) exhibit high open-circuit voltages (V_{OC}s) and short circuit currents (J_{SC}s), the FFs for a large fraction of them are relatively low due to the difference in the morphological and electrical properties between the polymer donor (high degree of polymerization) and the PSMA (low degree of polymerization). One of the solutions to this issue is optimizing the morphology and charge transport property, but for a given set of donor and acceptor materials, this can be difficult. From the experience of device engineering for organic photovoltaics in recent decades, another angle to tackle this problem, other than modifying the active layer directly, is interface engineering. So far, most high-performance all-PSCs utilize N,N′-Bis(N,N-dimethylpropan-1-amine oxide)perylene-3,4,9,10-tetracarboxylic diimide (PDINO), MoO$_x$, or poly [(9,9-bis(3′-(N,N-dimethylamino)propyl)2,7-fluorene)-alt-5,5′-bis(2,2′-thiophene)-2,6-naphthalene1,4,5,8-tetracaboxylic-N,N′-di(2-ethylhexyl)imide] (PNDIT-F3N) as the electron transport material (ETM). Among them, PNDIT-F3N and its bromide version PNDIT-F3N-Br, are the workhorse ETMs in small molecule-based OSCs. However, the FFs of PNDIT-F3N-based all-PSCs are generally not high (hardly more than 70%, see Figure 1), which limits the further improvement of device efficiency [23–36]. For instance, Ref. [37] studied PM6:PYF-T devices with an efficiency of 14.10% and an FF of 67.73%. Ref. [38] researched PM6:PY-IT-based all-PSCs and reported an efficiency of 15.15% and an FF of 67.70% [39]. In 2020, Li's group reported a new PDI-derived electron transport material, aliphatic amine-functionalized perylene-diimide (PDINN), which showed better contact with non-fullerenes active layers and better conductivity. The enhanced interfacial stability and higher conductivity, as well as the ability to reduce the work function of the metal cathode, make it more suitable for use as an electron transport material. In addition, PDINN is simple to synthesize and could be synthesized in large quantities by a one-step reaction. Therefore, PDINN is a low-cost alternative ETM for OSCs and has great promise for future large-scale production. Since its first report, PDINN has been most utilized in polymer:small molecule acceptor OSCs and proved effective in different cases, but it has not been adopted in all-PSCs with PCEs >16%.

In this work, we report all-PSCs with PM6 as the donor, PYF-T-*o* or PY-IT as the acceptor, and PNDIT-F3N or PDINN as the electron transport material. We systematically compare the performance of the all-PSC devices with PDINN and PNDIT-F3N, and we show that the device performance for PDINN-based devices is higher than those of PNDIT-F3N-based ones. Particularly, we show that the FF of the devices increases from 69.21% (PNDIT-F3N) to 72.05% (PDINN) for the PY-IT-based all-PSCs, with a corresponding efficiency increase from 15.47% to 16.41%, which is the highest power conversion efficiency (PCE) for PY-IT based binary all-PSCs. Through different characterizations, e.g., transient photocurrent and transient photovoltage, we find that the difference in their optical properties does not contribute much to the device performance variation. Instead, we attribute the main increase, i.e., the FF, of the PDINN-based devices, to the faster charge extraction and enhanced charge carrier lifetime, which are observed in both all-PSC systems we studied.

2. Experiments

2.1. Materials

Poly[(2,6-(4,8-bis(5-(2-ethylhexyl-3-fluoro)thiophen-2-yl)-benzo [1,2b:4,5-b′]dithiophene))-alt-(5,5-(1′,3′-di-2-thienyl-5′,7′-bis(2-ethylhexyl)benzo [1′,2′c:4′,5′c′] dithio-phene-4,8-dione)] (PM6) was purchased from Dongguan Volt Ampere Photo-electric Technology Co., Ltd. (Dongguan, China). Poly[(2,2′-((2Z,2′Z)-((12,13-bis(2-octyldodecyl)-3,9-diundecyl-12,13-dihydro [1,2,5]thiadiazolo [3,4e]thieno [2″,3″:4′,5′]thieno [2′,3′:4,5] pyrrolo [3,2-g]thieno [2′,3′:4,5]thieno [3,2-b]-in-dole-2,10-diyl)bis(methanylylidene))bis(5-methyl-3-oxo-2,3-dihydro-1Hindene-2,1-diyl-idene)) dimalononitrile-co-2,5-thiophene (PY-IT) and N,N′-Bis{3-[3-(Dimethylamino)propylamino]propyl}perylene-3,4,9,10-tetracarboxylic diimide (PDINN) were purchased from Solarmer Materials Inc. (Beijing, China). Poly[(9,9-bis(3′-(N,N-dimethylamino)pro-pyl)2,7-fluorene)-alt-5,5′-bis(2,2′-thiophene)-2,6-naphthalene1, 4,5,8-tetracabox-ylic-N,N′-di(2-ethylhexyl)imide] (PNDIT-F3N) and PYF-T-*o* were purchased from eFlexPV Limited. Poly(3,4-ethylenedioxythiophene) polystyrene sulfonate (PEDOT:PSS) (Clevios P VP 4083) were purchased from Heraeus Inc. (Hanau, Germany). All the other reagents and chemicals were purchased from Sigma Aldrich or Aladdin (Burlington, MA, USA and Shanghai, China) and used as received. Purity of solvents: chloroform (>99.8%), methanol (>99.5%), and acetic acid (>99.5%).

Solubility of PDINN and PNDIT-F3N:

PDINN: the solubility was 26.7 mg/mL in methanol without the assistance of any acid [40].

PNDIT-F3N: the solubility was >30 mg/mL in common organic solvents [41].

2.2. Device Fabrication

Organic solar cells were fabricated in a conventional device configuration of ITO(50 nm)/ PEDOT:PSS(30 nm)/active layer(100~150 nm)/ETL(PNDIT-F3N or PDINN)(5~10 nm)/ Ag(100 nm), as shown in the SEM image in Figure S3. The patterned indium tin oxide(ITO) glass was scrubbed with detergent and then sonicated with deionized water, acetone, and isopropanol sequentially and baked overnight in an oven. The glass substrate was treated whit UV-Ozone for 10 min before use. PEDOT:PSS solution was spin-casted onto them at 5200 rpm for 20 s, then dried at 150 °C for 10 min in air.

The different kinds of devices:

1. The PM6:PYF-T-*o* blend (1:1.2 weight ratio) was dissolved in chloroform (the concentration of donor was 6 mg mL^{-1} for all blends) with 1-chloronaphthalene (1% vol) as an additive and stirred overnight in a nitrogen-filled glove box. The blend solution was spin-casted at 2500 rpm for 30 s onto the PEDOT:PSS films, followed by a thermal annealing step at 95 °C for 5 min.

2. The PM6:PY-IT blend (1:1.2 weight ratio) was dissolved in chloroform (the concentration of donor was 6 mg mL^{-1} for all blends) with 1-chloronaphthalene (1% vol) as an additive and stirred overnight in a nitrogen-filled glove box. The blend solution was spin-casted at 2700 rpm for 30 s onto the PEDOT:PSS films, followed by a thermal annealing step at 95 °C for 5 min.

For both types of devices, either methanol with 0.5% vol acetic acid blend solution of PNDIT-F3N at a concentration of 0.5 mg mL^{-1} or a pure methanol solution of PDINN at a concentration of 1.0 mg mL^{-1} was spin-coated onto the active layer, respectively, at 2000 rpm for 30s and 3000 rpm for 30 s as the electron transport layer (ETL). Around 100 nm of Ag were evaporated under 1×10^{-4} Pa through a shadow mask. Then, encapsulation was carried out.

2.3. Characterization

The current density–voltage (*J-V*) curves of the PSCs were measured using a Keithley 2400 Source Meter under AM 1.5 G (100 mW cm^{-2}) using an Enlitech solar simulator. The light intensity was calibrated using a standard Si diode with a KG5 filter to bring spectral mismatch to unity. An optical microscope (Olympus BX51) was used to define the device area (7.2 mm^2) in a glove box filled with nitrogen (oxygen and water contents are smaller than 0.1 ppm). EQEs were measured using an Enlitech QE-S EQE system equipped with a standard single-crystal Si photovoltaic cell. Monochromatic light was generated from an Enlitech 300 W lamp source.

Transient photovoltage (TPV) and transient photocurrent (TPC) measurements: In TPV measurements, the devices were placed under background light bias enabled by a focused Quartz Tungsten-Halogen Lamp with an intensity of similar to working devices, i.e., the device voltage is close to the V_{OC} under solar illumination conditions. Photo-excitations were generated with 8 ns pulses from a laser system (Oriental Spectra, NLD520, Hyderabad, India). The wavelength for the excitation was tuned to 518 nm with a spectral width of 3 nm. A digital oscilloscope was used to acquire the TPV signal at the open-circuit condition. TPC signals were measured under short-circuit conditions under the same excitation wavelength without background light bias.

3. Results and Discussion

The chemical structural formulas of the donor PM6 and the acceptors PYF-T-*o* and PY-IT are shown in Figure 2a. The chemical structural formulas of the electron transport materials (PDINN and PNDIT-F3N) are shown in Figure 2b. Figure 2c shows the highest occupied molecular orbital (HOMO) and the lowest unoccupied molecular orbital (LUMO) energy levels of the active layer materials, as well as the energy level diagrams of the electron transport materials of PDINN and PNDIT-F3N.

To study the optical property of the two different interfacial layers, we measured the transmittance and absorption of PNDIT-F3N and PDINN. To mimic the thickness of PNDIT-F3N and PDINN used in devices, we used identical spin-coating parameters to prepare the PNDIT-F3N and PDINN films. From the absorption and transmittance curves (Figure 3a,b), we found that PNDIT-F3N showed absorption in the UV and visible regions. For instance, the absorption of the PNDIT-F3N film, despite being weak, peaked at ~390 nm. In addition, PNDIT-F3N also showed absorption in the range of ~570–650 nm. These absorptions of PNDIT-F3N can be reflected in the overall absorption spectra of active layer/PNDIT-F3N, which is shown in Figure 3c,d.

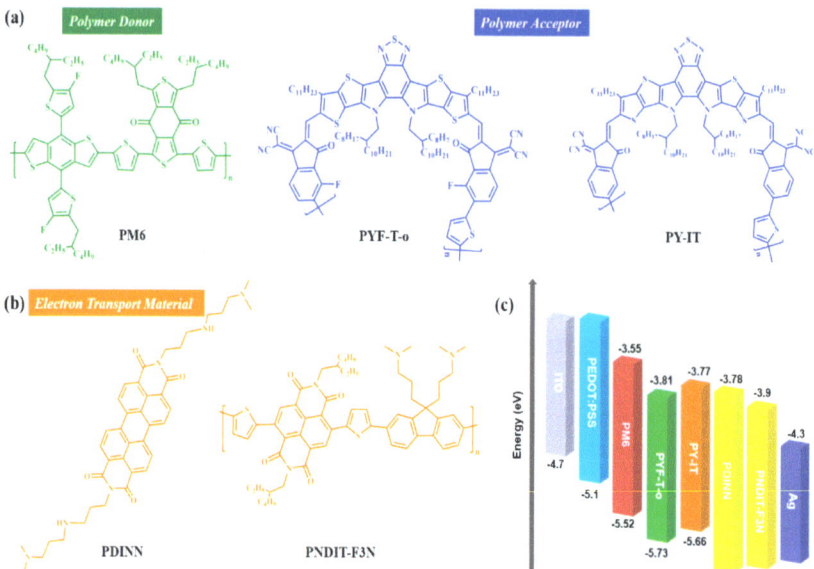

Figure 2. (**a**) Chemical structures of PM6, PYF-T-*o* and PY-IT. (**b**) Chemical structures of PDINN and PNDIT-F3N. (**c**) Energy diagram of materials (literature data: the energy levels of ITO, PEDOT: PSS, PM6, PYF-T-*o*, PNDIT-F3N, and Ag are taken from ref. [41], PY-IT from ref. [42], and PDINN from ref. [39]).

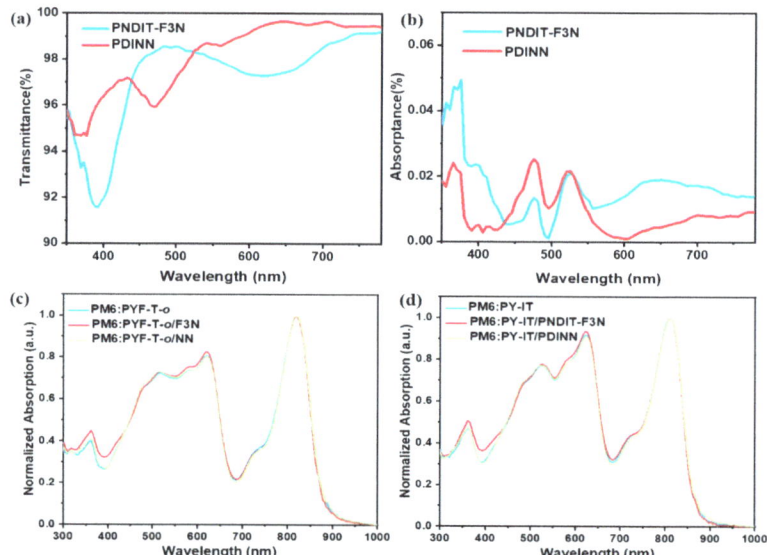

Figure 3. (**a**) UV-Vis transmittance spectra of PNDIT-F3N and PDINN. (**b**) UV-Vis absorption spectra of PNDIT-F3N and PDINN. (**c**,**d**) Absorption spectra of the active layers based on PYF-T-*o* and PY-IT, respectively.

For the PYF-T-*o*-based devices, the difference between V_{OC} and J_{SC} of PNDIT-F3N and PDINN devices was not significant. The main difference derived from the change in FF, where PDINN improved the filling factor of the device from 68.30% to 69.90% for PNDIT-F3N. For the PY-IT-based devices, PNDIT-F3N and PDINN had little effect on the J_{SC} of the device, and the main difference continued to come from the significant increase in the fill factor, for which PDINN increased from 69.21% of PNDIT-F3N to 72.05% of the device. As a result, these improvements significantly increased the efficiency of the PDINN-based devices by 16.41%. The corresponding specific device performance parameters are listed in Table 1 and Tables S2 and S3. We then conducted EQE, as shown in Figure 4b. From the EQE curves, we found that the current obtained from the EQE integration is consistent with the J_{SC} obtained from the J-V test (Figure 4a).

Table 1. Photovoltaic parameters of the solar cell devices based on PM6:PYF-T-o and PM6:PY-IT under AM 1.5 G illumination at 100 mW cm^{-2}.

Devices	V_{OC} (V)	J_{SC} (mA/cm^2)	FF (%)	PCEs (%)	S	$n_{id,l}$	$n_{id,d}$
PM6:PYF-T-o/ PNDIT-F3N	0.913 [a] (0.910 ± 0.002)	24.91 (24.9 ± 0.53)	68.30 (66.7 ± 0.96)	15.47 (15.1 ± 0.27)	0.925	1.21	1.82
PM6:PYF-T-o/ PDINN	0.908 (0.907 ± 0.003)	24.83 (24.8 ± 0.31)	69.90 (69.1 ± 0.78)	15.78 (15.5 ± 0.17)	0.930	1.21	1.71
PM6:PY-IT/ PNDIT-F3N	0.938 (0.938 ± 0.005)	23.85 (23.6 ± 0.47)	69.21 (67.8 ± 0.83)	15.47 (15.1 ± 0.29)	0.936	1.13	1.62
PM6:PY-IT/ PDINN	0.950 (0.951 ± 0.002)	23.95 (24.0 ± 0.17)	72.05 (70.5 ± 0.89)	16.41 (16.1 ± 0.17)	0.961	1.05	1.58

[a] Parameters for devices with the highest PCEs. Values in brackets are average values and standard deviations based on 10 independent devices.

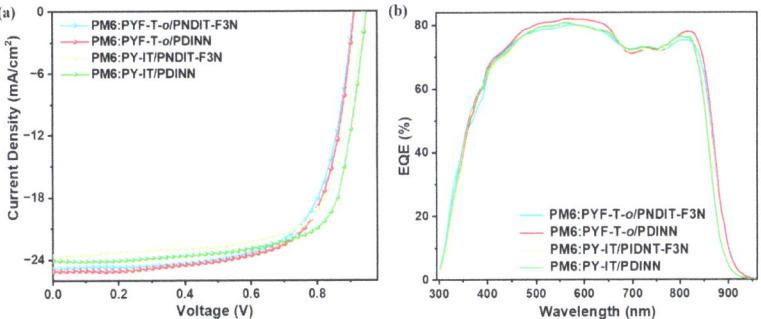

Figure 4. (**a**) J-V curves and (**b**) the corresponding EQE spectra of all-PSCs based on PM6 and PYF-T-o or PY-IT.

To investigate recombination, we first measured the J-V characteristics of the four devices under different light intensities (I). Figure 5a plots the relationship between J_{SC} and light intensity. By linearly fitting the J_{SC} versus light intensity data, we obtained the slope (S) for the four devices. The S values for the PM6:PYF-T-o/PNDIT-F3N, PM6:PYF-T-o/PDINN, PM6:PY-IT/PNDIT-F3N, and PM6:PY-IT/PDINN devices are 0.925, 0.930, 0.936, and 0.961, respectively. It is known that the closer S is to unity, the weaker the bimolecular recombination. Therefore, the smallest S for PM6:PY-IT/PDINN suggests the weakest bimolecular recombination in it, and among the four different devices, the trend of S is consistent with the trend of the FF of the devices.

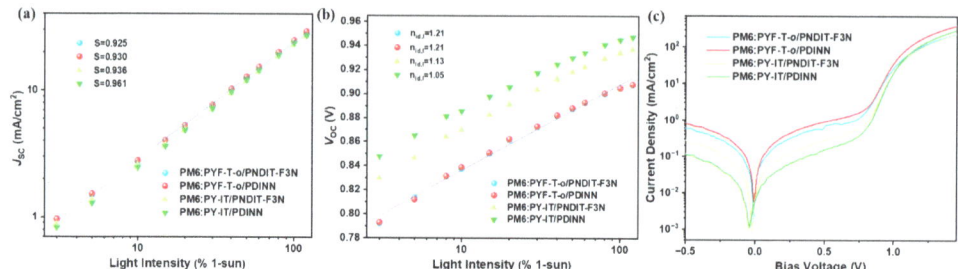

Figure 5. (**a**,**b**) are light intensity dependence of J_{SC} and V_{OC} of the OSCs based on PM6:PYF-T-*o* and PM6:PY-IT, respectively. (**c**) Dark *J-V* characteristics of the OSCs.

To access trap-assisted recombination, we plot the V_{OC} versus light intensity result in Figure 5b. By fitting the V_{OC} versus ln(*I*) curves, we obtained the ideality factor, $n_{id,l}$, from the equation $n_{id,l} = \frac{q}{kT}\frac{\partial V_{OC}}{\partial \ln(I)}$. The $n_{id,l}$s of the PM6:PYF-T-*o*/PNDIT-F3N, PM6:PYF-T-*o*/PDINN, PM6:PY-IT/PNDIT-F3N, and PM6:PY-IT/PDINN devices are 1.21, 1.21, 1.13, and 1.05, respectively. From diode theory, higher $n_{id,l}$ means that trap-assisted recombination is stronger. Therefore, PM6:PY-IT/PDINN has the weakest trap-assisted recombination ($n_{id,l}$ = 1.05) among the four devices, which agrees with its highest FF.

Another method to obtain the ideality factor is to fit the exponential region of the dark *J-V* curve. We measured the dark *J-V* curves for the devices, and the results are shown in Figure 5c. From the fitting, the $n_{id,d}$ of the PM6:PYF-T-*o*/PNDIT-F3N, PM6:PYF-T-*o*/PDINN, PM6:PY-IT/PNDIT-F3N, and PM6:PY-IT/PDINN devices are 1.823, 1.719, 1.642 and 1.582, respectively. The difference between the magnitude of $n_{id,l}$ and $n_{id,d}$ is detailed elsewhere [43], but the trend of the $n_{id,d}$ is overall consistent with that of $n_{id,l}$.

To further study the charge recombination and charge extraction, we performed TPV and TPC measurements. The details of the experimental setup for these measurements can be found in the experimental section. As shown in Figure 6a, we fitted the decay using a monoexponential function, which revealed that the decay constants have the following relationship, $\tau_{PDINN} > \tau_{PNDIT-F3N}$, indicating that the charge carrier lifetime is longer in devices based on PDINN than in PNDIT-F3N-based devices. This indicates that the recombination in the PDINN-based devices is weaker than that in the PNDIT-F3N-based devices. One hypothesis is that the PDINN has better/higher surface coverage than PNDIT-F3N so the contact between the active layer and cathode is reduced in PDINN-based devices, which reduces surface recombination and improves FF. In addition, from the TPC measurements (Figure 6b) for both PYF-T-*o* and PY-IT-based devices, the charge extraction in the PDINN-based devices is significantly faster than in the PNDIT-F3N-based devices. This indicates that the charge collection efficiency of PDINN is higher than that of PNDIT-F3N, which could be one of the determining factors of the high FF of the PNDIT-F3N devices.

By analyzing the results of AFM (Figure 7), the RMS of the PDINN film is 1.37 nm, and that of PNDIT-F3N is 1.87 nm. Leaving energetics alone, just from a morphological point of view, the smoother surface of PDINN is beneficial for obtaining better coverage on the active layer, which could then reduce the direct contact between the active layer material and the metal electrode. This could reduce surface recombination, protect the active layer from hot metal penetration or reaction during evaporation, increase device shunt resistance, and enhance the FF of the device.

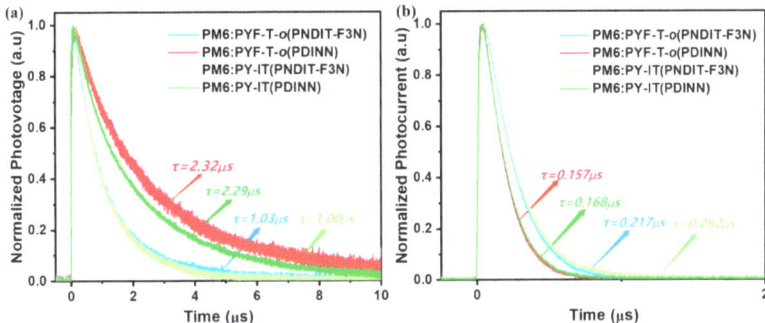

Figure 6. TPV (**a**) and TPC (**b**) decays of the PM6:PYF-T-*o* and PM6:PY-IT-based devices.

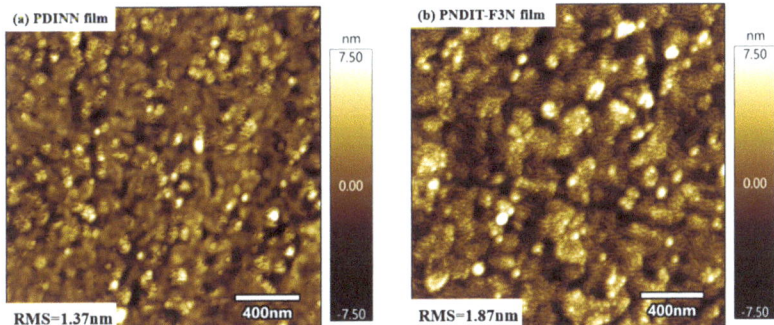

Figure 7. AFM images of PDINN and PNDIT–F3N.

4. Conclusions

In conclusion, we addressed the relatively low FF in PY-IT and PYF-T-*o*-based all-polymer solar cells through active layer-cathode interface engineering. Specifically, we used PDINN as an electron transport layer material in PM6:PYF-T-*o* and PM6:PY-IT-based devices, compared it with the widely employed, high-performance ETM, PNDIT-F3N, compared its photovoltaic performance, and investigated the charge extraction and recombination. It was found that the PDINN-based devices demonstrated faster charge extraction and longer charge carrier lifetime compared to PNDIT-F3N devices. Consequently, we demonstrated that PDINN could effectively promote the FF of the all-PSC devices studied in this work and thus improve the PCE of the devices. Particularly in the PM6:PY-IT-based device, the FF increased from 67.99% (PNDIT-F3N) to 72.05% (PDINN), and the PCEs increased from 15.47% to 16.41%, which is the highest efficiency reported to date for PY-IT-based binary all-polymer solar cells.

Supplementary Materials: The following supporting information can be downloaded at: https://www.mdpi.com/article/10.3390/polym14183835/s1.

Author Contributions: Data curation, G.Z. (Guoping Zhang), L.W. and C.Z.; Formal analysis, G.Z. (Guoping Zhang), L.W. and C.Z.; Funding acquisition, M.Q., S.L. and G.Z. (Guangye Zhang); Investigation, G.Z. (Guoping Zhang), C.Z., Y.W., R.H., J.C., S.H. and M.Q.; Methodology, L.W., L.C. and Z.L.; Project administration, W.C. and G.Z. (Guangye Zhang); Supervision, W.C., L.C., Z.L., S.L. and G.Z. (Guangye Zhang); Writing—original draft, W.C. and M.Q.; Writing—review & editing, L.C., Z.L., S.L. and G.Z. (Guangye Zhang). All authors have read and agreed to the published version of the manuscript.

Funding: The work was supported by the Guangdong Basic and Applied Basic Research Foundation (2022A1515010875), the Guangdong Basic and Applied Basic Research Foundation (2021A1515110017), the Natural Science Foundation of Top Talent of SZTU (grant no. 20200205), and the Project of Education Commission of Guangdong Province of China (2021KQNCX080). L.C. acknowledges support from the Natural Science Foundation of Top Talent of SZTU (grant no. 20200206). S.L. acknowledges support from the Education Department of Guangdong Province (2021KCXTD045).

Acknowledgments: Guoping Zhang, Lihong Wang, and Chaoyue Zhao contributed equally to this work. The work was supported by the Guangdong Basic and Applied Basic Research Foundation (2022A1515010875), the Guangdong Basic and Applied Basic Research Foundation (2021A1515110017), the Natural Science Foundation of Top Talent of SZTU (grant no. 20200205), and the Project of Education Commission of Guangdong Province of China (2021KQNCX080). L.C. acknowledges support from the Natural Science Foundation of Top Talent of SZTU (grant no. 20200206). S.L. acknowledges support from the Education Department of Guangdong Province (2021KCXTD045). We thank the Analysis and Testing Center of Shenzhen Technology University for their support on AFM and SEM measurements.

Conflicts of Interest: The authors declare no conflict of interest.

References

1. Xu, X.; Li, Y.; Peng, Q. Ternary Blend Organic Solar Cells: Understanding the Morphology from Recent Progress. *Adv. Mater.* **2021**, e2107476. [CrossRef] [PubMed]
2. Kan, B.; Ershad, F.; Rao, Z.; Yu, C. Flexible organic solar cells for biomedical devices. *Nano Res.* **2021**, *14*, 2891–2903. [CrossRef]
3. Liu, Y.; Liu, B.; Ma, C.-Q.; Huang, F.; Feng, G.; Chen, H.; Hou, J.; Yan, L.; Wei, Q.; Luo, Q.; et al. Recent progress in organic solar cells (Part I material science). *Sci. China Chem.* **2021**, *65*, 224–268. [CrossRef]
4. Song, Y.; Zhang, K.; Dong, S.; Xia, R.; Huang, F.; Cao, Y. Semitransparent Organic Solar Cells Enabled by a Sequentially Deposited Bilayer Structure. *ACS Appl. Mater. Interfaces* **2020**, *12*, 18473–18481. [CrossRef]
5. Yang, Y.; Feng, E.; Li, H.; Shen, Z.; Liu, W.; Guo, J.; Luo, Q.; Zhang, J.; Lu, G.; Ma, C.; et al. Layer-by-layer slot-die coated high-efficiency organic solar cells processed using twin boiling point solvents under ambient condition. *Nano Res.* **2021**, *14*, 4236–4242. [CrossRef]
6. Heydari Gharahcheshmeh, M.; Tavakoli, M.M.; Gleason, E.F.; Robinson, M.T.; Kong, J.; Gleason, K.K. Tuning, optimization, and perovskite solar cell device integration of ultrathin poly(3,4-ethylene dioxythiophene) films via a single-step all-dry process. *Sci. Adv.* **2019**, *5*, eaay0414. [CrossRef]
7. Wang, X.; Ishwara, T.; Gong, W.; Campoy-Quiles, M.; Nelson, J.; Bradley, D.D.C. High-Performance Metal-Free Solar Cells Using Stamp Transfer Printed Vapor Phase Polymerized Poly(3,4-Ethylenedioxythiophene) Top Anodes. *Adv. Funct. Mater.* **2012**, *22*, 1454–1460. [CrossRef]
8. Ma, R.; Li, G.; Li, D.; Liu, T.; Luo, Z.; Zhang, G.; Zhang, M.; Wang, Z.; Luo, S.; Yang, T.; et al. Understanding the Effect of End Group Halogenation in Tuning Miscibility and Morphology of High-Performance Small Molecular Acceptors. *Sol. RRL* **2020**, *4*, 2000250. [CrossRef]
9. Lin, Y.; Li, Y.; Zhan, X. Small molecule semiconductors for high-efficiency organic photovoltaics. *Chem. Soc. Rev.* **2012**, *41*, 4245–4272. [CrossRef]
10. Wu, J.; Li, G.; Fang, J.; Guo, X.; Zhu, L.; Guo, B.; Wang, Y.; Zhang, G.; Arunagiri, L.; Liu, F.; et al. Random terpolymer based on thiophene-thiazolothiazole unit enabling efficient non-fullerene organic solar cells. *Nat. Commun.* **2020**, *11*, 4612. [CrossRef]
11. Zhang, G.; Zhao, J.; Chow, P.C.Y.; Jiang, K.; Zhang, J.; Zhu, Z.; Zhang, J.; Huang, F.; Yan, H. Nonfullerene Acceptor Molecules for Bulk Heterojunction Organic Solar Cells. *Chem. Rev.* **2018**, *118*, 3447–3507. [CrossRef] [PubMed]
12. Ma, R.; Yu, J.; Liu, T.; Zhang, G.; Xiao, Y.; Luo, Z.; Chai, G.; Chen, Y.; Fan, Q.; Su, W.; et al. All-polymer solar cells with over 16% efficiency and enhanced stability enabled by compatible solvent and polymer additives. *Aggregate* **2021**, *3*, e58. [CrossRef]
13. Ma, R.; Yang, T.; Xiao, Y.; Liu, T.; Zhang, G.; Luo, Z.; Li, G.; Lu, X.; Yan, H.; Tang, B. Air-Processed Efficient Organic Solar Cells from Aromatic Hydrocarbon Solvent without Solvent Additive or Post-Treatment: Insights into Solvent Effect on Morphology. *Energy Environ. Mater.* **2021**, *5*, 977–985. [CrossRef]
14. Luo, Z.; Liu, T.; Ma, R.; Xiao, Y.; Zhan, L.; Zhang, G.; Sun, H.; Ni, F.; Chai, G.; Wang, J.; et al. Precisely Controlling the Position of Bromine on the End Group Enables Well-Regular Polymer Acceptors for All-Polymer Solar Cells with Efficiencies over 15. *Adv. Mater.* **2020**, *32*, e2005942. [CrossRef]
15. Li, D.; Song, L.; Fang, H.; Teng, Y.; Cui, H.; Li, Y.-Y.; Liu, R.; Niu, Q. Optimization of Biomethane Production in Mono-Cardboard Digestion: Key Parameters Influence, Batch Test Kinetic Evaluation, and DOM Indicators Variation. *Energy Fuels* **2019**, *33*, 4340–4351. [CrossRef]
16. Lee, W.H.; Liu, B.T.; Lee, R.H. Difluorobenzothiadiazole based two-dimensional conjugated polymers with triphenylamine substituted moieties as pendants for bulk heterojunction solar cells. *Express Polym. Lett.* **2017**, *11*, 910–923. [CrossRef]
17. Shiau, S.-Y.; Chang, C.-H.; Chen, W.-J.; Wang, H.-J.; Jeng, R.-J.; Lee, R.-H. Star-shaped organic semiconductors with planar triazine core and diketopyrrolopyrrole branches for solution-processed small-molecule organic solar cells. *Dye. Pigment.* **2015**, *115*, 35–49. [CrossRef]

18. Luo, Z.; Ma, R.; Yu, J.; Liu, H.; Liu, T.; Ni, F.; Hu, J.; Zou, Y.; Zeng, A.; Su, C.J.; et al. Heteroheptacene-based acceptors with thieno[3,2-b]pyrrole yield high-performance polymer solar cells. *Natl. Sci. Rev.* **2022**, *9*, nwac076. [CrossRef]
19. Ans, M.; Ayub, K.; Bhatti, I.A.; Iqbal, J. Designing indacenodithiophene based non-fullerene acceptors with a donor-acceptor combined bridge for organic solar cells. *RSC Adv.* **2019**, *9*, 3605–3617. [CrossRef]
20. Fan, Q.; Su, W.; Chen, S.; Kim, W.; Chen, X.; Lee, B.; Liu, T.; Méndez-Romero, U.A.; Ma, R.; Yang, T.; et al. Mechanically Robust All-Polymer Solar Cells from Narrow Band Gap Acceptors with Hetero-Bridging Atoms. *Joule* **2020**, *4*, 658–672. [CrossRef]
21. Sun, H.; Guo, X.; Facchetti, A. High-Performance n-Type Polymer Semiconductors: Applications, Recent Development, and Challenges. *Chem* **2020**, *6*, 1310–1326. [CrossRef]
22. Halls, J.J.M.; Walsh, C.A.; Greenham, N.C.; Marseglia, E.A.; Friend, R.H.; Moratti, S.C.; Holmes, A.B. Efficient photodiodes from interpenetrating polymer networks. *Nature* **1995**, *376*, 498–500. [CrossRef]
23. Wang, W.; Wu, Q.; Sun, R.; Guo, J.; Wu, Y.; Shi, M.; Yang, W.; Li, H.; Min, J. Controlling Molecular Mass of Low-Band-Gap Polymer Acceptors for High-Performance All-Polymer Solar Cells. *Joule* **2020**, *4*, 1070–1086. [CrossRef]
24. Du, J.; Hu, K.; Meng, L.; Angunawela, I.; Zhang, J.; Qin, S.; Liebman-Pelaez, A.; Zhu, C.; Zhang, Z.; Ade, H.; et al. High-Performance All-Polymer Solar Cells: Synthesis of Polymer Acceptor by a Random Ternary Copolymerization Strategy. *Angew. Chem. Int. Ed. Engl.* **2020**, *59*, 15181–15185. [CrossRef] [PubMed]
25. Shi, S.; Chen, P.; Chen, Y.; Feng, K.; Liu, B.; Chen, J.; Liao, Q.; Tu, B.; Luo, J.; Su, M.; et al. A Narrow-Bandgap n-Type Polymer Semiconductor Enabling Efficient All Polymer Solar Cells. *Adv. Mater.* **2019**, *31*, e1905161. [CrossRef]
26. Zhao, R.; Wang, N.; Yu, Y.; Liu, J. Organoboron Polymer for 10% Efficiency All-Polymer Solar Cells. *Chem. Mater.* **2020**, *32*, 1308–1314. [CrossRef]
27. Sun, H.; Tang, Y.; Koh, C.W.; Ling, S.; Wang, R.; Yang, K.; Yu, J.; Shi, Y.; Wang, Y.; Woo, H.Y.; et al. High-Performance All-Polymer Solar Cells Enabled by an n-Type Polymer Based on a Fluorinated Imide-Functionalized Arene. *Adv. Mater.* **2019**, *31*, e1807220. [CrossRef]
28. Shi, Y.; Guo, H.; Huang, J.; Zhang, X.; Wu, Z.; Yang, K.; Zhang, Y.; Feng, K.; Woo, H.Y.; Ortiz, R.P.; et al. Distannylated Bithiophene Imide: Enabling High-Performance n-Type Polymer Semiconductors with an Acceptor-Acceptor Backbone. *Angew. Chem. Int. Ed. Engl.* **2020**, *59*, 14449–14457. [CrossRef]
29. Wang, Y.; Yan, Z.; Uddin, M.A.; Zhou, X.; Yang, K.; Tang, Y.; Liu, B.; Shi, Y.; Sun, H.; Deng, A.; et al. Triimide-Functionalized n-Type Polymer Semiconductors Enabling All-Polymer Solar Cells with Power Conversion Efficiencies Approaching 9%. *Sol. RRL* **2019**, *3*, 1900107. [CrossRef]
30. Kolhe, N.B.; Tran, D.K.; Lee, H.; Kuzuhara, D.; Yoshimoto, N.; Koganezawa, T.; Jenekhe, S.A. New Random Copolymer Acceptors Enable Additive-Free Processing of 10.1% Efficient All-Polymer Solar Cells with Near-Unity Internal Quantum Efficiency. *ACS Energy Lett.* **2019**, *4*, 1162–1170. [CrossRef]
31. Kolhe, N.B.; Lee, H.; Kuzuhara, D.; Yoshimoto, N.; Koganezawa, T.; Jenekhe, S.A. All-Polymer Solar Cells with 9.4% Efficiency from Naphthalene Diimide-Biselenophene Copolymer Acceptor. *Chem. Mater.* **2018**, *30*, 6540–6548. [CrossRef]
32. Fan, Q.; Ma, R.; Liu, T.; Yu, J.; Xiao, Y.; Su, W.; Cai, G.; Li, Y.; Peng, W.; Guo, T.; et al. High-performance all-polymer solar cells enabled by a novel low bandgap non-fully conjugated polymer acceptor. *Sci. China Chem.* **2021**, *64*, 1380–1388. [CrossRef]
33. Tang, A.; Li, J.; Zhang, B.; Peng, J.; Zhou, E. Low-Bandgap n-Type Polymer Based on a Fused-DAD-Type Heptacyclic Ring for All-Polymer Solar Cell Application with a Power Conversion Efficiency of 10.7. *ACS Macro Lett.* **2020**, *9*, 706–712. [CrossRef] [PubMed]
34. Lee, J.-W.; Choi, N.; Kim, D.; Phan, T.N.-L.; Kang, H.; Kim, T.-S.; Kim, B.J. Side Chain Engineered Naphthalene Diimide-Based Terpolymer for Efficient and Mechanically Robust All-Polymer Solar Cells. *Chem. Mater.* **2021**, *33*, 1070–1081. [CrossRef]
35. Su, N.; Ma, R.; Li, G.; Liu, T.; Feng, L.-W.; Lin, C.; Chen, J.; Song, J.; Xiao, Y.; Qu, J.; et al. High-Efficiency All-Polymer Solar Cells with Poly-Small-Molecule Acceptors Having π-Extended Units with Broad Near-IR Absorption. *ACS Energy Lett.* **2021**, *6*, 728–738. [CrossRef]
36. Fu, H.; Li, Y.; Yu, J.; Wu, Z.; Fan, Q.; Lin, F.; Woo, H.Y.; Gao, F.; Zhu, Z.; Jen, A.K. High Efficiency (15.8%) All-Polymer Solar Cells Enabled by a Regioregular Narrow Bandgap Polymer Acceptor. *J. Am. Chem. Soc.* **2021**, *143*, 2665–2670. [CrossRef]
37. Yu, H.; Qi, Z.; Yu, J.; Xiao, Y.; Sun, R.; Luo, Z.; Cheung, A.M.H.; Zhang, J.; Sun, H.; Zhou, W.; et al. Fluorinated End Group Enables High-Performance All-Polymer Solar Cells with Near-Infrared Absorption and Enhanced Device Efficiency over 14%. *Adv. Energy Mater.* **2020**, *11*, 2003171. [CrossRef]
38. Zhao, C.; Huang, H.; Wang, L.; Zhang, G.; Lu, G.; Yu, H.; Lu, G.; Han, Y.; Qiu, M.; Li, S.; et al. Efficient All-Polymer Solar Cells with Sequentially Processed Active Layers. *Polymer* **2022**, *14*, 2058. [CrossRef]
39. Yao, J.; Qiu, B.; Zhang, Z.G.; Xue, L.; Wang, R.; Zhang, C.; Chen, S.; Zhou, Q.; Sun, C.; Yang, C.; et al. Cathode engineering with perylene-diimide interlayer enabling over 17% efficiency single-junction organic solar cells. *Nat. Commun.* **2020**, *11*, 2726. [CrossRef]
40. Wu, Z.; Sun, C.; Dong, S.; Jiang, X.F.; Wu, S.; Wu, H.; Yip, H.L.; Huang, F.; Cao, Y. n-Type Water/Alcohol-Soluble Naphthalene Diimide-Based Conjugated Polymers for High-Performance Polymer Solar Cells. *J. Am. Chem. Soc.* **2016**, *138*, 2004–2013. [CrossRef]
41. Zhang, W.; Sun, C.; Angunawela, I.; Meng, L.; Qin, S.; Zhou, L.; Li, S.; Zhuo, H.; Yang, G.; Zhang, Z.G.; et al. 16.52% Efficiency All-Polymer Solar Cells with High Tolerance of the Photoactive Layer Thickness. *Adv. Mater.* **2022**, *34*, e2108749. [CrossRef] [PubMed]

42. Yu, H.; Pan, M.; Sun, R.; Agunawela, I.; Zhang, J.; Li, Y.; Qi, Z.; Han, H.; Zou, X.; Zhou, W.; et al. Regio-Regular Polymer Acceptors Enabled by Determined Fluorination on End Groups for All-Polymer Solar Cells with 15.2% Efficiency. *Angew. Chem. Int. Ed. Engl.* **2021**, *60*, 10137–10146. [CrossRef] [PubMed]
43. Kirchartz, T.; Deledalle, F.; Tuladhar, P.S.; Durrant, J.R.; Nelson, J. On the Differences between Dark and Light Ideality Factor in Polymer:Fullerene Solar Cells. *J. Phys. Chem. Lett.* **2013**, *4*, 2371–2376. [CrossRef]

Review

Defect Passivation Scheme toward High-Performance Halide Perovskite Solar Cells

Bin Du [1,*], Kun He [1], Xiaoliang Zhao [1] and Bixin Li [2,3,*]

[1] School of Materials Science and Engineering, Xi'an Polytechnic University, Xi'an 710048, China
[2] School of Physics and Chemistry, Hunan First Normal University, Changsha 410205, China
[3] Shaanxi Institute of Flexible Electronics (SIFE), Northwestern Polytechnical University (NPU), Xi'an 710072, China
* Correspondence: dubin@xpu.edu.cn (B.D.); jkylbxin@hnfnu.edu.cn (B.L.)

Abstract: Organic-inorganic halide perovskite solar cells (PSCs) have attracted much attention in recent years due to their simple manufacturing process, low cost, and high efficiency. So far, all efficient organic-inorganic halide PSCs are mainly made of polycrystalline perovskite films. There are transmission barriers and high-density defects on the surface, interface, and grain boundary of the films. Among them, the deep-level traps caused by specific charged defects are the main non-radiative recombination centers, which is the most important factor in limiting the photoelectric conversion efficiency of PSCs devices to the Shockley-Queisser (S-Q) theoretical efficiency limit. Therefore, it is imperative to select appropriate passivation materials and passivation strategies to effectively eliminate defects in perovskite films to improve their photovoltaic performance and stability. There are various passivation strategies for different components of PSCs, including interface engineering, additive engineering, antisolvent engineering, dopant engineering, etc. In this review, we summarize a large number of defect passivation work to illustrate the latest progress of different types of passivators in regulating the morphology, grain boundary, grain size, charge recombination, and defect density of states of perovskite films. In addition, we discuss the inherent defects of key materials in carrier transporting layers and the corresponding passivation strategies to further optimize PSCs components. Finally, some perspectives on the opportunities and challenges of PSCs in future development are highlighted.

Keywords: perovskite solar cell; defect passivation; interface engineering; surface treatment; dopant passivation

Citation: Du, B.; He, K.; Zhao, X.; Li, B. Defect Passivation Scheme toward High-Performance Halide Perovskite Solar Cells. *Polymers* **2023**, *15*, 2010. https://doi.org/10.3390/polym15092010

Academic Editor: Rong-Ho Lee

Received: 29 March 2023
Revised: 20 April 2023
Accepted: 20 April 2023
Published: 24 April 2023

Copyright: © 2023 by the authors. Licensee MDPI, Basel, Switzerland. This article is an open access article distributed under the terms and conditions of the Creative Commons Attribution (CC BY) license (https:// creativecommons.org/licenses/by/ 4.0/).

1. Introduction

Traditional energy, which was mainly used for power generation in the past, is facing the problems of resource shortage and ecological impact. Therefore, green and renewable new energy has attracted much attention in recent years. Thereinto, the photovoltaic (PV) technology that converts light energy into electrical energy is particularly impressive due to the advantages of simple preparation of collection equipment (solar cells) and easy mass production [1–3]. Solar cells have been developed for three generations since the birth of Bell Labs in 1954. Although silicon-based solar cells (the first generation of solar cells) and thin film compound solar cells (the second generation of solar cells) have reached more than 23% power conversion efficiency (PCE), various factors limit their further development, such as the need for high-temperature processing, high-purity material requirements, and high manufacturing costs of photovoltaic module systems. In particular, the highly toxic raw materials required for thin film compound solar cells are not conducive to industrial mass production [4–7]. Recently developed quantum dots and dye-sensitized solar cells (third-generation solar cells) also have the disadvantages of a slow development process, a complex preparation process, and low PCE, which has led to the search for new photovoltaic materials to improve the most advanced photovoltaic technology to further develop solar cells.

Perovskite solar cells (PSCs) are the latest third-generation solar cells. After just over a decade of development, the highest PCE has soared to 25.8% [8]. The light-absorbing layer material of high-efficiency and stable PSCs is mainly composed of perovskite with ABX_3 structure. The A-site is a monovalent organic or inorganic cation with a larger radius, including MA^+ (methylamine ion, $CH_3NH_3^+$), FA^+ (formamidine ion, $HC(NH_2)_2^+$), Cs^+, etc., the B-site is a divalent metal cation, usually Pb^{2+}, Sn^{2+}, etc., and the X-site is a halogen/pseudohalogen anion, such as I^-, Cl^-, SCN^- or their mixture [9]. The A-position cation will be surrounded by BX_6 octahedral with 12-fold coordination symmetry in the ideal perovskite structure, which makes the perovskite possess an excellent absorbance coefficient (~10^5 cm^{-1}) and long carrier diffusion length (>1 μm), which greatly reduces the processing cost and material cost of perovskite photovoltaic devices, making commercialization possible [10,11]. At the same time, by adjusting the A-site cation composition in the perovskite crystal, the perovskite has a band gap adjustability (1.2–3.0 eV) beyond the reach of other photovoltaic devices, which also makes the Shockley-Queisser (S-Q) theoretical efficiency limit of perovskite/silicon tandem solar cells and perovskite/perovskite tandem solar cells exceed 30%, far more than that of single-junction solar cells [12,13]. In 2009, a pioneering work by Miyasaka et al. reported a 3.8% PCE PSCs based on liquid electrolyte using the halide perovskite material $MAPbI_3$ [14]. In 2012, Park et al. first used a solid-state semiconductor 2,2′,7,7′-tetrakis[N,N-di(4-methoxyphenyl)amine]-9,9′-spirobifluorene (spiro-OMeTAD) as a hole transport layer material to successfully prepare the first all-solid-state PSCs with a PCE of 9.7% and no encapsulation [15]. Since then, PSCs based on organic-inorganic halide perovskite absorbing layers have been getting more and more attention from researchers, and the certified PCE is increasing year by year [16–18]. In 2021, Seok et al. obtained a PCE of 25.5% by interfacial modulation, which was a world record that year [19]. In 2022, You et al. eventually obtained a 25.6% champion-certified PCE by introducing the additive RbCl to stabilize the perovskite phase [20].

Planar PSCs devices typically consist of transparent conductive oxide (TCO) electrodes (indium tin oxide (ITO) or fluorine-doped tin oxide (FTO)), electron transport layer (ETL), perovskite light-absorbing layer, hole transport layer (HTL), and metal electrodes (Au, Ag, Cu). Planar PSCs can be divided into n-i-p planar structures and p-i-n planar structures according to the different locations of carrier transport layer distribution, where TCO electrodes/ETL/perovskite absorbing layer/HTL/metal electrodes device structure is n-i-p planar structures, TCO electrodes/HTL/perovskite absorbing layer/ETL/metal electrodes device structure is p-i-n planar structures. Although these two PSCs have different structures, they both convert light energy into electrical energy by using the photovoltaic effect of the semiconductor PN junction. In PSCs, the perovskite light-absorbing layer (intrinsic semiconductor) forms a PN junction with adjacent ETL (n-type semiconductor) and HTL (p-type semiconductor). The PN junction can absorb a large number of photons to generate excitons under the irradiation of sunlight, and the perovskite with high light absorption coefficient can absorb a large number of photons to generate excitons. The exciton binding energy is low, and it is easy to dissociate into electrons and holes. These carriers (electrons and holes) will move under the action of the PN junction and reach the two ends of the electrode through different transport layers to form a photovoltage. Once the electrode is connected, it will form a current loop to convert light energy into electrical energy [21].

The ideal crystal structure is in a state where each atom is in its corresponding position, but the actual crystal structure will be more or less affected by the defects introduced during crystal growth and the post-treatment process. The spin-coating preparation process and post-annealing process of PSCs devices will form various defects on the surface or grain boundaries of polycrystalline perovskite crystals, including (1) insufficiently coordinated anions or cations; (2) inherent point defects caused by the preparation process, such as halide vacancies, cation vacancies, and Pb-I antisites; (3) ion migration at grain boundaries; and (4) exogenous impurities [22]. These defects with positive or negative charges will introduce transition energy levels in the band gap. When the transition energy level is located in one-third of the band gap, deep-level defects will be formed, which is the source of the notorious trap-assisted recombination (also known as Shockley–Reid–Hall recombination; SRH). SRH recombination is not conducive to the extraction and migration of carriers in perovskite films and affects the lifetime of carriers, which has been identified as an important reason for limiting open circuit voltage (V_{OC}), short circuit current density (J_{SC}), and fill factor (FF) in PSCs [23,24]. In addition, there are more defects in the carrier transport interfaces, so it is necessary to passivate the defects of each component of the PSCs to improve the efficiency and stability of the device (Figure 1).

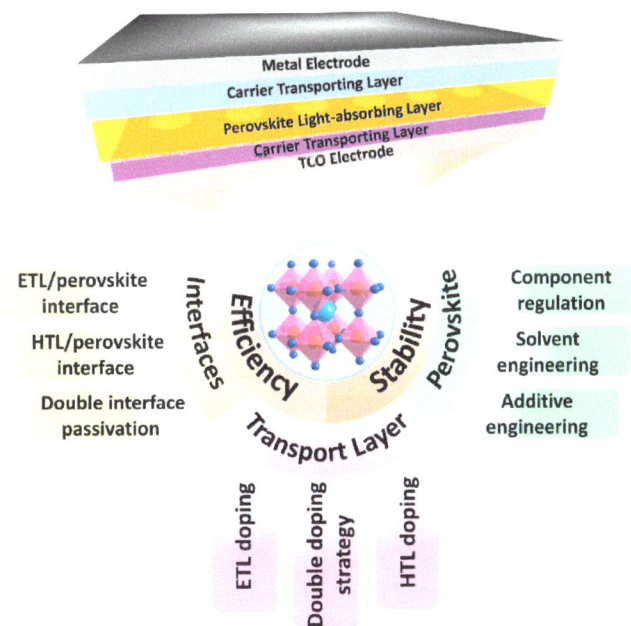

Figure 1. Planar PSCs device structures and passivation strategies based on different components of PSCs to improve device efficiency and stability.

This review focuses on the progress and challenges of a series of passivation strategies that can effectively eliminate the main defects in PSC devices. Based on the passivation strategies for different components of PSCs, we highlight the latest research on interface engineering, perovskite surface treatment, and dopant passivation, including ETL (HTL)/perovskite interface passivation, the solvent component, precursor additive engineering, anti-solvent engineering, and ETL (HTL) doping. We summarized the important role of different functional groups in the defect passivation process. Finally, some perspectives on the opportunities and challenges of PSCs in future development are put forward.

2. Interface Engineering

It can be seen from the working principle of PSCs that the generation, transmission, and extraction of free carriers will go through multiple interfaces, including TCO electrode/ETL interface, ETL/perovskite interface, perovskite/HTL interface, and HTL/metal electrode interface. However, there will be a severe interface recombination process at the ETL (HTL)/perovskite interface. On the one hand, the recombination is caused by the ineffective matching of energy levels between interfaces. On the other hand, the interface defects from the carrier transport layer and the perovskite layer itself are also the main recombination centers [25,26]. The interface engineering can simply and effectively adjust the interface energy level mismatch to overcome the interface loss and also optimize the interface morphology for carrier transport [19]. In view of this, the passivation strategy of improving V_{OC} and FF by using interface engineering to minimize interface recombination has been widely studied in recent years.

2.1. ETL/Perovskite Interface Passivation

ETL in PSCs is generally an n-type semiconductor material, which is used to extract the electrons formed by dissociation in the perovskite layer and transport them to the TCO electrode (n-i-p planar structures) or the metal electrode (p-i-n planar structures) while blocking the holes in the perovskite layer to avoid carrier recombination [27]. It has been reported that the ETL/perovskite interface contains a large number of deep defects limiting the efficiency and stability of PSCs devices, which is about 100 times greater than that of the perovskite layer defects [28]. Meanwhile, the energy level matching between ETL and perovskite layer is crucial to improve the carrier extraction and collection efficiency and the V_{OC} of the device. TiO_2 is the earliest ETL material used in various n-i-p planar structures PSCs. Loo et al. found that the β-$CsPbI_3$ lattice near the TiO_2/perovskite interface will undergo polymorphic transformation under illumination, which is due to the strain at the TiO_2/$CsPbI_3$ interface [29]. In order to suppress this lattice distortion, three alkyltrimethoxysilane derivatives (C0 = methyltrimethoxysilane, C2 = propyltrimethoxysilane, and C3 = butyltrimethoxysilane) with different alkyl chain lengths were inserted between the TiO_2/$CsPbI_3$ interface as strain release layers (SRL). The results of depth-resolved grazing-incidence wide-angle X-ray scattering (GIWAXS) show that the C3 with the longest alkyl chain provides a more flexible interface, which can effectively reduce the thermal expansion mismatch between the TiO_2/$CsPbI_3$ interface, reduce the interface stress and improve the phase stability and device stability of the interface (Figure 2a,b). Compared with the original TiO_2/$CsPbI_3$ device, the PCE of TiO_2/C3/$CsPbI_3$ device increased from 15.7% to 20.1%. Wu et al. introduced cystamine dihydrochloride (CMDR) with double amino groups between dense TiO_2 and perovskite layer to modify the interface [30]. The diamino group of CMDR can not only form Ti-N bond with TiO_2 but also form hydrogen bond with I^- in perovskite, which effectively inhibits the generation of excessive metal reduction of lead (Pb^0) defects in PbI_2. This bilateral synergistic passivation strategy effectively reduces the contact resistance, trap state density and non-radiative recombination of carriers between the interfaces of perovskite films and finally increases the PCE from 18.41% to 20.63%.

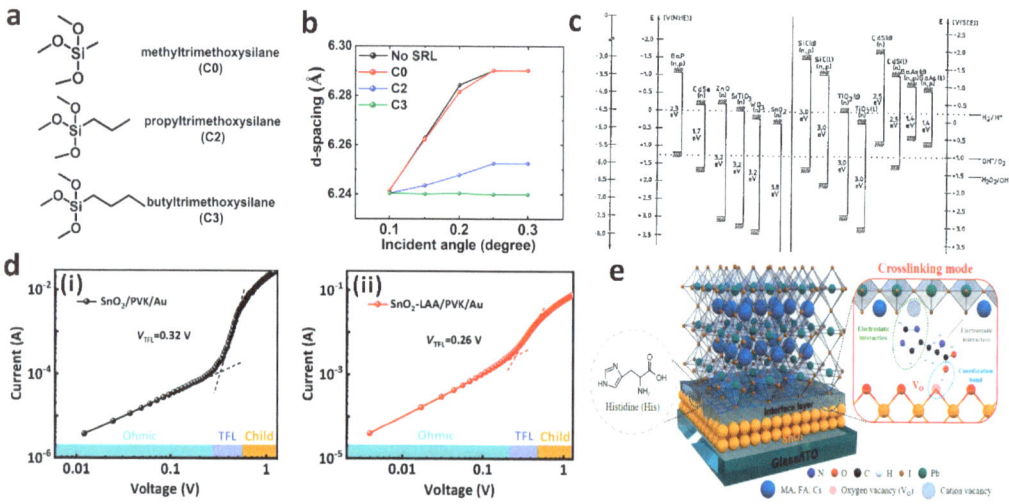

Figure 2. *Cont.*

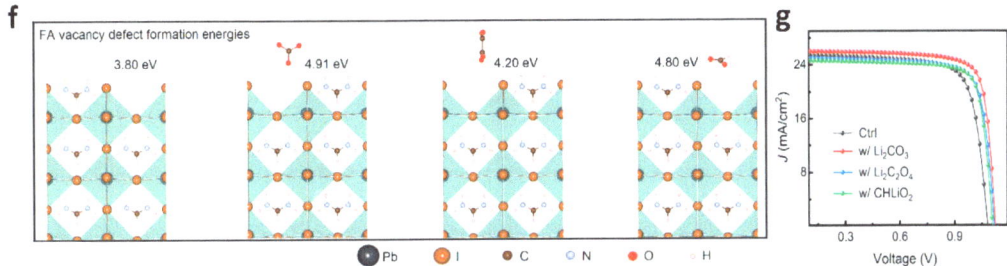

Figure 2. (**a**) Chemical structure of methoxysilane compound used as SRL. (**b**) GIWAXS was used to measure the (110) interplanar spacing of CsPbI$_3$ deposited on different SRLs at a critical angle of 0.24°. Reproduced with permission [29]. Copyright 2022, American Chemical Society. (**c**) The band position of the commonly used electron transport layer. Reproduced with permission [31]. Copyright 1996, American Chemical Society. (**d**) Dark I-V curves of devices with different structures: (i) FTO/SnO$_2$/PVK/Au, (ii) FTO/SnO$_2$-LAA/PVK/Au. Reproduced with permission [32]. Copyright 2022, Wiley-VCH. (**e**) A schematic diagram of the formation of His molecules between the SnO$_2$ layer and the perovskite layer. Reproduced with permission [33]. Copyright 2022, American Chemical Society. (**f**) Calculation of the energy (from left to right) of formation of FA vacancy defects before and after adsorption of CO$_3^{2-}$, C$_2$O$_4^{2-}$ and HCOO$^-$ ions on the FAPbI$_3$ (001) surface. (**g**) J-V curves of PSCs with and without lithium salt modification. Reproduced with permission [34]. Copyright 2022, American Chemical Society.

Compared with dense TiO$_2$, SnO$_2$ has higher electron mobility, wider band gap, lower photocatalytic activity and preparation temperature, and, most importantly, better matching with the energy level of perovskite. Therefore SnO$_2$ is one of the most ideal ETL materials for planar PSCs (Figure 2c) [31,35]. Nevertheless, the process of preparing the device will inevitably lead to the film having obvious pinholes or traps, which will lead to serious carrier recombination at the interface, resulting in a sharp decrease in device performance, so the post-processing of the SnO$_2$ layer is particularly important [36]. To solve the problem that oxygen vacancies (V$_O$) and hydroxyl defects on SnO$_2$ ETL damage perovskite films during the preparation of PSCs, Zhang et al. used multifunctional amino acid L-aspartic acid (LAA) to "match" the SnO$_2$/perovskite interface [32]. The -COOH in LAA can coordinate the mismatched Sn^{4+} in SnO$_2$, thereby reducing the V$_O$ defect of SnO$_2$ and also neutralizing the alkalinity of the hydroxyl group on the side of SnO$_2$. The amino group of LAA itself connects the perovskite layer through hydrogen bonds to improve the quality of the perovskite film. LAA forms a "channel" between SnO$_2$/perovskite through bilateral synergistic passivation, accelerates electron transfer at the interface, and reduces the density of trap states at the interface (Figure 2d). When the V$_{OC}$ is 1.15 V, the PCE of SnO$_2$/LAA-based PSCs is up to 22.73%, which is much higher than 20.02% of SnO$_2$-based PSCs. Similarly, Tao et al. also attempted to use the designed multifunctional histidine (His) as a cross-linking agent for the SnO$_2$/perovskite interface [33]. This close cross-linking of SnO$_2$ with perovskite facilitates the extraction and transfer of electrons, improves the quality of perovskite films, and reduces non-radiative recombination between interfaces (Figure 2e). Crosslinking agents can also effectively adjust the interface level and accelerate electron transfer. Finally, the PSC device based on His-modified SnO$_2$ produced a champion PCE of 22.91%, V$_{OC}$ of 1.17 V, J$_{SC}$ of 24.21 mA cm^{-2}, and FF of 80.9%.

In addition to these monomolecular layers that act as interface crosslinking agents, salt molecules containing both anions and cations show great potential in regulating the ETL/perovskite interface energy level, and if necessary, the ETL/perovskite interface layer is modified by customizing salt molecules of anions (cations) with specific passivation functions. By studying the effects of a series of lithium salt anions (CO$_3^{2-}$, C$_2$O$_4^{2-}$, and HCOO$^-$) on the SnO$_2$ layer, FAPbI$_3$ layer, and SnO$_2$/FAPbI$_3$ interface, Bi et al. developed a strategy for Li salt molecules to passivate the SnO$_2$/FAPbI$_3$ interface [34]. They found that when the C-O group and C=O group of these anions are in optimal configurations, they will coordinate with the uncoordinated Sn^{4+} and FA^+ on both sides of the interface to form stronger bonds. The results of density functional theory (DFT) calculations (Figure 2f) show that compared with $C_2O_4^{2-}$ and $HCOO^-$, the binding energy between CO_3^{2-} and FA^+ is the strongest, which increases the formation energy of V_{FA} defects and releases the residual stress of $FAPbI_3$ lattice. The modification of Li_2CO_3 greatly promoted the charge transfer between the $SnO_2/FAPbI_3$ interface, effectively reduced the carrier recombination between the interfaces, optimized the crystallinity of the perovskite film, and finally obtained $FAPbI_3$ PSCs with a PCE of 23.5% (Figure 2g). In addition to Li^+, guanidine cations (GA^+) are also widely used in $SnO_2/FAPbI_3$ interface defect passivation to prepare high-quality perovskite films. Zang et al. prepared different guanidinium salts (GASCN, GASO4, GAAc, and GACl) by using a series of anions (SCN^-, SO_4^{2-}, Ac^- and Cl^-) that have a positive effect on the surface defect passivation, interface energy level matching, and the crystallization of PbI_2 and perovskite layers [37]. The effects of different anions on the interfacial chemical interaction strength, trap density, film crystallinity, and energy level on the device performance were systematically studied. The results show that all interface passivators can effectively passivate the defects on the surface of SnO_2 and perovskite, adjust the interface band alignment, and promote the crystallization of perovskite. The passivation effect of anions from SCN^-, SO_4^{2-}, and Ac^- to Cl^- is getting better. Compared with 21.84% of the control device, the device champion PCE modified by GASCN, GA_2SO_4, GAAc, and GACl was 22.76%, 23.43%, 23.57%, and 23.74%, respectively. In addition, the thermal stability and environmental stability of the modified equipment were also improved. Sun et al. used biguanide hydrochloride (BGCl) as a multifunctional interfacial modifier to simultaneously optimize charge extraction and transport at the SnO_2/perovskite interface and promote the growth of perovskite crystals, confirming the synergistic passivation effect of BGCl (Figure 3a) [38]. Firstly, The N in BGCl is coordinated with the uncoordinated Sn^{4+} in SnO_2 by Lewis coupling in an alkaline environment. With the injection of electrons on the surface of SnO_2, the electron extraction and transport capacity at the SnO_2/perovskite interface is significantly enhanced. In addition, Cl^- ions in BGCl occupy the oxygen vacancy (V_O) through electrostatic coupling, thereby reducing the V_O density on SnO_2. Secondly, $-NH_2/-NH_3^+$ in BGCl can be anchored to I/I^- in PbI_2 by hydrogen bonding to achieve uniform perovskite crystal growth. After BGCl modification, the interface defects of SnO_2/perovskite are effectively passivated, and the quality of perovskite film is improved. Finally, 24.4% certified PCE is achieved. This work provides an effective method for the selection and design of interface modification molecules.

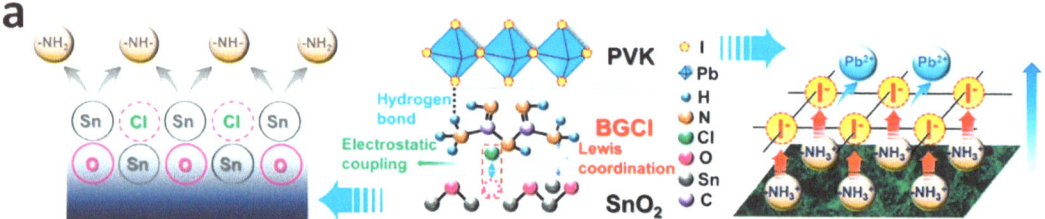

Figure 3. Cont.

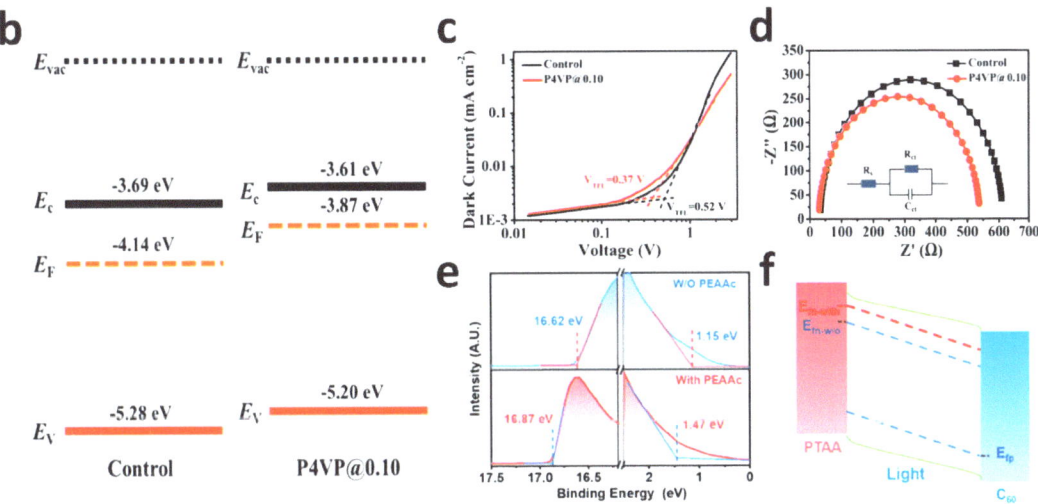

Figure 3. (a) A schematic diagram of the passivation mechanism of BGCl at the ETL/perovskite interface. Reproduced with permission [38]. Copyright 2022, Wiley-VCH. (b) Control and P4VP modified perovskite film energy level diagram. (c) Dark J-V curve of a pure electronic device with a device structure of ITO/SnO$_2$/MAPbI$_3$/P4VP/PCBM/Al. (d) In dark conditions, the EIS of the controlled and P4VP-modified device is measured at 640 mV, and the illustration shows the equivalent circuit for fitting the impedance spectrum data. Reproduced with permission [39]. Copyright 2020, The Royal Society of Chemistry. (e) UPS spectra of perovskite films with and without PEAAc treatment. (f) Schematic diagram of Fermi level splitting under illumination. Reproduced with permission [40]. Copyright 2022, American Chemical Society.

The interface recombination at the ETL/perovskite interface is also important in p-i-n PSCs. Liu et al. introduced poly-4-vinylpyridine (P4VP) as an intermediate film between the perovskite/[6,6]-phenyl-C61-butyric acid methyl ester (PCBM) interface to passivate defects existing on the surface and grain boundaries [39]. The results show that P4VP effectively adjusts the energy level matching of perovskite/PCBM, which is conducive to efficient charge extraction between interfaces and inhibits hole transfer. At the same time, the space-charge-limited current (SCLC) and electrochemical impedance spectroscopy (EIS) characterization results further confirmed that P4VP could effectively passivate surface defects (Figure 3b–d). Finally, the PCE of PSCs modified by P4VP increased from 17.46% to 20.02%. Based on the introduction of the ethanediamine dihydroiodide (EDAI$_2$) interface passivation layer, Ding et al. introduced the hexamethylene diisocyanate (HDI) interface layer to further treat the perovskite/PCBM interface [41]. Their results show that the interface recombination after EDAI$_2$/HDI passivation is significantly inhibited, and a very low non-radiative V_{OC} loss of 0.10 V is obtained. They verified that the isocyanate group in the HDI molecule could easily cross-link with the amine group in EDAI$_2$ even at room temperature, and the cross-linking molecule is formed on the surface of the perovskite, which helps to hinder the diffusion of EDA^{2+} cations into the perovskite, making EDAI$_2$/HDI passivated PSCs have excellent thermal stability. Xu et al. introduced an ultra-thin interface layer of phenylethylammonium acetate (PEAAc) at the wide band gap perovskite/C60 interface, which effectively alleviated the surface defects of perovskite coordination and greatly reduced the non-radiative recombination loss [40]. Through ultraviolet photoelectron spectroscopy (UPS), they found that the Fermi level of perovskite moved upward after PEAAc treatment, indicating that there were more n-type perovskite crystal planes on the surface of perovskite, which was conducive to electron extraction and hole blocking at the perovskite/C60 interface. More importantly, the higher Fermi level of the PEAAc-treated perovskite film can lead to a larger splitting between the electron quasi-Fermi level and the hole quasi-Fermi level under illumination, which contributes to V_{OC} enhancement (Figure 3e,f). Therefore, the PEAAc-modified wide-bandgap (1.68 eV) PSCs device achieved a champion PCE of 20.66% with a high V_{OC} of 1.25 V, which is one of the highest V_{OC} values reported in wide-bandgap perovskite devices in recent years.

2.2. HTL/Perovskite Interface Passivation

HTL is usually a p-type semiconductor material. Except for blocking electrons and improving hole mobility, the most important role of HTL is to improve the stability of PSC devices. It has been reported that the presence of HTL in PSCs can increase stability by 90% [42]. The HTL/perovskite interface passivation strategy is also as important as ETL/perovskite interface passivation. For n-i-p planar structures PSCs, the first thing to consider is that the interface passivation material is directly deposited on the perovskite film, so the solvent for dissolving the interface passivation material must be an inert solvent that cannot destroy the perovskite film. The appropriate interface passivator must have the ability to passivate the surface defects of the perovskite film, increase the hole transfer rate and match the energy level between the HTL and the perovskite layer [43–45]. Halide anions or pseudohalide anions have been shown to react chemically with anion vacancies or cation defects on the surface of perovskite films through ionic bonds or hydrogen bonds, thereby increasing the crystallinity of perovskite films. Kong et al. introduced a tetrabutylammonium chloride (TBAC) monolayer at the perovskite/spiro-OMeTAD interface by a simple solution method [46]. When TBAC is deposited on the perovskite film, the Cl^- in TBAC will enter the perovskite lattice by occupying the I^- vacancy in the film or acting as a gap, which makes TBAC have a strong interface dipole to promote the built-in electric field and reduces the contact barrier for hole extraction. They studied the self-assembly behavior of the TBAC interface layer on the perovskite surface by Kelvin probe force microscopy (KPFM) and capacitance-voltage (C-V) tests and found that the built-in electric field induced by the TBAC dipole layer was significantly enhanced, which was further confirmed by EIS results (Figure 4a–c). Finally, the n-i-p PSCs based on the ITO/SnO2/perovskite/TBAC/spiro-OMeTAD/Au structure achieved a 23.5% champion PCE. Song et al. used phenethylammonium fluoride (PEAF) deposited on perovskite films by rapid thermal evaporation as an interfacial passivator for PSCs [47]. Fluoride anions in PEAF have a small ionic radius and high Lewis basicity, which can form strong hydrogen bonds with organic cations in perovskite N-H—F and uncoordinated Pb^{2+} to form strong ionic bonds. The quality of the perovskite film modified by PEAF is significantly improved. At the same time, the lifetime of the treated perovskite film is significantly longer (153 ns) than that of the original perovskite film (20 ns), which indicates that PEAF effectively passivates perovskite defects and reduces non-radiative recombination. Finally, the PSCs treated with PEAF had a high PCE of 23.2%, and the stability was significantly enhanced. Inspired by this work, Pan et al. innovatively synthesized hexadecyltrimethylammonium hexafluorophosphate (HTAP) by using pseudo-halide anion PF_6^- and hexadecyltrimethylammonium cation (HTA^+) and coated it on the top of perovskite layer to achieve a terminal sealing strategy [48]. This strategy can provide a good "channel" for hole extraction and provide a defect passivation layer for enhancing V_{OC} and FF. PF_6^- can fill the halide anion vacancies on the perovskite film and anchor the uncoordinated Pb^{2+}, helping to improve the crystallization and morphology of the perovskite film. In addition, HTA with an ultra-long chain can prevent the erosion of water molecules and effectively enhance the resistance of the device to environmental erosion. They showed the macroscopic color changes of the devices after modification under light immersion. The results revealed that the control devices were seriously discolored and decomposed while the modified devices did not change significantly, indicating that HTAP sealing can effectively improve the optical stability of PSCs. The optimal device modified by HTAP obtained 23.14% of the champion PCE, and the lead leakage was effectively alleviated. Similar to PF_6^-, Li et al. studied the role of pseudohalide anion SCN^- in the passivation of perovskite/HTL interface to obtain high-quality perovskite films. They synthesized a new bifunctional material acetamidine thiocyanate (AASCN), which showed the synergistic passivation effect of polar cations and pseudohalide anions [49]. After AASCN enters the perovskite film, the polar AA⁺ containing four rotation-restricted C-N bonds can improve the stability of the perovskite, and the N-H bond of AA⁺ can effectively passivate the film by forming hydrogen bonds N-H—I with iodine vacancies in the perovskite. SCN⁻, with a small size and high structural freedom, can interact with Pb-I octahedron, and then through the Ostwald ripening process, the crystallization of perovskite films can be improved during the secondary crystal growth process to obtain higher quality perovskite films. Therefore, the PCE of $FA_{0.25}MA_{0.75}PbI_3$ PSCs increased from 21.43% to 23.17%, and the V_{OC} increased from 1.095 V to 1.167 V.

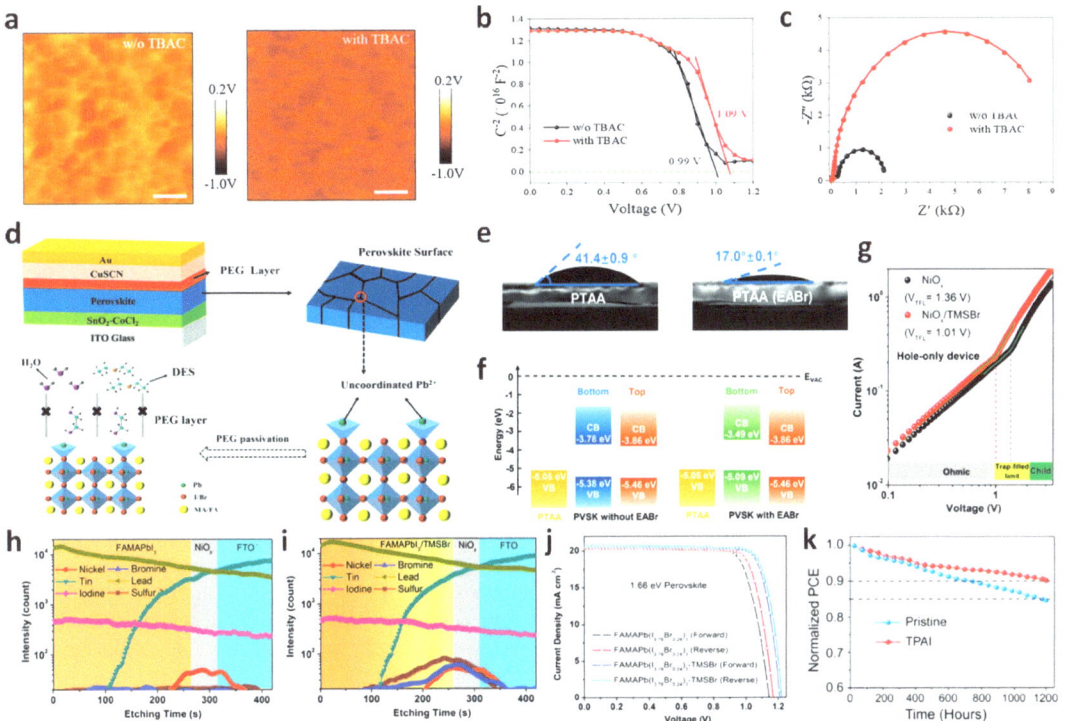

Figure 4. (a) Control (left) and TBAC-modified (right) KPFM images of perovskite films. (b) C-V characterization of the Mott-Schottky diagram and (c) EIS spectrum Nyquist diagram of the control device and the TBAC-PSCs device. Reproduced with permission [46]. Copyright 2022, Elsevier. (d) The schematic diagram of CuSCN-based PSCs device, the chemical structure of PEG and the coordination diagram of PEG and perovskite. Reproduced with permission [50]. Copyright 2022, Wiley-VCH. (e) Contact angles of perovskite precursor solution droplets on original PTAA (left) and EABr modified PTAA (right). (f) Energy level diagrams of PTAA and two perovskite films. Reproduced with permission [51]. Copyright 2022, Wiley-VCH. (g) Dark C-V curves of pure hole devices with and without TMSBr treatment, from left to right, are the low bias voltage region (Ohmic region), trap filling state region, and SCLC state region. ToF-SIMS curves of $FAMAPbI_3$ perovskite films on different substrates: (h) FTO/NiO$_x$ and (i) FTO/NiO$_x$/TMSBr. (j) J-V curves of $FAMAPb(I_{0.76}Br_{0.24})_3$ PSCs before and after TMSBr modification. Reproduced with permission [52]. Copyright 2022, The Royal Society of Chemistry. (k) The operating stability curves of the original and TPAI-modified unpackaged devices under an ambient atmosphere. Reproduced with permission [53]. Copyright 2022, The Royal Society of Chemistry.

In addition to the most common spiro-OMeTAD, inorganic p-type semiconductor cuprous thiocyanate (CuSCN) has attracted much attention as a new type of HTL. It has good transparency in the entire visible and infrared regions, good chemical stability, and higher hole mobility than spiro-OMeTAD [54]. However, devices based on CuSCN HTL have suffered from serious interface recombination problems. In order to solve the problem of interface instability, Long et al. added polyethylene glycol (PEG) as an intermediate film at the perovskite/CuSCN interface, effectively avoiding direct contact between perovskite and CuSCN to prevent SCN^- from destroying the perovskite crystal structure [50]. PEG can anchor MA^+ and I^- in perovskite to inhibit ion migration, improve the poor contact between the perovskite and CuSCN interface, enhance the hole mobility of the perovskite film, and passivate the uncoordinated Pb^{2+} in the perovskite film to inhibit the Pb^0 defects (Figure 4d). At the same time, the unique hygroscopicity of PEG molecules can form a water barrier around the perovskite film, effectively enhancing the environmental stability of PSCs. Through this new interface engineering, an excellent PCE of 19.20% was finally achieved. This is one of the most efficient standards reported to date in CuSCN-based PSCs. This work broadens the prospects for the commercialization of efficient and stable CuSCN-based PSCs.

The HTL/perovskite interface passivation of p-i-n planar structures PSCs has been rarely studied in the past. However, similar to the HTL/perovskite interface passivation in the n-i-p structure, the HTL/perovskite interface passivation in the p-i-n structure will also greatly affect the final device performance. Poly(bis{4-phenyl}{2,4,6-trimethylphenyl}amine) (PTAA) is one of the most commonly used HTL semiconductors for p-i-n PSCs. However, the large surface tension and incomplete surface coverage between PTAA and perovskite films are severe challenges for the preparation of high-performance PSCs. In this regard, Wu et al. introduced an inorganic potassium fluoride (KF) interfacial buffer layer onto the PTAA substrate to adjust the surface energy level difference between PTAA and perovskite [55]. KF can effectively reduce the valence band maximum (VBM) of PTAA, which is beneficial to the extraction of holes. In addition, the introduction of the KF layer significantly increases the composite resistance of the PTAA/perovskite interface, thereby inhibiting the carrier recombination between the interfaces. Finally, the PSCs modified by KF showed a PCE of 21.51%, J_{SC} of 23.95 mA cm^{-2}, V_{OC} of 1.09 V, and FF of 82.4%. Xing et al. introduced ethylammonium bromide (EABr) into the bottom interface of perovskite films to study its passivation effect [51]. The water contacts angle of the EABr-modified perovskite film decreased from the original 42° to 17°, indicating that the wettability of the EABr-modified PTAA substrate was improved, which was beneficial to the growth of the perovskite film. In addition, the ammonium group in EABr can significantly reduce the unreacted PbI_2 crystals at the PTAA/perovskite interface. These crystals have been reported to be the main defect sources and main degradation sites of perovskite films. The introduction of EABr also moves the VBM at the bottom of the perovskite upward by 0.29 eV, which improves the energy level alignment between perovskite and PTAA and promotes the extraction of holes (Figure 4e,f). Finally, the PSCs modified based on the EABr interface layer achieved a V_{OC} of 1.20 V, the champion PCE also increased from 20.41% to 21.06%, and the stability was also improved. Another work of Xing et al. is the introduction of donor-acceptor-donor organic molecule 4,4′,4″-(1-hexyl-1H-dithieno [3′,2′:3,4; 2″,3″:5,6] benzo [1,2-d] imidazole-2,5,8-triyl) tris (N,N-bis(4-methoxyphenyl) aniline (M2) to try to alleviate the inherent hydrophobicity of PTAA, which hinders the production of high-quality perovskite films on PTAA substrates [56]. The wettability of PTAA after M2 modification is greatly improved, and the crystallinity of the perovskite film formed on this substrate is significantly enhanced. More importantly, due to the excellent hole extraction and transport properties of M2, PTAA/M2 also exhibits higher hole mobility and conductivity than the original PTAA. The introduction of the M2 layer can also reduce the highest occupied molecular orbital (HOMO) energy level gap between PTAA and perovskite, thereby reducing the Voc loss. Therefore, the champion PCE of p-i-n PSCs based on PTAA/M2 increased from 18.67% to 20.23%, and the operational stability and light stability were enhanced.

Due to its high carrier mobility and high transmittance, nickel oxide (NiO_x) has become a common HTL in p-i-n PSCs in addition to PTAA. However, recent reports have pointed out that the photo-induced degradation of NiO_x-perovskite heterojunction is the main factor limiting the life of NiO_x-based PSCs devices [57]. For this reason, Qi et al. used vapor deposition to introduce a trimethylsulfonium bromide (TMSBr) buffer layer between the NiO_x/perovskite interface to eliminate the multi-step photodegradation of the NiO_x-perovskite heterojunction accompanying the device preparation process [52]. Time-of-flight secondary ion mass spectrometry (ToF-SIMS) results confirmed the penetration of TMS^+ and Br into the perovskite layer. The TMSBr buffer layer can eliminate the deprotonation and redox reaction between the organic iodide in the perovskite precursor and the Ni^{3+} in the NiO_x layer, which greatly improves device efficiency and stability. At the same time, the TMSBr buffer layer also has lattice parameters matching with perovskite crystals and strong trap passivation ability. TMS^+ in TMSBr can also significantly delay the proton transfer process at the NiO_x/perovskite interface. Finally, the p-i-n PSCs with TMSBr buffer layer achieved a champion PCE of 22.1%, and the time to reduce the efficiency to 80% of its initial value under AM 1.5G illumination was 2310 h, which is one of the highest service life reported by NiO_x-based PSCs (Figure 4g–j). In addition, the mismatch of the thermal expansion coefficient leads to the residual strain caused by the interface between NiO_x and perovskite, which accelerates the degradation of perovskite film and reduces the stability of the device. In this regard, Yang et al. introduced a tetrapentylammonium iodide (TPAI) buffer interface by sequential deposition method to prepare strain-free hybrid perovskite film [53]. Due to the low interaction energy between its flexibility and perovskite, the TPAI buffer layer can effectively release the in-plane tensile stress of PbI_2, thereby expanding the PbI_2 layer spacing, which is beneficial to release the residual stress between NiO_x and PbI_2 films in the subsequent perovskite phase transformation and accelerate the transformation of PbI_2 to perovskite. The TPAI buffer layer can also passivate interface defects, increase hole mobility, and improve device stability (Figure 4k). The champion PCE of IPAI-modified $MAPbI_3$ PSCs is 22.14%, with an FF of 84.6%. This work paves a new way to fabricate strain-free hybrid PSCs.

3. Perovskite Surface Treatment

As the light-absorbing layer, perovskite thin films are the most important part of PSCs devices. Perovskite thin films with high phase purity, low structural defects, excellent morphology, and high crystallinity are the key factors in obtaining high-efficiency PSCs. However, the perovskite crystal structure will be more or less affected by crystal growth and the post-treatment process, resulting in defects in the crystal [58–60]. To achieve high PCE and stability, it is very important to further improve the quality of perovskite films. It is urgent to eliminate SRH recombination in perovskite films. Generally, the most effective passivation methods are the solvent component, precursor additive engineering, and anti-solvent engineering methods.

3.1. The Solvent Component

With the rapid development of PSCs, the spin-coating process has become the most effective film manufacturing method, and various modified deposition based on the spin-coating process has achieved good performance [61,62]. Perovskite precursor solution plays an important role in the crystallinity, morphology, and stoichiometry of perovskite films. The solvent in the precursor solution can control perovskite nucleation and crystal growth to achieve uniform, pinhole-free high-quality perovskite films [63]. Precursor solutions contain inorganic and organic precursors with different properties, allowing only a limited selection of common solvents with sufficient solubility for the vast majority of mixtures [64]. Since the advent of PSCs, finding solvents or solvent mixtures with the appropriate properties to significantly increase the performance and stability of PSCs devices has been a major goal for many researchers.

In the early research stage of PSCs, the preparation of perovskite films mainly focused on the use of a single solvent perovskite precursor solution. At this time, a unipolar solvent such as N,N-dimethylformamid (DMF) or γ-butyrolactone (GBL) is typically used to prepare a perovskite precursor solution. The solution is spin-coated onto the substrate to form a wet perovskite precursor film, which is then converted into a perovskite film by annealing to remove the solvent [65,66]. Since the solubility of PbI_2 in DMF or GBL is relatively poor, and the weak coordination between PbI_2 and DMF or GBL makes it very easy for PbI_2 to crystallize preferentially from the precursor solution during the spin coating process, the morphology of the prepared perovskite film is always poor. To overcome this problem, Han et al. prepared an $MAPbI_3$ precursor solution by using dimethyl sulfoxide (DMSO) as an alternative precursor solvent [67]. The results show that DMSO can effectively delay the rapid crystallization of PbI_2, thus overcoming the problem of incomplete conversion of PbI_2. A high PCE planar PSC of 13.5% was prepared. Nevertheless, DMSO without high viscosity cannot be used as a single precursor solvent in one-step deposition methods. Further studies confirm that the use of a single solvent for perovskite film deposition is not the best choice for precise control to achieve high-quality films with the disadvantage of low nucleation rate and fast crystal growth, resulting in the formation of needle-like crystals in the film. The incomplete coverage of the film in the deposition area not only reduces the active area of the light-absorbing layer and induces direct contact between the interfaces but also leads to the intensification of perovskite SRH recombination. To overcome this shortcoming, Seok et al. first proposed the use of a mixed solvent of GBL and DMSO to prepare a perovskite precursor solution [68]. Through the PbI_2/MAI@DMSO mesophase, a very uniform and dense perovskite film is formed, and PSCs with a champion PCE of 16.2% can be fabricated. Subsequently, a series of mixed solvents, including DMF, DMSO, N-methyl-2-pyrrolidone (NMP), and GBL, were used as perovskite precursor solvents. So far, DMF/DMSO mixed solvent strategy has been widely used in the precursor solution of high-efficiency PSCs [69,70].

With the advancement of PSCs commercialization and the increasing emphasis on environmental protection and experimental safety issues, the toxicity of a series of polar aprotic solvents (DMF, DMSO, NMP, etc.), which are most widely used as perovskite precursor solvents, has been gradually discussed. More and more green, non-toxic solvents with controlled lattice growth have been introduced into PSCs systems [71]. Chen et al. used tin oxide nanorods (SnO_2-NRs) as ETL substrates and used green solvent triethyl phosphate (TEP) as the main solvent of the perovskite precursor to prepare perovskite films [72]. SnO_2-NRs can promote the nucleation process and delay the perovskite crystallization rate by providing a large number of heterogeneous nucleation sites with reduced Gibbs free energy. The strong interaction between the green solvent TEP and PbI_2 can slow down the crystal growth rate. A perovskite film with uniform morphology and large grain size was prepared, and the perovskite defects were effectively passivated. Gao et al. first reported the use of a green, non-toxic Lewis base solvent N-formylmorpholine (NFM) to replace toxic DMF [73]. The interaction between NFM and PbI_2 is stronger than that of DMF, which is beneficial to inhibiting the rapid crystallization of PbI_2 and delaying crystal growth. Besides, NFM has a higher viscosity than DMF, and the slow evaporation rate can lead to a wider anti-solvent drop window for crystal growth, which provides favorable conditions for the formation of dense and smooth high-quality perovskite films (Figure 5a). The trap state density of PbI_2@NFM-based PSCs was significantly reduced, and the trap recombination and non-radiative recombination of perovskite were effectively suppressed. The final champion PCE reached 22.78%, while the PCE based on PbI_2@DMF solvent was 21.97%. In addition, the humidity stability of PbI_2@NFM-based PSCs is greatly enhanced, and it still maintains more than 90% of their initial efficiency after aging for more than 30 days at a relative humidity of ~35% in ambient air.

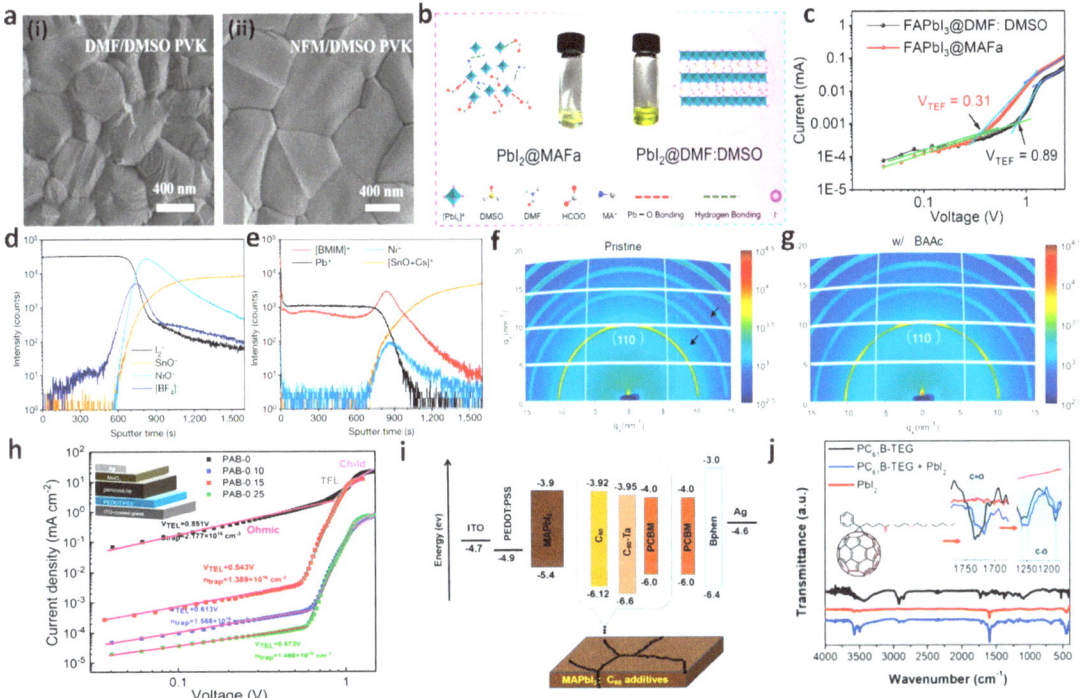

Figure 5. (**a**) SEM images of perovskite films with different precursor solvents: (**i**) DMF/DMSO and (**ii**) NFM/DMSO. Reproduced with permission [73]. Copyright 2022, American Chemical Society. (**b**) PbI$_2$@MAFa and PbI$_2$@DMF/DMSO solution image and solution interaction diagram. (**c**) The J-V characteristic curves of pure electronic devices based on FAPbI$_3$@DMF/DMSO perovskite film (prepared in an N$_2$-filled glove box) and FAPbI$_3$@MAFa perovskite film (prepared at 70–90% humidity). Reproduced with permission [74]. Copyright 2021, American Association for the Advancement of Science. The ToF-SIMS depth profiles of perovskite films on FTO/NiO/BMIMBF$_4$ substrates are measured with (**d**) negative and (**e**) positive polarity. Reproduced with permission [75]. Copyright 2019, Springer Nature. (**f**) Control and (**g**) 2D GIWAXS images of BAAc-treated perovskite films, where the diffraction rings at about 11 nm^{-1} and 16 nm^{-1} are marked with black arrows. Reproduced with permission [76]. Copyright 2022, Wiley-VCH. (**h**) Dark J-V curves of pure hole devices with different contents of PAB-modified perovskite layers [77]. Copyright 2022, Elsevier. (**i**) Energy level diagram of PSCs modified by different fullerene derivative materials. Reproduced with permission [78]. Copyright 2020, Elsevier. (**j**) FTIR spectra of PbI$_2$ interacting with PC$_{61}$B-TEG. Reproduced with permission [79]. Copyright 2022, Wiley-VCH.

In recent years, ionic liquids (ILs) have attracted much attention in the emerging solvents of perovskite precursors. ILs are a kind of low melting point salt ($T_m < 100\ ^\circ$C). It has excellent physical and chemical properties and has good compatibility with PSCs. ILs have an excellent liquid range and thermal working range, which can reach 300 $^\circ$C from $-90\ ^\circ$C in some cases [80,81]. The treatment of perovskite precursor solutions usually requires a wide temperature range, so ILs work well in the solvent engineering of PSCs. In addition to the interactions that exist in conventional organic solvents (hydrogen bonding, Van der Waals interactions, etc.), ILs also have specific ionic interactions (electrostatic attraction or repulsion of charged particles), which allow ILs to be mixed with a wide range of polar substances and dissolve both organic and inorganic substances. The cations in ILs have alkyl chains of different lengths, which is beneficial to improve their solubility in less polar fluids [82,83]. The solubility of the solvent in the perovskite precursor has an important influence on the crystallization process and quality of the prepared film, so ILs can be used as a perovskite precursor solvent.

In the past, the preparation of high-quality perovskite films must be carried out in an inert atmosphere, and the temperature and humidity must be strictly controlled. Huang et al. first reported the use of ionic liquid methylamine acetate (MAAc) to replace the traditional solvent DMF as the perovskite precursor to prepare high-quality MAPbI$_3$ films in ambient air [84]. Unlike DMF, MAAc dissolves perovskite precursors by forming Pb-O strong chelates and N-H—I hydrogen bonds with PbI$_2$, so it has a stronger ability to induce directional crystallization and chemically passivate grain boundaries. The unique molecular structure of MAAc ultimately improves the quality of perovskite films and the performance and stability of devices. The PSCs based on MAAc solvent achieve a champion efficiency of 21.18%. In another work, Huang et al. synthesized stable black α-FAPbI$_3$ in ambient air using methylamine formate (MAFa) as a precursor solvent (Figure 5b) [74]. During the formation of PbI$_2$ thin films, the strong chelation between the C=O group and Pb^{2+} leads to the regular arrangement of PbI$_2$ crystals, forming a series of PbI$_2$ crystal structures with nanoscale "ion channels" and growing perpendicularly to the substrate. These channels accelerate the entry of FAI into the interior of PbI$_2$ thin films and react with PbI$_2$ to form stable black phase α-FAPbI$_3$ perovskite thin films. Meanwhile, the formate ions remaining at the crystallization site can anchor the defects in situ, reducing the possibility of film defect formation, and the surface roughness of the perovskite film decreases from 20.5 nm to 10.1 nm. In addition, the trap state density of FAPbI$_3$@MAFa perovskite film is also significantly lower than that of FAPbI$_3$@DMF/DMSO perovskite film, which is more favorable to reduce the probability of carrier recombination (Figure 5c). Finally, the PSCs device based on MAFa solvent has 24.1% PCE, which is much higher than 22.1% based on DMF/DMSO solvent.

3.2. Precursor Additive Engineering

In addition to precursor solvent engineering, additive engineering that can increase grain size, passivate defects, and improve carrier extraction and transport to suppress SRH non-radiative recombination is also an important passivation strategy. The additives in the perovskite precursor solution can regulate the crystallization of perovskite, stabilize the phase state of perovskite, passivate the defects of perovskite, and optimize the interface morphology as well as the energy level of perovskite [85–87]. At present, there are many kinds of additives used in PSCs, such as ILs, polymers, and small organic molecules. The diversity of available additives is mainly due to the good coordination ability of anions and cations in halide perovskites, which is the basis for the solution processing of halide PSCs.

Taima et al. first used ILs as the precursor additive of MAPbI$_3$ perovskite by adding 1 wt% 1-hexyl-3-methylimidazolium chloride (HMImCl) [88]. Compared with the original MAPbI$_3$ film, the film after ILs treatment is smooth and uniform, which provides a new idea for the production of high-quality perovskite films. Inspired by this work, Snaith et al. presented an inverted mixed cation PSCs with 1-butyl-3-methylimidazolium tetrafluoroborate (BMIMBF$_4$) as an additive [75]. They were surprised to find that in the BMIMBF$_4$-modified perovskite film, BF$_4$ is mainly located in the embedded interface, while BMIM exists in the whole bulk film and accumulates in the embedded interface (Figure 5d,e). This result indicates that [BMIM]$^+$ and [BF$_4$]$^-$ ions accumulate at the perovskite/NiO$_x$ interface. The improvement of PSCs performance is mainly due to the presence of BMIM, and BF$_4$ ensures that the introduction of ILs does not negatively affect film performance and device performance. The final champion PCE reached 19.80%, and the environmental stability of the device was greatly improved. The PCE was only reduced by 14% after 100 h of aging under full-spectrum sunlight at 60–65 °C.

The protonated amine carboxylic acid ILs mentioned in the previous section have also received extensive attention in additive engineering due to their unique molecular structure and high solubility. Zhang et al. innovatively reported the crystallization kinetics control of MAPbI$_3$ perovskite precursor additive MAAc in carbon-based mesoporous PSCs [89]. The crystallinity of MAPbI$_3$ film modified by MAAc increases obviously, and the defect density decreases. In addition, they further elucidated the effect of MAAc on crystal growth kinetics by Fourier transform infrared (FTIR). The results show that MAAc has an effective coordination effect on non-coordinated Pb^{2+} defects, which is beneficial to inhibit the non-radiative recombination of carriers and promote charge transfer in the device. Similar to MAAc, butylammonium acetate (BAAc) is also a protonated amine carboxylic acid ILs containing acetate anions (Ac$^-$). Recently, Yang et al. added BAAc as an additive in the PbI$_2$@DMF solution to adjust perovskite crystallization by strong bonding interaction with the PbI$_2$ precursor solution and obtained high-quality perovskite films with significantly increased grain size [76]. The results of GIWAXS verify that the BAAc perovskite film has a high diffraction intensity along the (110) ring at q = 10 nm^{-1} (Figure 5f,g). The diffraction rings at 11 nm^{-1} and 16 nm^{-1} in the perovskite film doped with BAAc are significantly suppressed. These phenomena indicate that the chemical bond between BAAc and [PbI$_6$]$^{4-}$ skeleton forms directional crystallization, which further indicates that BAAc has the effect of regulating the crystallization kinetics of perovskite. In addition, they found that the defects of perovskite films prepared by doping BAAc were significantly reduced and the performance and stability of the devices were improved. Finally, the best device prepared by BAAc has a PCE of 20.1%, V$_{OC}$ of 1.12 V, FF of 79%, and J$_{SC}$ of 22.7 mA cm^{-2}.

Polymers have become one of the most effective passivation additives for PSCs due to their special functional groups. Some atoms (S and N) in the polymer can react with Pb^{2+} in the perovskite to stabilize the perovskite structure and improve the crystallinity and morphology of the perovskite film. In addition, due to its excellent hydrophobicity, thermoplasticity, electrical conductivity, and mechanical stability, the addition of polymers can effectively reduce the sensitivity of perovskite materials to water, oxygen, temperature, and ultraviolet radiation, which helps to improve the stability of the device [90]. Therefore, polymers as indispensable PSCs additives have been extensively explored in regulating the nucleation and crystallization processes of perovskite films and improving device performance.

Su et al. first used PEG as a perovskite precursor additive to prepare high-quality perovskite films [91]. They found that the morphology of PEG-modified films was greatly improved, the surface was smoother, no obvious holes and the roughness was significantly reduced. This is mainly because PEG can slow down the growth and aggregation of perovskite crystals during nucleation and reduce the gap between perovskite grain boundaries during phase transformation. The optimized perovskite film has higher absorption to promote charge transfer, which greatly improves V$_{OC}$ and J$_{SC}$. Cheng et al. introduced polyvinyl alcohol (PVA), poly (methyl acrylate) (PMA), and polyacrylic acid (PAA) as additives into MAPbI$_3$ PSCs to explain the role of different functional groups (-OH in PVA, -C=O in PMA and -COOH in PAA) in the passivation process of additives [92]. The FTIR spectra of MAPbI$_3$ films doped with three different additives showed that the -OH peak of PVA-MAPbI$_3$ and the -C=H peak of PMA-MAPbI$_3$ had a red shift. They explained that -OH in PVA and MA$^+$ in MAPbI$_3$ formed hydrogen bonds, and -C=O in PMA complexed with uncoordinated Pb^{2+} in MAPbI$_3$. In addition, the shift of -OH and -C=H in PAA-MAPbI$_3$ is more obvious than that in PVA-MAPbI$_3$ and PMA-MAPbI$_3$. They believe that the -COOH of PVA can not only selectively interact with MA$^+$ and I$^-$ through hydrogen bonds but also complex with uncoordinated Pb^{2+} to more effectively passivate defects. Finally, PAA-modified MAPbI$_3$ PSCs achieved a champion PCE of 20.29% and a V$_{OC}$ of 1.13 V in all modified devices.

Most polymers contain one or two passivation functional groups. Due to the complex synthesis process and harsh experimental conditions, polymers with three or more passivation functional groups are rarely reported. Polyamide derivatives (PAB) is a rare polymer containing three functional groups (hydroxyl, secondary amine, and carboxyl), which is synthesized by a novel multicomponent reaction between benzoxazine-isocyanide chemistry (BIC). Ling et al. first used phenolic hydroxyl substituted PAB as precursor solution additives to passivate the perovskite active layer [77]. They found that the hydroxyl and carboxyl groups in PAB can act as Lewis bases to react strongly with Pb^{2+} in perovskite, thereby passivating defects. At the same time, the N atom in the secondary amine can coordinate with I^- due to its power supply characteristics. The interaction of these functional groups with the perovskite material effectively suppresses the non-radiative recombination of the carriers (Figure 5h), ultimately increasing the champion efficiency of PSCs from 19.45% to 21.13%.

Unlike large-sized polymers, small organic molecules have attracted much attention in PSCs additive engineering due to their small size and ability to enter perovskite lattices for passivation. Fullerene was first discovered by Smalley et al. in 1985. Because of its unique physical and chemical properties, it has attracted wide attention from the scientific community, including the photovoltaics industry [93]. Fullerenes and its derivatives are very suitable for PSCs due to their unique high electron mobility and surprisingly small recombination energy [94]. Nowadays, fullerene-based materials have been widely used as electron transfer layer materials and interface defect passivation materials for PSCs [95,96]. The role of fullerenes and their derivatives in PSCs additive engineering is also crucial. After entering the perovskite lattice, these small molecules can completely cover the surface of the perovskite grain boundary and act on ions attempting to move along the grain boundary by physical blocking [97]. Wu et al. systematically compared and analyzed a series of fullerene derivative additives, such as C_{60}, PCBM, and C_{60}-taurine (C_{60}-Ta), and added them to perovskite precursors to construct perovskite-fullerene heterojunction PSCs [78]. Compared with PCBM, the energy levels of C_{60} and C_{60}-Ta are more matched with the energy levels of perovskite (Figure 5i), thereby enhancing the electron transfer of perovskite, inhibiting the carrier recombination and prolonging the carrier lifetime of perovskite. Finally, compared with the control device with a PCE of 14.87%, the efficiency of all fullerene modification devices was improved. The PCE of PSCs with C_{60}-Ta was 16.46%, which was slightly lower than 16.59% of C_{60} PSCs but higher than 15.94% of PCBM-based PSCs. In addition, they also studied the effect of chemical composition in fullerene derivatives on the performance of PSCs device parameters, trying to explain the specific role of the C_{60} cage and grafted side chain in fullerene and its derivative additives. C_{60} has higher carrier mobility than C_{60}-Ta, but the grafted side chain of C_{60}-Ta can more effectively improve crystal quality and reduce defects, thereby further improving device stability. It is worth noting that the grafting side chain of C_{60}-Ta has a negative effect on carrier migration, resulting in a lower final efficiency of C_{60}-Ta-modified devices than that of C_{60}-modified devices. Therefore, it is very important to select a suitable C_{60} grafted side chain to simultaneously reduce the perovskite defect state and improve carrier transport to balance the stability and PCE of PSCs. Jeon et al. innovatively proposed a new method to introduce [6,6]-phenyl-C_{61}-butyric acid 2-[2-(2-methoxyethoxy)ethoxy]ethyl ester (PC_{61}B-TEG) into perovskite devices and induce favorable vertical gradients [79]. Because the TEG on fullerene can significantly improve its solubility in the polar solvent of perovskite precursor, the charge transfer ability and grain defect passivation ability of the modified perovskite film is significantly enhanced. FTIR results show that when the fullerene derivative is mixed with PbI_2 (Figure 5j), the peaks of C=O and C-O in PC_{61}B-TEG move downward, indicating that the additive effectively passivates Pb^{2+}, indicating that the vertical gradient PC_{61}B-TEG additive can effectively passivate the defect sites of perovskite, and the coated PC_{61}B-TEG interface also significantly enhances the carrier transport capacity. Finally, it is found that devices based on different perovskites exhibit higher performance parameters than conventional devices. For $MAPbI_3$-based devices, the PCE of devices with PC_{61}B-TEG added increased from 17% to 19.5%. The device based on $FA_{0.65}MA_{0.35}PbI_{3-x}Cl_x$ has more significant performance improvement. Compared with the traditional device with a V_{OC} of 1.14 V, J_{SC} of 24.97 mA cm^{-2}, FF of 79%, and PCE of 21.88%, the improved device has a V_{OC} of 1.13 V, J_{SC} of 25.42 mA cm^{-2}, FF of 81% and PCE of 23.34%, which is also the highest certification efficiency for PSCs prepared by fullerene derivative additives so far.

3.3. Anti-Solvent Engineering

Anti-solvent engineering is another way to introduce additives into the perovskite light-absorbing layer for component regulation. Anti-solvents, such as chlorobenzene (CB), are a type of non-polar solvents that are miscible with the deposition solvent in the perovskite precursor solution and insoluble with perovskite salts. It plays an important role in the surface morphology and crystallization properties of perovskite films. After the perovskite precursor solution is spin-coated on the substrate for a specific time, an anti-solvent is added dropwise to the rotating precursor solution to prepare a perovskite film. The film produced by this method is smoother and of higher quality than the film prepared by spin-coating without adding a solvent dropwise [98]. Anti-solvent engineering provides a practical way to passivate the carrier non-radiative recombination problem and suppress its defects as much as possible. But dropping anti-solvent on the perovskite layer results in a fast and uncontrollable crystallization process and produces a large number of grain boundaries and surface defects. Wang et al. added polyvinyl butyral (PVB) as an additive to the anti-solvent CB when preparing MAPbI$_3$ films by one-step spin-coating process, which was added to the film surface before the end of spin-coating to help the perovskite crystal growth to improve film quality (Figure 6a) [99]. They found that the grain size of the modified perovskite film increased significantly to 600–700 nm, and the number of grain boundaries decreased significantly, which was beneficial to the reduction of the defect density of the film. Gao et al. added poly{4,8-bis[(2-ethylhexyl)oxy] benzo [1,2-b:4,5-b']dithiophene-2,6-diyl-alt-3-fluoro-2-[(2-ethylhexyl)carbonyl]thieno[3,4-b]thiophene-4,6-diyl} (PTB7) as an additive to the anti-solvent CB of perovskite film to fully study the passivation mechanism of PTB7 [100]. They found that the introduction of PTB7 significantly reduced defects in perovskite films and increased crystallinity. The synchrotron radiation grazing incidence X-ray diffraction (GIXRD) results of perovskite films before and after adding PTB7 at different grazing incidence angles and different detection depths also confirm this conclusion (Figure 6b). Gao et al. explained that this phenomenon is due to the large size of the perovskite surface that does not enter the perovskite lattice and only exists at the grain boundary or on the surface of the film. The PTB7 molecule interacts with the Pb atom in the precursor to form a Lewis base coordination bond, thereby slowing down the crystallization kinetics of the film nucleation and increasing the crystallinity. Based on the starting point of green environmental protection, Wang et al. added non-toxic polymer polyvinylpyrrolidone (PVP) as an additive to the green anti-solvent isopropanol to passivate perovskite films [101]. Time-resolved photoluminescence (TRPL) spectra show that the non-radiative linear recombination of carriers in the passivated device is significantly reduced, mainly due to the lone pair electrons in PVP can share C=O bonds with uncoordinated Pb^{2+} in perovskite to inhibit ion migration and stabilize perovskite crystal structure (Figure 6c).

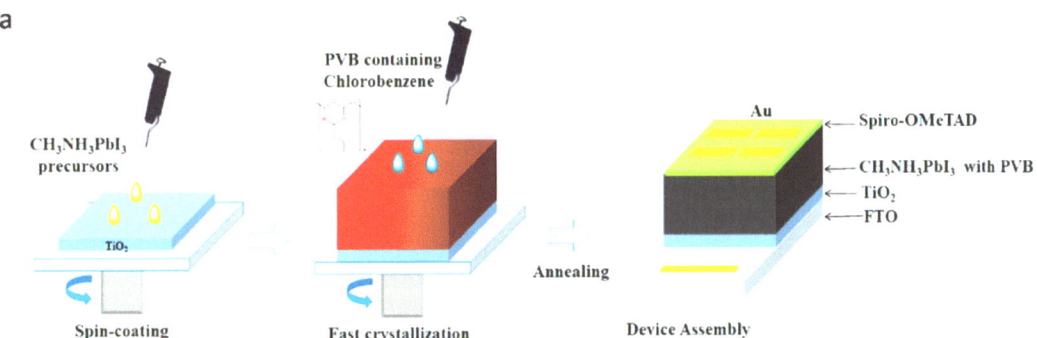

Figure 6. *Cont.*

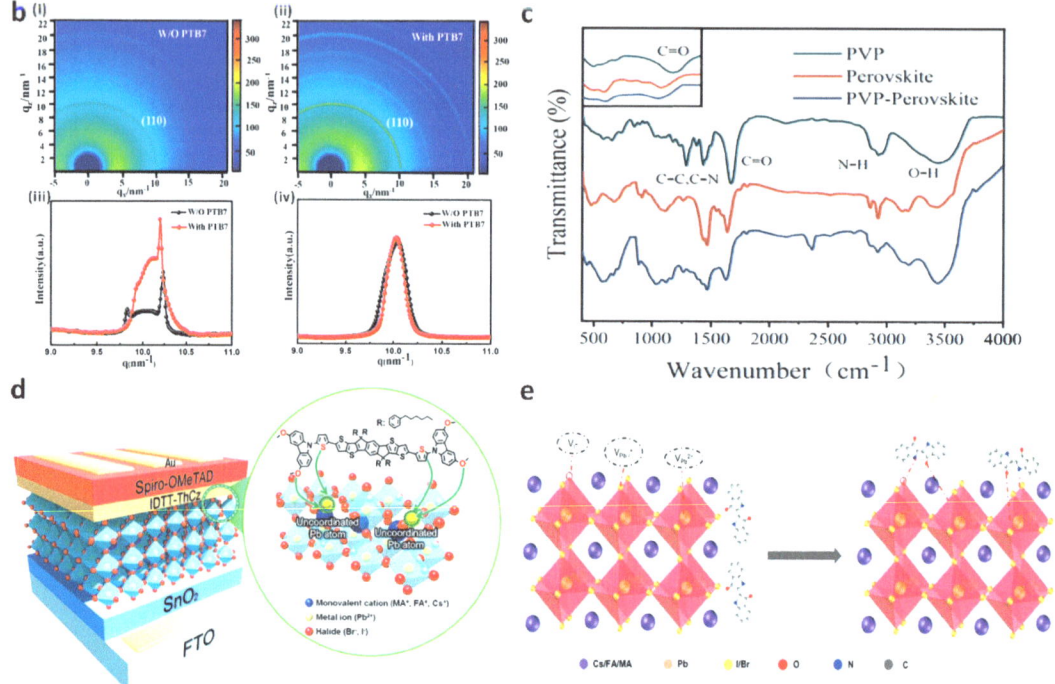

Figure 6. (**a**) One-step preparation of PSCs schematic, in which PVB is introduced into the perovskite layer by anti-solvent engineering. Reproduced with permission [99]. Copyright 2021, Elsevier. (**b**) 2D-GIXRD diagram of (**i**) original and (**ii**) PTB7 modified perovskite films at a grazing incidence angle of 0.06° and azimuthal integral intensity diagram of (**iii**) perovskite (110) plane and (**iv**) perovskite (110) diffraction peak. Reproduced with permission [100]. Copyright 2022, Elsevier. (**c**) FTIR spectra of PVP, perovskite and PVP-modified perovskite films. Reproduced with permission [101]. Copyright 2022, Elsevier. (**d**) The interaction diagram of IDTT-ThCz molecule with perovskite layer. Reproduced with permission [102]. Copyright 2021, Wiley-VCH. (**e**) Indigo and mixed hybrid perovskite passivation mechanism diagram. Reproduced with permission [103]. Copyright 2022, Wiley-VCH.

In addition to polymers, small organic molecules are also used in anti-solvent engineering. Song et al. first synthesized an indacenodithieno[3,2-b]thiophene-based small molecule (IDTT-ThCz) and introduced it as an additive anti-solvent into perovskite to assist crystallization (Figure 6d) [102]. Finally, PSCs with high PCE and obvious thermal stability were prepared. The FF of the modified device is as high as 80.4%, and the PCE is as high as 22.5%. Simultaneously, 95% of the initial PCE can be retained after 500 h storage under thermal conditions (85 °C). As they explained, the Lewis atoms in IDTT-ThCz can react with Pb^{2+} in the perovskite precursor, passivate the electronic defect state and effectively inhibit the degradation of the perovskite layer. Due to the unique p-type semiconductor characteristics of IDTT-ThCz, the charge extraction capability of PSCs has also been significantly improved. Ma et al. first reported a natural small organic dye molecule Indigo as a passivator for the design and preparation of high-quality hybrid perovskite films through anti-solvent engineering [103]. They treated the $Cs_{0.05}FA_{0.85}MA_{0.10}Pb(I_{0.90}Br_{0.10})_3$ perovskite film by dissolving the Lewis base indigo molecule in CB at an optimal concentration and proved that the presence of the C=O/-NH functional group has a significant effect on the passivation of the original perovskite film defects. The carbonyl group (electron-pair donor) in the indigo molecule can interact with the uncoordinated Lewis acid Pb^{2+} on the perovskite surface and the Pb-I antisite defect, and the amino group can interact with the I-site. In addition, the hydrogen bond between indigo molecules and the perovskite surface can inhibit ion migration and further passivate perovskite film defects (Figure 6e). Therefore, the champion PCE of PSCs passivated by indigo increased from 20.18% to 23.22%.

4. Dopant Passivation

Although introducing additives directly into a perovskite precursor solution can reduce the trap-state density of the film and suppress non-radiative recombination of carriers, this passivation strategy carries the risk of introducing impurities into the perovskite crystal that affects the long-range ordered structure of the perovskite crystal [104]. Therefore, the researchers turned the passivation target to the carrier transport layer adjacent to the perovskite layer. The elemental doping dopant passivation project on the carrier transport layer can promote the carrier transport rate and adjust the energy level barrier between the interfaces, which will further passivate the perovskite film defects, control the crystallization process of the perovskite film and increase the crystallinity.

4.1. ETL Doping

The biggest disadvantage of TiO_2 is the low electron mobility, which greatly limits the performance parameters of the device. By doping alkali metals or transition metals to passivate the dense TiO_2 layer and/or mesoporous TiO_2 layer, the electronic band structure and the trap state of TiO_2 will be changed. It is very helpful to improve the charge transport performance of the device. Inspired by previous studies, Chu et al. used metal Li ions doped into TiO_2 (Li-TiO_2) as a new ETL for carbon-based $CsPbIBr_2$ PSCs [105]. They found that the optical band gap of TiO_2 did not change after Li doping, but the crystallinity of TiO_2 films could be effectively improved (Figure 7a). At the same time, the carrier recombination at the Li-TiO_2/$CsPbIBr_2$ interface is suppressed, which greatly improves the efficiency and stability of inorganic $CsPbIBr_2$ PSCs. Liu et al. developed a fast one-step laser-assisted doping process to incorporate the transition metal tantalum (Ta) into the TiO_2 ETL (Ta-TiO_2), inducing the crystallization of the TiO_2 film from its amorphous precursor to the anatase phase [106]. The conductivity and electron transport capacity of the TiO_2 film treated by the best laser process is improved, and the high concentration of Ti^{3+} defects on the surface of the film is effectively suppressed. The perovskite film with Ta-TiO_2 ETL as the substrate has good coverage and crystallinity while reducing the non-radiative recombination of carriers. The $MA_{0.1}FA_{0.9}PbI_3$ PSCs device based on Ta-TiO_2 ETL finally achieved a champion PCE of 18.34%, mainly due to a significant increase in FF from 73% to 76.5% (Figure 7b).

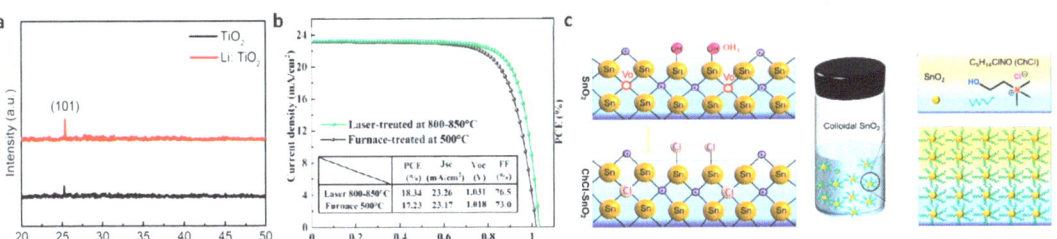

Figure 7. Cont.

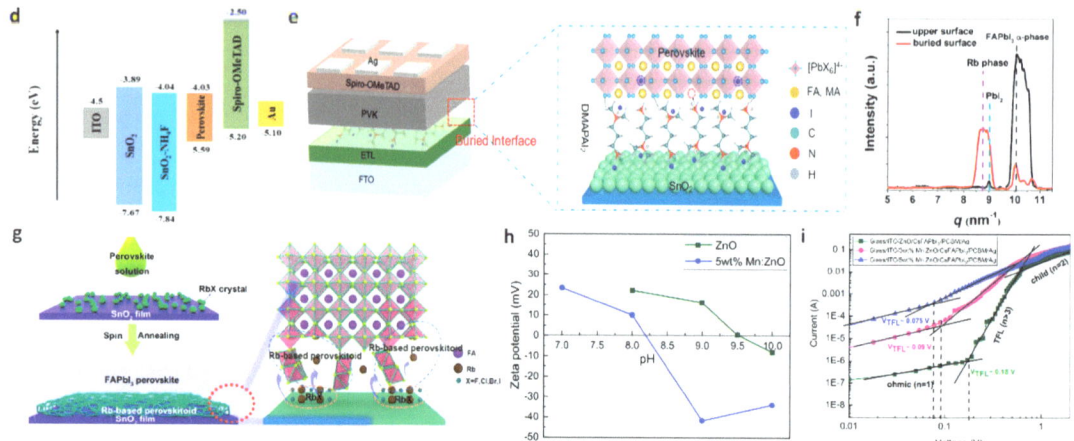

Figure 7. (**a**) X-ray diffraction (XRD) patterns of TiO$_2$ and Li: TiO$_2$ samples. Reproduced with permission [105]. Copyright 2023, Elsevier. (**b**) J-V curves of Cs$_{0.1}$FA$_{0.9}$PbI$_3$ devices based on furnace-treated and optimized laser-processed Ta-TiO$_2$ films. Reproduced with permission [106]. Copyright 2022, American Chemical Society. (**c**) The surface defect sites of SnO$_2$ and the schematic diagram of ChCl-SnO$_2$ nanoparticle solution. Reproduced with permission [107]. Copyright 2022, Wiley-VCH. (**d**) Energy level diagram of PSCs before and after NH$_4$F modification. Reproduced with permission [108]. Copyright 2022, American Chemical Society. (**e**) PSCs device structure diagram and DMAPAI$_2$ embedded interface diagram. Reproduced with permission [109]. Copyright 2022, Elsevier. (**f**) Basic integral intensity corresponding to GIWAXS data of RbCl modified film upper surface and buried surface. (**g**) A schematic diagram of RbX crystallized on the surface of SnO$_2$ and Rb-based perovskite formed on the buried interface of perovskite film. Reproduced with permission [110]. Copyright 2022, American Chemical Society. (**h**) IEP measurements of ZnO and Mn:ZnO powders. (**i**) SCLC measurements for ZnO and Mn:ZnO-based electron-only devices. Reproduced with permission [111]. Copyright 2022, American Chemical Society.

As an emerging ETL, SnO$_2$ has attracted much attention due to its low-temperature solution treatment and high electron mobility (100–200 cm^2 V^{-1} s^{-1}). However, because of the absence of high-temperature sintering, there are a large number of oxygen vacancies on the SnO$_2$ film. SnO$_2$ nanoparticles easily form agglomerates in the solution state, which often leads to a large number of intrinsic defects, resulting in poor film uniformity, crystallinity, and then obvious leakage current. Meanwhile, the uncoordinated Sn will hang on the surface of SnO$_2$ and become a trap state of electrons in the conduction band, forming a potential barrier to hinder electron transport [112]. It is necessary to use dopants with different characteristics to modify SnO$_2$ to passivate surface defects so that ETL can form effective contact with perovskite and reduce interface carrier recombination. Ammonium salt is a commonly used SnO$_2$ dopant. Zhang et al. developed a molecular bridge strategy to change the properties of the buried interface in n-i-p PSCs by introducing a multifunctional dopant 2-Hydroxyethyl trimethylammonium chloride (ChCl) into SnO$_2$ ETL (Figure 7c) [107]. The multifunctional molecular structure (NH$_4$$^+$, Cl$^-$, -OH) in the dopant ChCl can be used as a molecular bridge to passivate defects in colloidal SnO$_2$ and simultaneously regulate perovskite crystallization. Therefore, the perovskite film has larger grains, high uniformity and low defects, which is beneficial for suppressing non-radiative recombination and reducing voltage loss. At the same time, the embedded ChCl-SnO$_2$ ETL also exhibits reduced defect state density, matched energy levels and high conductivity. Therefore, at a significant V$_{OC}$ of up to 1.193 V, the device PCE increased significantly from 20.0% to 23.07%. Similar to this work, Liu et al. proposed a modification of ETL by incorporating organic ammonium salt propylammonium chloride (PACl) into SnO$_2$

colloidal solution to study the interaction mechanism between organic salts and alkaline colloidal solution [113]. PACl can passivate perovskite layer defects and enhance the crystallization of perovskite films. The main reason is that Cl^- and PA^+ will be introduced after PACl is incorporated into SnO_2 colloidal alkaline solution, and Cl^- will diffuse to the PbI_2 layer to promote perovskite nucleation and increase perovskite grain size. PA^+ can passivate grain boundaries and reduce perovskite film defects. Therefore, the overall performance of the device based on SnO_2-PACl ETL has been significantly improved. Champion device PCE was 22.27%. The incorporation of PACl also significantly improved the stability of PSCs, and the PCE remained 85% of the original value after 800 h in air. Chang et al. introduced a method to effectively passivate the surface defects of SnO_2 thin films by doping ammonium fluoride (NH_4F) into SnO_2 precursor [108]. The F in NH_4F can repair the terminal hydroxyl defects on the surface of SnO_2 and reduce the defects on the surface of SnO_2 and perovskite. The terminal hydroxyl groups on the surface of SnO_2 have been confirmed to act as defect sites to introduce deep level defects into the band gap, thus causing carrier recombination between interfaces. The doping of NH_4F makes the energy level configuration of the device more conducive to electron extraction (Figure 7d). The results show that the PSCs based on SnO_2-NH_4F ETL reaches 22.12% PCE, and the V_{OC} is 70 mV higher than that of the control device.

In addition to chloride ammonium salt and fluoride ammonium salt, iodide ammonium salt is also a kind of commonly used organic ammonium salt. Shi et al. reported a strategy to passivate the SnO_2 layer using an asymmetric diammonium salt N, N-dimethyl-1,3-propanediamine dihydroiodide ($DMAPAI_2$) (Figure 7e) [109]. The I^- in $DMAPAI_2$ can passivate V_O on the surface of SnO_2 by electrostatic coupling, thereby enhancing the electron mobility of SnO_2 and adjusting the energy level structure. The ammonium cations on the surface of $DMAPAI_2$-SnO_2 can interact with iodides in perovskite precursors through ionic bonds and/or hydrogen bonds to slow down the growth process of perovskite, which is conducive to promoting uniform nucleation and growth. Based on this strategy, the PCE of PSCs increased significantly from 20.78% to 23.20%.

Inorganic halide salts are also effective ETL dopants for preparing efficient PSCs. Wu et al. incorporated rubidium chloride (RbCl) into the SnO_2 precursor solution and crystallized an island pattern on the surface of SnO_2 based on the "rigid skeleton" structural properties of Rb [110]. In the process of perovskite crystal growth, RbCl crystal can be used as a nucleation center to act as a "scaffold" to anchor the uncoordinated atoms on the surface of the perovskite film to reduce defects (Figure 7f,g). They found that the grain size and crystallization strength of SnO_2-RbCl-based perovskite films were greatly improved. The simulation results also confirmed that the perovskite termination layer based on SnO_2-RbCl could significantly inhibit the formation of surface iodide vacancies, which contributes to the crystallization and passivation of perovskite films. The enhancement of defects leads to a slower carrier recombination rate in the perovskite film, which further reduces the non-radiative recombination of the perovskite surface and improves device stability. Finally, PSCs prepared by the RbCl dopant showed 25.14% champion PCE.

ZnO with high electron mobility (120 $cm^2 V^{-1} s^{-1}$) is also an excellent ETL material for planar p-i-n structure PSCs, which has higher light transmittance and better conduction band offset similar to TiO_2 and SnO_2 [114]. However, the high alkalinity of ZnO and the higher isoelectric point (IEP) than other metal oxides make MA^+ and FA^+ in the perovskite film rapidly deprotonated, thereby increasing more defect sites and result in higher charge recombination at the interface. In this concern, Krishnamoorthy et al. first doped solution-treated Mn into ZnO to adjust its IEP [111]. They found that the IEP (~8.2) of Mn: ZnO was significantly lower than that of the original ZnO (~9.5). X-ray photoelectron spectroscopy (XPS) analysis also shows that the formation of OH peaks and oxygen vacancies in modified ZnO is relatively low. SCLC measurement shows that the trap state density of PSCs based on Mn: ZnO is significantly reduced, which indicates that the doping of Mn reduces the defects of ZnO and perovskite films, thus ensuring better electron transport between perovskite and ZnO (Figure 7h,i). Finally, the PCE of Mn: ZnO-based PSCs increased from

11.7% to 13.6%, which was about 15% higher than that of the original ZnO-based PSCs. Akram et al. doped Al into ZnO to improve carrier mobility to suppress the generation of deep-level defects [115]. When 1% or 2% Al is doped in ZnO, the lattice shrinkage leads to an increase in grain size, the uniformity, and the smoothness of the film. The surface defects are effectively passivated. Because Al^{3+} replaces Zn^{2+} in the lattice site, the carrier concentration increases, indicating that the ETL/perovskite interface can achieve more effective charge extraction.

4.2. HTL Doping

Spiro-OMeTAD doped with lithium bis(trifluoromethanesulfonyl)imide (Li-TFSI) and 4-*tert*-butylpyrimidine (tBP) is considered to be the most effective HTL material for planar n-i-p structure PSCs with many record PCE [116–119]. However, Li-TFSI/tBP doped spiro-OMeTAD has poor environmental conductivity, which cannot effectively passivate the perovskite/HTL interface and reduce perovskite crystal defects. The unstable HTL compositions and iodide salts will cause serious degradation of the device, resulting in unstable device performance. Overcoming these shortcomings of the spiro-OMeTAD layer can effectively improve the performance and stability of PSCs. Chen et al. designed a passivation strategy to incorporate multi-walled carbon nanotubes (MWCNT:NiO_x) modified with multifunctional NiO_x quantum dots (QD) into spiro-OMeTAD [120]. Due to the strong interaction between O in MWCNT:NiO_x and H in spiro-OMeTAD, the conductivity of modified HTL is improved. The anchoring effect of the Li-O bond in MWCNT:NiO_x on Li-TFSI effectively limits the migration of Li^+ ions. MWCNT:NiO_x also passivates perovskite crystal defects by forming Ni-I bonds with perovskite. The interface defects can be reduced, and the extraction and transfer of holes are promoted. The PSCs device fabricated by this passivation strategy has a PCE of up to 22.73%, which is 1.2 times greater than that of the original spiro-OMeTAD-based PSCs. The environmental, thermal, and light stability are improved significantly. To further enhance the stability of spiro-OMeTAD, Li et al. designed N2, N2′, N7, N7′-tetrakis (4-((2-methoxyethoxy) methyl) phenyl) -tetra(yridine-4-yl)-9,9′-spirobi[fluorene]-2,2′,7,7′-tetraamine (spiro-BD-2OEG) composed of a main chain of spirobifluorene (spiro), a terminal group of phenylpyridine-4-amine (BD) and oligo (ethyl-eneglycol) (OEG) side chain [121]. They found that spiro-BD-2OEG provides a strong π-π interaction between the easily reduced benzene ring and the pyridine group of tBP, thereby further inhibiting the volatilization of tBP compared with spiro-OMeTAD. The lone pair electrons of the pyridine part of spiro-BD-2OEG combine with Li^+ to accelerate its dissolution, which is beneficial in inhibiting morphological defects and stabilizing the composition. Spiro-BD-2OEG doped spiro-OMeTAD film has long-range ordered molecular order and low roughness, which helps to form strong electronic contact with perovskite film and improve stability. The photoluminescence (PL) mapping strength of the perovskite modified by spiro-BD-2OEG was significantly reduced, indicating that the surface defects of the perovskite were effectively passivated and the non-radiative recombination at the interface was inhibited (Figure 8a,b). Finally, the PSCs device based on piro-BD-2OEG doped spiro-OMeTAD achieved an excellent PCE of 24.19%.

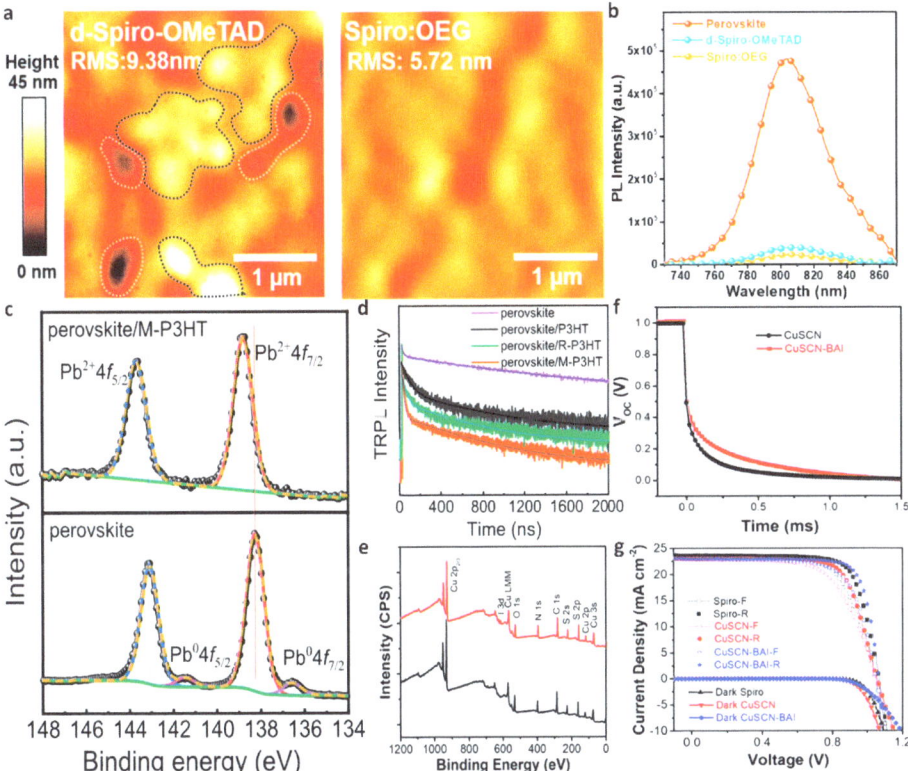

Figure 8. (**a**) AFM images of Spiro-OMeTAD and Spiro:OEG films coated on perovskite, where black dashed lines indicate aggregation of Li-TFSI and white dashed lines indicate pinholes. (**b**) PL spectra of perovskite, perovskite and Spiro-OMeTAD or Spiro:OEG HTL. Reproduced with permission [121]. Copyright 2022, Wiley-VCH. (**c**) High-resolution XPS spectra of Pb 4f spectra and (**d**) TRPL spectra of M-P3HT modified perovskite films. Reproduced with permission [122]. Copyright 2022, Springer Nature. (**e**) XPS full measurement spectra of CuSCN and CuSCN-BAI films. (**f**) Transient photovoltage decay (TPV) curves and (**g**) dark and light J−V curves of PSCs based on CuSCN and CuSCN-BAI films. Reproduced with permission [123]. Copyright 2022, American Chemical Society.

Compared with the classic spiro-OMeTAD, poly (3-hexylthiophene) (P3HT) is a lower-cost and more stable HTL material in planar n-i-p structure PSCs. However, the alkyl side chain of P3HT will directly contact the perovskite film, resulting in poor electron contact at the P3HT/perovskite interface, which exacerbates the non-radiative recombination of PSCs and makes the PCE of pure P3HT-based PSCs generally low [124]. To solve this problem, Gao et al. used 2-((7-(4-(bis(4-methoxyphenyl)amino)phenyl)−10-(2-(2-ethoxyethoxy)ethyl)−10H-phenoxazin-3-yl)methylene)-malononitrile (MDN) to modify P3HT to improve the bad contact between perovskite and P3HT [122]. The N atom in the malononitrile group in the MDN can be electrostatically coupled with the uncoordinated Pb on the perovskite surface, effectively suppressing the generation of Pb0 defects. In addition, the triphenylamine group in the MDN can form a π-π stacking with P3HT to establish a charge transport path between the perovskite/HTL. They found that the PL intensity of P3HT-modified perovskite decreased, indicating that the density of trap states at the perovskite/HTL interface decreased, and the non-radiative recombination was alleviated (Figure 8c,d). Finally, 22.87% PCE was achieved using MDN-doped P3HT as HTM, which was much higher than 12.48% of the control device.

The doping passivation engineering of HTL in planar p-i-n structure PSCs has also attracted much attention in recent years. NiO_x is considered to be one of the most promising HTMs for p-i-n inverted PSCs due to its high stability, high mobility, and low cost. In addition, it has a wide band gap with high transmittance and proper VB alignment. However, compared with organic HTL materials, the biggest problem faced by NiO_x-based PSCs is the lower V_{OC}. In this regard, Park et al. reported the device parameters of PSCs based on modified NiO_x by incorporating ammonium salt into NiO_x precursor solution to study the inherent properties of modified NiO_x [125]. They found that the morphology of the NiO_x film doped with ammonium salt was improved, the defects were reduced, and the crystallinity was higher. The energy level and hole conductivity of NiO_x are optimized, which is beneficial to hole transport. In addition, the strong interaction between ammonium salt and perovskite also optimizes the quality of perovskite films, enhances the interfacial properties between NiO_x and perovskite layers, and reduces trap-assisted recombination. The $MAPbI_3$ PSCs based on this new NiO_x obtained a 19.91% champion PCE and an extremely high V_{OC} of 1.13 V. In addition to NiO_x, CuSCN is also one of the low-cost and highly stable HTL materials in p-i-n inverted PSCs, which have attracted much attention recently [126]. The conductivity of the solution-treated CuSCN HTL is low, which is not conducive to the extraction and transmission of holes in PSCs devices. To improve the conductivity of CuSCN, Ye et al. doped n-butylammonium iodide (BAI) into the CuSCN precursor to optimize its p-conductivity [123]. BAI can effectively complex with Cu^{2+} in CuSCN to achieve complete coverage of the perovskite active layer. In addition, the complexation helps to generate more Cu vacancies in the CuSCN HTL, resulting in a significant increase in hole concentration and p conductivity of the CuSCN film. At the same time, the high hole extraction rate of the modified CuSCN inhibits the non-radiative recombination at the HTL/perovskite interface and achieves high device stability. Finally, the modified PSCs achieved 19.24% PCE, showing better stability than the control device in the air environment (Figure 8e–g).

5. Summary and Prospect

Among the third-generation solar cells, PSCs have attracted much attention due to their simple manufacturing process, low cost, and fast development. The surface defect degree of the perovskite film is one of the main factors affecting the PCE of the device, so we can improve efficiency by passivating defects. Here, we provide an in-depth review of the passivation strategies reported so far and list the detailed photovoltaic parameters of the highly efficient PSCs achieved over the past year using different passivation strategies in Table 1. Although single-junction PSCs have achieved a high PCE of 25.8%, there is still large room for further improvement from the S-Q theoretical limit of 1.6 eV band gap (30.5%). The J_{SC} of most reported high-efficiency PSCs is close to the theoretical value, while V_{OC} and FF are still lower than the theoretical value, so strategies to improve V_{OC} and FF should be reasonably formulated to further enhance PCE.

Table 1. The passivators used in different PSC passivation strategies in the past year and the performance comparison before and after passivation.

Strategy	Passivator	Processing	V_{OC} [V]	J_{SC} [mA cm^{-2}]	FF [%]	PCE [%]	N_{trap} [cm^{-3}]	Ref.
Interface Engineering	TFPhFACl	Control	1.08	25.49	79.77	21.90	3.58×10^{15}	[127]
		Passivated	1.16	25.42	81.26	24.00	1.44×10^{14}	
Interface Engineering	BTACl	Control	1.08	23.78	71.74	18.43	2.74×10^{16}	[128]
		Passivated	1.19	23.85	76.54	21.72	1.41×10^{16}	
Interface Engineering	I-TFBA	Control	1.13	21.80	79.20	19.50	1.43×10^{16}	[129]
		Passivated	1.18	22.80	82.20	22.02	1.14×10^{16}	
Interface Engineering	CsF	Control	1.15	25.28	75.32	21.93	9.67×10^{15}	[130]
		Passivated	1.18	25.47	77.31	23.13	4.83×10^{15}	
Interface Engineering	LDA-Cl	Control	1.09	25.30	79.47	21.91	5.51×10^{16}	[131]
		Passivated	1.13	25.46	80.69	23.28	5.09×10^{16}	

Table 1. Cont.

Strategy	Passivator	Processing	V_{OC} [V]	J_{SC} [mA cm^{-2}]	FF [%]	PCE [%]	N_{trap} [cm^{-3}]	Ref.
Interface Engineering	BDDAB	Control	1.06	23.96	75.95	19.39	3.41×10^{15}	[132]
		Passivated	1.10	24.98	80.10	22.08	1.15×10^{15}	
Perovskite Surface Treatment	Taurine	Control	1.07	23.94	77.00	19.73	1.13×10^{15}	[133]
		Passivated	1.13	24.72	80.20	22.54	4.70×10^{14}	
Perovskite Surface Treatment	AHPD	Control	0.69	30.68	74.18	15.72	1.40×10^{13}	[134]
		Passivated	0.81	30.03	79.10	19.18	7.80×10^{12}	
Perovskite Surface Treatment	L-arginine	Control	1.13	22.98	78.56	20.37	9.98×10^{13}	[135]
		Passivated	1.18	23.57	82.36	22.96	3.58×10^{15}	
Perovskite Surface Treatment	FID	Control	1.08	25.31	76.80	21.07	1.29×10^{16}	[136]
		Passivated	1.14	25.42	81.00	23.44	8.85×10^{15}	
Perovskite Surface Treatment	ORO	Control	1.05	24.55	75.36	19.47	1.90×10^{16}	[137]
		Passivated	1.09	24.66	76.47	20.62	1.20×10^{16}	
Perovskite Surface Treatment	L-Theanine	Control	1.15	25.10	76.99	22.29	1.73×10^{16}	[138]
		Passivated	1.19	25.13	81.87	24.58	0.73×10^{16}	
Perovskite Surface Treatment	DA	Control	1.07	22.33	79.59	19.04	8.50×10^{15}	[139]
		Passivated	1.13	23.36	83.92	22.15	3.80×10^{15}	
Perovskite Surface Treatment	PB	Control	0.98	18.68	74.00	13.55	4.04×10^{16}	[140]
		Passivated	1.09	21.02	79.00	18.01	2.04×10^{16}	
Perovskite Surface Treatment	VC	Control	1.16	24.95	78.46	22.67	3.27×10^{16}	[141]
		Passivated	1.16	25.00	79.34	23.01	1.55×10^{16}	
Dopant Passivation	F4-TCNQ	Control	1.09	22.55	73.49	18.06	3.99×10^{15}	[142]
		Passivated	1.11	23.70	78.58	20.67	1.16×10^{15}	
Dopant Passivation	GV	Control	1.06	22.28	68.28	16.25	9.76×10^{15}	[143]
		Passivated	1.09	23.33	75.73	19.20	7.17×10^{15}	
Dopant Passivation	IMBF$_4$	Control	1.09	24.66	75.07	20.18	1.96×10^{15}	[144]
		Passivated	1.15	24.90	80.50	23.05	1.59×10^{15}	
Dopant Passivation	LiOH	Control	1.15	22.90	73.27	19.26	3.54×10^{15}	[145]
		Passivated	1.15	24.20	76.26	21.31	2.83×10^{15}	
Dopant Passivation	g-C$_3$N$_5$	Control	1.16	22.51	73.84	19.28	2.04×10^{16}	[146]
		Passivated	1.18	23.97	78.98	22.34	6.82×10^{15}	

We convince that the passivation of deep defect-induced traps plays an irreplaceable role in the latest progress of PSCs device performance and stability. However, the complete understanding of the passivation mechanism has not been fully resolved, which may be largely due to the versatility of some passivators, the overall complexity of the discussed system, and limited experimental techniques. It is difficult to achieve efficient passivation in perovskite film manufacturing processes if various types of defects and their concentrations and trap depths cannot be accurately identified. This may be the biggest challenge limiting defect passivation of PSCs. Looking ahead, a better understanding of the passivation mechanism is needed to guide the selection, design, and combination of passivators to produce synergistic passivation, which is crucial for further improving the efficiency and stability of PSCs.

In addition, although lead-based PSCs show superior performance, the toxicity of lead is a common concern for all developers and consumers because lead leakage that may occur during manufacturing, installation, or disposal can seriously pollute the environment and endanger humans. Since the lattice disorder and trap density of perovskite films are proportional to the size of perovskite devices, there are still problems in the manufacture of large-area devices for commercialization. Moreover, due to the structural disorder caused by interface and grain boundary defects, PSCs are difficult to maintain good efficiency and stability in large-scale production, so much effort should be spared to developing large-scale manufacturing technologies for perovskite photovoltaics. Considering the continuous efforts of researchers and the improvement of the performance and stability of perovskite devices, the actual commercialization of PSCs seems to be fully achievable in the near future.

Author Contributions: Funding acquisition, writing-original draft preparation, B.D.; writing-review and editing, B.D. and K.H.; validation and supervision, B.L.; formal analysis and investigation X.Z. All authors have read and agreed to the published version of the manuscript.

Funding: This work was financially supported by the Doctoral research start-up project of Xi'an Polytechnic University (107020579), Shaanxi Natural Science Foundation of China (2023-JC-QN-0680), China Postdoctoral Science Foundation (2021M690127), the Natural Science Foundation of Hunan Province (Grant No. 2021JJ40141), the Scientific Research Fund of Hunan Provincial Education Department (20B121), and the Training Project of Changsha City for Distinguished Young Scholars (kq2107023).

Institutional Review Board Statement: Not applicable.

Informed Consent Statement: Not applicable.

Data Availability Statement: Not applicable.

Conflicts of Interest: The authors declare no competing financial interest.

References

1. Xiang, H.; Liu, P.; Wang, W.; Ran, R.; Zhou, W.; Shao, Z. Towards highly stable and efficient planar perovskite solar cells: Materials development, defect control and interfacial engineering. *Chem. Eng. J.* **2021**, *420*, 127599. [CrossRef]
2. Kim, D.H.; Whitaker, J.B.; Li, Z.; van Hest, M.F.A.M.; Zhu, K. Outlook and Challenges of Perovskite Solar Cells toward Terawatt-Scale Photovoltaic Module Technology. *Joule* **2018**, *2*, 1437–1451. [CrossRef]
3. Roy, P.; Kumar Sinha, N.; Tiwari, S.; Khare, A. A review on perovskite solar cells: Evolution of architecture, fabrication techniques, commercialization issues and status. *Sol. Energy* **2020**, *198*, 665–688. [CrossRef]
4. Chapin, D.M.; Fuller, C.S.; Pearson, G.L. A New Silicon p-n Junction Photocell for Converting Solar Radiation into Electrical Power. *J. Appl. Phys.* **1954**, *25*, 676–677. [CrossRef]
5. Green, M.A.; Dunlop, E.D.; Hohl-Ebinger, J.; Siefer, G.; Yoshita, M.; Kopidakis, N.; Bothe, K.; Hao, X. Solar cell efficiency tables (version 61). *Prog. Photovolt.* **2023**, *31*, 3–16. [CrossRef]
6. Yoshikawa, K.; Kawasaki, H.; Yoshida, W.; Irie, T.; Konishi, K.; Nakano, K.; Uto, T.; Adachi, D.; Kanematsu, M.; Uzu, H.; et al. Silicon heterojunction solar cell with interdigitated back contacts for a photoconversion efficiency over 26%. *Nat. Energy* **2017**, *2*, 17032. [CrossRef]
7. Liu, W.; Li, H.; Qiao, B.; Zhao, S.; Xu, Z.; Song, D. Highly efficient CIGS solar cells based on a new CIGS bandgap gradient design characterized by numerical simulation. *Sol. Energy* **2022**, *233*, 337–344. [CrossRef]
8. NREL Efficiency Chart. Available online: https://www.nrel.gov/pv/cell-efficiency.html (accessed on 10 November 2022).
9. Wu, P.; Wang, S.; Li, X.; Zhang, F. Advances in SnO_2-based perovskite solar cells: From preparation to photovoltaic applications. *J. Mater. Chem. A* **2021**, *9*, 19554–19588. [CrossRef]
10. Huang, J.; Yuan, Y.; Shao, Y.; Yan, Y. Understanding the physical properties of hybrid perovskites for photovoltaic applications. *Nat. Rev. Mater.* **2017**, *2*, 17042. [CrossRef]
11. Huang, J.; Shao, Y.; Dong, Q. Organometal Trihalide Perovskite Single Crystals: A Next Wave of Materials for 25% Efficiency Photovoltaics and Applications Beyond? *J. Phys. Chem. Lett.* **2015**, *6*, 3218–3227. [CrossRef]
12. Bush, K.A.; Palmstrom, A.F.; Yu, Z.J.; Boccard, M.; Cheacharoen, R.; Mailoa, J.P.; McMeekin, D.P.; Hoye, R.L.Z.; Bailie, C.D.; Leijtens, T.; et al. 23.6%-efficient monolithic perovskite/silicon tandem solar cells with improved stability. *Nat. Energy* **2017**, *2*, 17009. [CrossRef]
13. Sahli, F.; Werner, J.; Kamino, B.A.; Brauninger, M.; Monnard, R.; Paviet-Salomon, B.; Barraud, L.; Ding, L.; Diaz Leon, J.J.; Sacchetto, D.; et al. Fully textured monolithic perovskite/silicon tandem solar cells with 25.2% power conversion efficiency. *Nat. Mater.* **2018**, *17*, 820–826. [CrossRef] [PubMed]
14. Kojima, A.; Teshima, K.; Shirai, Y.; Miyasaka, T. Organometal Halide Perovskites as Visible-Light Sensitizers for Photovoltaic Cells. *J. Am. Chem. Soc.* **2009**, *131*, 6050–6051. [CrossRef] [PubMed]
15. Kim, H.S.; Lee, C.R.; Im, J.H.; Lee, K.B.; Moehl, T.; Marchioro, A.; Moon, S.J.; Humphry-Baker, R.; Yum, J.H.; Moser, J.E.; et al. Lead iodide perovskite sensitized all-solid-state submicron thin film mesoscopic solar cell with efficiency exceeding 9%. *Sci. Rep.* **2012**, *2*, 591. [CrossRef] [PubMed]
16. Min, H.; Kim, M.; Lee, S.-U.; Kim, H.; Kim, G.; Choi, K.; Lee, J.H.; Seok, S.I. Efficient, stable solar cells by using inherent bandgap of alpha-phase formamidinium lead iodide. *Science* **2019**, *366*, 749–753. [CrossRef]
17. Jeong, M.; Choi, I.W.; Go, E.M.; Cho, Y.; Kim, M.; Lee, B.; Jeong, S.; Jo, Y.; Choi, H.W.; Lee, J.; et al. Stable perovskite solar cells with efficiency exceeding 24.8% and 0.3-V voltage loss. *Science* **2020**, *369*, 1615–1620. [CrossRef]
18. Yoo, J.J.; Seo, G.; Chua, M.R.; Park, T.G.; Lu, Y.; Rotermund, F.; Kim, Y.K.; Moon, C.S.; Jeon, N.J.; Correa-Baena, J.P.; et al. Efficient perovskite solar cells via improved carrier management. *Nature* **2021**, *590*, 587–593. [CrossRef] [PubMed]
19. Min, H.; Lee, D.Y.; Kim, J.; Kim, G.; Lee, K.S.; Kim, J.; Paik, M.J.; Kim, Y.K.; Kim, K.S.; Kim, M.G.; et al. Perovskite solar cells with atomically coherent interlayers on SnO_2 electrodes. *Nature* **2021**, *598*, 444–450. [CrossRef]

20. Zhao, Y.; Ma, F.; Qu, Z.; Yu, S.; Shen, T.; Deng, H.-X.; Chu, X.; Peng, X.; Yuan, Y.; Zhang, X.; et al. Inactive $(PbI_2)_2$RbCl stabilizes perovskite films for efficient solar cells. *Science* **2022**, *377*, 531–534. [CrossRef]
21. Tan, S.; Huang, T.; Yavuz, I.; Wang, R.; Yoon, T.W.; Xu, M.; Xing, Q.; Park, K.; Lee, D.K.; Chen, C.H.; et al. Stability-limiting heterointerfaces of perovskite photovoltaics. *Nature* **2022**, *605*, 268–273. [CrossRef]
22. Chen, B.; Rudd, P.N.; Yang, S.; Yuan, Y.; Huang, J. Imperfections and their passivation in halide perovskite solar cells. *Chem. Soc. Rev.* **2019**, *48*, 3842–3867. [CrossRef]
23. Shockley, W.; Read, W.T. Statistics of the Recombinations of Holes and Electrons. *Phys. Rev.* **1952**, *87*, 835–842. [CrossRef]
24. Ono, L.K.; Liu, S.F.; Qi, Y. Reducing Detrimental Defects for High-Performance Metal Halide Perovskite Solar Cells. *Angew. Chem. Int. Ed.* **2020**, *59*, 6676–6698. [CrossRef] [PubMed]
25. Chen, J.; Park, N.G. Causes and Solutions of Recombination in Perovskite Solar Cells. *Adv. Mater.* **2019**, *31*, 1803019. [CrossRef] [PubMed]
26. Park, K.; Lee, J.-H.; Lee, J.-W. Surface Defect Engineering of Metal Halide Perovskites for Photovoltaic Applications. *ACS Energy Lett.* **2022**, *7*, 1230–1239. [CrossRef]
27. Schulz, P.; Cahen, D.; Kahn, A. Halide Perovskites: Is It All about the Interfaces? *Chem. Rev.* **2019**, *119*, 3349–3417. [CrossRef]
28. Ni, Z.; Bao, C.; Liu, Y.; Jiang, Q.; Wu, W.-Q.; Chen, S.; Dai, X.; Chen, B.; Hartweg, B.; Yu, Z.; et al. Resolving spatial and energetic distributions of trap states in metal halide perovskite solar cells. *Science* **2020**, *367*, 1352–1358. [CrossRef] [PubMed]
29. Liu, T.; Zhao, X.; Zhong, X.; Burlingame, Q.C.; Kahn, A.; Loo, Y.-L. Improved Absorber Phase Stability, Performance, and Lifetime in Inorganic Perovskite Solar Cells with Alkyltrimethoxysilane Strain-Release Layers at the Perovskite/TiO_2 Interface. *ACS Energy Lett.* **2022**, *7*, 3531–3538. [CrossRef]
30. Wei, Y.; Rong, B.; Chen, X.; Chen, Y.; Liu, H.; Ye, X.; Huang, Y.; Fan, L.; Wu, J. Efficiency improvement of perovskite solar cell utilizing cystamine dihydrochloride for interface modification. *Mater. Res. Bull.* **2022**, *155*, 111949. [CrossRef]
31. Nozik, A.J.; Memming, R. Physical Chemistry of Semiconductor−Liquid Interfaces. *J. Phys. Chem.* **1996**, *100*, 13061–13078. [CrossRef]
32. Geng, Q.; Jia, X.; He, Z.; Hu, Y.; Gao, Y.; Yang, S.; Yao, C.; Zhang, S. Interface Engineering via Amino Acid for Efficient and Stable Perovskite Solar Cells. *Adv. Mater. Interfaces* **2022**, *9*, 2201641. [CrossRef]
33. Li, Y.; Li, S.; Shen, Y.; Han, X.; Li, Y.; Yu, Y.; Huang, M.; Tao, X. Multifunctional Histidine Cross-Linked Interface toward Efficient Planar Perovskite Solar Cells. *ACS Appl. Mater. Interfaces* **2022**, *14*, 47872–47881. [CrossRef] [PubMed]
34. Zhang, Y.; Kong, T.; Xie, H.; Song, J.; Li, Y.; Ai, Y.; Han, Y.; Bi, D. Molecularly Tailored SnO_2/Perovskite Interface Enabling Efficient and Stable $FAPbI_3$ Solar Cells. *ACS Energy Lett.* **2022**, *7*, 929–938. [CrossRef]
35. Jiang, Q.; Zhang, X.; You, J. SnO_2: A Wonderful Electron Transport Layer for Perovskite Solar Cells. *Small* **2018**, *14*, 1801154. [CrossRef]
36. Park, S.Y.; Zhu, K. Advances in SnO_2 for Efficient and Stable n-i-p Perovskite Solar Cells. *Adv. Mater.* **2022**, *34*, 2110438. [CrossRef] [PubMed]
37. Zhuang, Q.; Zhang, C.; Gong, C.; Li, H.; Li, H.; Zhang, Z.; Yang, H.; Chen, J.; Zang, Z. Tailoring multifunctional anion modifiers to modulate interfacial chemical interactions for efficient and stable perovskite solar cells. *Nano Energy* **2022**, *102*, 107747. [CrossRef]
38. Xiong, Z.; Chen, X.; Zhang, B.; Odunmbaku, G.O.; Ou, Z.; Guo, B.; Yang, K.; Kan, Z.; Lu, S.; Chen, S.; et al. Simultaneous Interfacial Modification and Crystallization Control by Biguanide Hydrochloride for Stable Perovskite Solar Cells with PCE of 24.4%. *Adv. Mater.* **2022**, *34*, 2106118. [CrossRef]
39. Zhang, Q.; Xiong, S.; Ali, J.; Qian, K.; Li, Y.; Feng, W.; Hu, H.; Song, J.; Liu, F. Polymer interface engineering enabling high-performance perovskite solar cells with improved fill factors of over 82%. *J. Mater. Chem. C* **2020**, *8*, 5467–5475. [CrossRef]
40. Chen, J.; Wang, D.; Chen, S.; Hu, H.; Li, Y.; Huang, Y.; Zhang, Z.; Jiang, Z.; Xu, J.; Sun, X.; et al. Dually Modified Wide-Bandgap Perovskites by Phenylethylammonium Acetate toward Highly Efficient Solar Cells with Low Photovoltage Loss. *ACS Appl. Mater. Interfaces* **2022**, *14*, 43246–43256. [CrossRef]
41. Wan, F.; Ke, L.; Yuan, Y.; Ding, L. Passivation with crosslinkable diamine yields 0.1 V non-radiative Voc loss in inverted perovskite solar cells. *Sci. Bull.* **2021**, *66*, 417–420. [CrossRef]
42. Kamat, P.V. Evolution of Perovskite Photovoltaics and Decrease in Energy Payback Time. *J. Phys. Chem. Lett.* **2013**, *4*, 3733–3734. [CrossRef]
43. Wang, H.; Ye, F.; Liang, J.; Liu, Y.; Hu, X.; Zhou, S.; Chen, C.; Ke, W.; Tao, C.; Fang, G. Pre-annealing treatment for high-efficiency perovskite solar cells via sequential deposition. *Joule* **2022**, *6*, 2869–2884. [CrossRef]
44. Cai, W.; Wang, Y.; Shang, W.; Liu, J.; Wang, M.; Dong, Q.; Han, Y.; Li, W.; Ma, H.; Wang, P.; et al. Lewis base governing superfacial proton behavior of hybrid perovskite: Basicity dependent passivation strategy. *Chem. Eng. J.* **2022**, *446*, 137033. [CrossRef]
45. Shini, F.; Thambidurai, M.; Dewi, H.A.; Jamaludin, N.F.; Bruno, A.; Kanwat, A.; Mathews, N.; Dang, C.; Nguyen, H.D. Interfacial passivation with 4-chlorobenzene sulfonyl chloride for stable and efficient planar perovskite solar cells. *J. Mater. Chem. C* **2022**, *10*, 9044–9051. [CrossRef]
46. Zhong, H.; Jia, Z.; Shen, J.; Yu, Z.; Yin, S.; Liu, X.; Fu, G.; Chen, S.; Yang, S.; Kong, W. Surface treatment of the perovskite via self-assembled dipole layer enabling enhanced efficiency and stability for perovskite solar cells. *Appl. Surf. Sci.* **2022**, *602*, 154365. [CrossRef]
47. Gu, W.-M.; Zhang, Y.; Jiang, K.-J.; Yu, G.; Xu, Y.; Huang, J.-H.; Zhang, Y.; Wang, F.; Li, Y.; Lin, Y.; et al. Surface fluoride management for enhanced stability and efficiency of halide perovskite solar cells via a thermal evaporation method. *J. Mater. Chem. A* **2022**, *10*, 12882–12889. [CrossRef]
48. Xu, S.; Zhang, L.; Liu, B.; Liang, Z.; Xu, H.; Zhang, H.; Ye, J.; Ma, H.; Liu, G.; Pan, X. Constructing of superhydrophobic and intact crystal terminal: Interface sealing strategy for stable perovskite solar cells with efficiency over 23%. *Chem. Eng. J.* **2023**, *453*, 139808. [CrossRef]
49. Wang, X.; Huang, H.; Du, S.; Cui, P.; Lan, Z.; Yang, Y.; Yan, L.; Ji, J.; Liu, B.; Qu, S.; et al. Facile Synthesized Acetamidine Thiocyanate with Synergistic Passivation and Crystallization for Efficient Perovskite Solar Cells. *Sol. RRL* **2022**, *6*, 2200717. [CrossRef]

50. Liu, W.; Dong, H.; Li, X.; Xia, T.; Zheng, G.; Tian, N.; Mo, S.; Peng, Y.; Yang, Y.; Yao, D.; et al. Interfacial Engineering Based on Polyethylene Glycol and Cuprous Thiocyanate Layers for Efficient and Stable Perovskite Solar Cells. *Adv. Mater. Interfaces* **2022**, *9*, 2201437. [CrossRef]
51. Ren, J.; Liu, T.; He, B.; Wu, G.; Gu, H.; Wang, B.; Li, J.; Mao, Y.; Chen, S.; Xing, G. Passivating Defects at the Bottom Interface of Perovskite by Ethylammonium to Improve the Performance of Perovskite Solar Cells. *Small* **2022**, *18*, 2203536. [CrossRef]
52. Wu, T.; Ono, L.K.; Yoshioka, R.; Ding, C.; Zhang, C.; Mariotti, S.; Zhang, J.; Mitrofanov, K.; Liu, X.; Segawa, H.; et al. Elimination of light-induced degradation at the nickel oxide-perovskite heterojunction by aprotic sulfonium layers towards long-term operationally stable inverted perovskite solar cells. *Energy Environ. Sci.* **2022**, *15*, 4612–4624. [CrossRef]
53. Liu, D.; Chen, M.; Wei, Z.; Zou, C.; Liu, X.; Xie, J.; Li, Q.; Yang, S.; Hou, Y.; Yang, H.G. Strain-free hybrid perovskite films based on a molecular buffer interface for efficient solar cells. *J. Mater. Chem. A* **2022**, *10*, 10865–10871. [CrossRef]
54. Chen, J.; Seo, J.Y.; Park, N.G. Simultaneous Improvement of Photovoltaic Performance and Stability by In Situ Formation of 2D Perovskite at (FAPbI$_3$)$_{0.88}$(CsPbBr$_3$)$_{0.12}$/CuSCN Interface. *Adv. Energy Mater.* **2018**, *8*, 1702714. [CrossRef]
55. Xu, J.; Dai, J.; Dong, H.; Li, P.; Chen, J.; Zhu, X.; Wang, Z.; Jiao, B.; Hou, X.; Li, J.; et al. Surface-tension release in PTAA-based inverted perovskite solar cells. *Org. Electron.* **2022**, *100*, 106378. [CrossRef]
56. Li, Y.; Wang, B.; Liu, T.; Zeng, Q.; Cao, D.; Pan, H.; Xing, G. Interfacial Engineering of PTAA/Perovskites for Improved Crystallinity and Hole Extraction in Inverted Perovskite Solar Cells. *ACS Appl. Mater. Interfaces* **2022**, *14*, 3284–3292. [CrossRef]
57. Boyd, C.C.; Shallcross, R.C.; Moot, T.; Kerner, R.; Bertoluzzi, L.; Onno, A.; Kavadiya, S.; Chosy, C.; Wolf, E.J.; Werner, J.; et al. Overcoming Redox Reactions at Perovskite-Nickel Oxide Interfaces to Boost Voltages in Perovskite Solar Cells. *Joule* **2020**, *4*, 1759–1775. [CrossRef]
58. Mohd Yusoff, A.R.b.; Vasilopoulou, M.; Georgiadou, D.G.; Palilis, L.C.; Abate, A.; Nazeeruddin, M.K. Passivation and process engineering approaches of halide perovskite films for high efficiency and stability perovskite solar cells. *Energy Environ. Sci.* **2021**, *14*, 2906–2953. [CrossRef]
59. Tang, Y.; Gu, Z.; Fu, C.; Xiao, Q.; Zhang, S.; Zhang, Y.; Song, Y. FAPbI$_3$ Perovskite Solar Cells: From Film Morphology Regulation to Device Optimization. *Sol. RRL* **2022**, *6*, 2200120. [CrossRef]
60. Tong, Y.; Najar, A.; Wang, L.; Liu, L.; Du, M.; Yang, J.; Li, J.; Wang, K.; Liu, S.F. Wide-Bandgap Organic-Inorganic Lead Halide Perovskite Solar Cells. *Adv. Sci.* **2022**, *9*, 2105085. [CrossRef]
61. Huang, K.; Feng, X.; Li, H.; Long, C.; Liu, B.; Shi, J.; Meng, Q.; Weber, K.; Duong, T.; Yang, J. Manipulating the Migration of Iodine Ions via Reverse-Biasing for Boosting Photovoltaic Performance of Perovskite Solar Cells. *Adv. Sci.* **2022**, *9*, 2204163. [CrossRef] [PubMed]
62. Zhao, Y.; Yavuz, I.; Wang, M.; Weber, M.H.; Xu, M.; Lee, J.H.; Tan, S.; Huang, T.; Meng, D.; Wang, R.; et al. Suppressing ion migration in metal halide perovskite via interstitial doping with a trace amount of multivalent cations. *Nat. Mater.* **2022**, *21*, 1396–1402. [CrossRef] [PubMed]
63. Rezaee, E.; Zhang, W.; Silva, S.R.P. Solvent Engineering as a Vehicle for High Quality Thin Films of Perovskites and Their Device Fabrication. *Small* **2021**, *17*, 2008145. [CrossRef] [PubMed]
64. Jung, M.; Ji, S.G.; Kim, G.; Seok, S.I. Perovskite precursor solution chemistry: From fundamentals to photovoltaic applications. *Chem. Soc. Rev.* **2019**, *48*, 2011–2038. [CrossRef] [PubMed]
65. Im, J.H.; Lee, C.R.; Lee, J.W.; Park, S.W.; Park, N.G. 6.5% efficient perovskite quantum-dot-sensitized solar cell. *Nanoscale* **2011**, *3*, 4088–4093. [CrossRef]
66. Lee, M.M.; Teuscher, J.; Miyasaka, T.; Murakami, T.N.; Snaith, H.J. Efficient Hybrid Solar Cells Based on Meso-Superstructured Organometal Halide Perovskites. *Science* **2012**, *338*, 643–647. [CrossRef]
67. Wu, Y.; Islam, A.; Yang, X.; Qin, C.; Liu, J.; Zhang, K.; Peng, W.; Han, L. Retarding the crystallization of PbI$_2$ for highly reproducible planar-structured perovskite solar cells via sequential deposition. *Energy Environ. Sci.* **2014**, *7*, 2934–2938. [CrossRef]
68. Jeon, N.J.; Noh, J.H.; Kim, Y.C.; Yang, W.S.; Ryu, S.; Seok, S.I. Solvent engineering for high-performance inorganic-organic hybrid perovskite solar cells. *Nat. Mater.* **2014**, *13*, 897–903. [CrossRef]
69. Hui, W.; Yang, Y.; Xu, Q.; Gu, H.; Feng, S.; Su, Z.; Zhang, M.; Wang, J.; Li, X.; Fang, J.; et al. Red-Carbon-Quantum-Dot-Doped SnO$_2$ Composite with Enhanced Electron Mobility for Efficient and Stable Perovskite Solar Cells. *Adv. Mater.* **2020**, *32*, 1906374. [CrossRef]
70. Yu, R.; Wu, G.; Shi, R.; Ma, Z.; Dang, Q.; Qing, Y.; Zhang, C.; Xu, K.; Tan, Z.A. Multidentate Coordination Induced Crystal Growth Regulation and Trap Passivation Enables over 24% Efficiency in Perovskite Solar Cells. *Adv. Energy Mater.* **2023**, *13*, 2203127. [CrossRef]
71. Chao, L.; Niu, T.; Gao, W.; Ran, C.; Song, L.; Chen, Y.; Huang, W. Solvent Engineering of the Precursor Solution toward Large-Area Production of Perovskite Solar Cells. *Adv. Mater.* **2021**, *33*, 2005410. [CrossRef]
72. Wu, X.; Zheng, Y.; Liang, J.; Zhang, Z.; Tian, C.; Zhang, Z.; Hu, Y.; Sun, A.; Wang, C.; Wang, J.; et al. Green-solvent-processed formamidinium-based perovskite solar cells with uniform grain growth and strengthened interfacial contact via a nanostructured tin oxide layer. *Mater. Horiz.* **2023**, *10*, 122–135. [CrossRef] [PubMed]
73. Fang, Y.; Jiang, Y.; Yang, Z.; Xu, Z.; Wang, Z.; Lu, X.; Gao, X.; Zhou, G.; Liu, J.M.; Gao, J. A Nontoxic NFM Solvent for High-Efficiency Perovskite Solar Cells with a Widened Processing Window. *ACS Appl. Mater. Interfaces* **2022**, *14*, 47758–47764. [CrossRef] [PubMed]
74. Hui, W.; Chao, L.; Lu, H.; Xia, F.; Wei, Q.; Su, Z.; Niu, T.; Tao, L.; Du, B.; Li, D.; et al. Stabilizing black-phase formamidinium perovskite formation at room temperature and high humidity. *Science* **2021**, *371*, 1359–1364. [CrossRef] [PubMed]
75. Bai, S.; Da, P.; Li, C.; Wang, Z.; Yuan, Z.; Fu, F.; Kawecki, M.; Liu, X.; Sakai, N.; Wang, J.T.; et al. Planar perovskite solar cells with long-term stability using ionic liquid additives. *Nature* **2019**, *571*, 245–250. [CrossRef] [PubMed]

76. Ran, J.; Wang, H.; Deng, W.; Xie, H.; Gao, Y.; Yuan, Y.; Yang, Y.; Ning, Z.; Yang, B. Ionic Liquid-Tuned Crystallization for Stable and Efficient Perovskite Solar Cells. *Sol. RRL* **2022**, *6*, 2200176. [CrossRef]
77. Li, L.; Tu, S.; You, G.; Cao, J.; Wu, D.; Yao, L.; Zhou, Z.; Shi, W.; Wang, W.; Zhen, H.; et al. Enhancing performance and stability of perovskite solar cells through defect passivation with a polyamide derivative obtained from benzoxazine-isocyanide chemistry. *Chem. Eng. J.* **2022**, *431*, 133951. [CrossRef]
78. Hu, L.; Li, S.; Zhang, L.; Liu, Y.; Zhang, C.; Wu, S.; Sun, Q.; Cui, Y.; Zhu, F.; Hao, Y.; et al. Unravelling the role of C_{60} derivatives as additives into active layers for achieving high-efficiency planar perovskite solar cells. *Carbon* **2020**, *167*, 160–168. [CrossRef]
79. Kim, K.; Wu, Z.; Han, J.; Ma, Y.; Lee, S.; Jung, S.K.; Lee, J.W.; Woo, H.Y.; Jeon, I. Homogeneously Miscible Fullerene inducing Vertical Gradient in Perovskite Thin-Film toward Highly Efficient Solar Cells. *Adv. Energy Mater.* **2022**, *12*, 2200877. [CrossRef]
80. Ahmed, E.; Breternitz, J.; Groh, M.F.; Ruck, M. Ionic liquids as crystallisation media for inorganic materials. *CrystEngComm* **2012**, *14*, 4874. [CrossRef]
81. Marsh, K.N.; Boxall, J.A.; Lichtenthaler, R. Room temperature ionic liquids and their mixtures—A review. *Fluid Phase Equilib.* **2004**, *219*, 93–98. [CrossRef]
82. Shahiduzzaman, M.; Muslih, E.Y.; Hasan, A.K.M.; Wang, L.; Fukaya, S.; Nakano, M.; Karakawa, M.; Takahashi, K.; Akhtaruzzaman, M.; Nunzi, J.-M.; et al. The benefits of ionic liquids for the fabrication of efficient and stable perovskite photovoltaics. *Chem. Eng. J.* **2021**, *411*, 128461. [CrossRef]
83. Niu, T.; Chao, L.; Gao, W.; Ran, C.; Song, L.; Chen, Y.; Fu, L.; Huang, W. Ionic Liquids-Enabled Efficient and Stable Perovskite Photovoltaics: Progress and Challenges. *ACS Energy Lett.* **2021**, *6*, 1453 1179. [CrossRef]
84. Chao, L.; Niu, T.; Gu, H.; Yang, Y.; Wei, Q.; Xia, Y.; Hui, W.; Zuo, S.; Zhu, Z.; Pei, C.; et al. Origin of High Efficiency and Long-Term Stability in Ionic Liquid Perovskite Photovoltaic. *Research* **2020**, *2020*, 2616345. [CrossRef] [PubMed]
85. He, Z.; Xu, C.; Li, L.; Liu, A.; Ma, T.; Gao, L. Highly efficient and stable perovskite solar cells induced by novel bulk organosulfur ammonium. *Mater. Today Energy* **2022**, *26*, 101004. [CrossRef]
86. Tian, J.; Zhang, K.; Xie, Z.; Peng, Z.; Zhang, J.; Osvet, A.; Lüer, L.; Kirchartz, T.; Rau, U.; Li, N.; et al. Quantifying the Energy Losses in $CsPbI_2Br$ Perovskite Solar Cells with an Open-Circuit Voltage of up to 1.45 V. *ACS Energy Lett.* **2022**, *7*, 4071–4080. [CrossRef]
87. Li, D.; Xia, T.; Liu, W.; Zheng, G.; Tian, N.; Yao, D.; Yang, Y.; Wang, H.; Long, F. Methylammonium thiocyanate seeds assisted heterogeneous nucleation for achieving high-performance perovskite solar cells. *Appl. Surf. Sci.* **2022**, *592*, 153206. [CrossRef]
88. Shahiduzzaman, M.; Yamamoto, K.; Furumoto, Y.; Kuwabara, T.; Takahashi, K.; Taima, T. Ionic liquid-assisted growth of methylammonium lead iodide spherical nanoparticles by a simple spin-coating method and photovoltaic properties of perovskite solar cells. *RSC Adv.* **2015**, *5*, 77495–77500. [CrossRef]
89. Wang, D.; Zhang, Z.; Huang, T.; She, B.; Liu, B.; Chen, Y.; Wang, L.; Wu, C.; Xiong, J.; Huang, Y.; et al. Crystallization Kinetics Control Enabled by a Green Ionic Liquid Additive toward Efficient and Stable Carbon-Based Mesoscopic Perovskite Solar Cells. *ACS Appl. Mater. Interfaces* **2022**, *14*, 9161–9171. [CrossRef]
90. Li, T.; Pan, Y.; Wang, Z.; Xia, Y.; Chen, Y.; Huang, W. Additive engineering for highly efficient organic–inorganic halide perovskite solar cells: Recent advances and perspectives. *J. Mater. Chem. A* **2017**, *5*, 12602–12652. [CrossRef]
91. Chang, C.Y.; Chu, C.Y.; Huang, Y.C.; Huang, C.W.; Chang, S.Y.; Chen, C.A.; Chao, C.Y.; Su, W.F. Tuning perovskite morphology by polymer additive for high efficiency solar cell. *ACS Appl. Mater. Interfaces* **2015**, *7*, 4955–4961. [CrossRef]
92. Li, X.; Sheng, W.; Duan, X.; Lin, Z.; Yang, J.; Tan, L.; Chen, Y. Defect Passivation Effect of Chemical Groups on Perovskite Solar Cells. *ACS Appl. Mater. Interfaces* **2022**, *14*, 34161–34170. [CrossRef] [PubMed]
93. Kroto, H.W.; Heath, J.R.; O'Brien, S.C.; Curl, R.F.; Smalley, R.E. C_{60}: Buckminsterfullerene. *Nature* **1985**, *318*, 162–163. [CrossRef]
94. Li, C.-Z.; Chueh, C.-C.; Yip, H.-L.; Zou, J.; Chen, W.-C.; Jen, A.K.Y. Evaluation of structure–property relationships of solution-processible fullerene acceptors and their n-channel field-effect transistor performance. *J. Mater. Chem.* **2012**, *22*, 14976. [CrossRef]
95. Menzel, D.; Al-Ashouri, A.; Tejada, A.; Levine, I.; Guerra, J.A.; Rech, B.; Albrecht, S.; Korte, L. Field Effect Passivation in Perovskite Solar Cells by a LiF Interlayer. *Adv. Energy Mater.* **2022**, *12*, 2201109. [CrossRef]
96. Liu, K.; Chen, S.; Wu, J.; Zhang, H.; Qin, M.; Lu, X.; Tu, Y.; Meng, Q.; Zhan, X. Fullerene derivative anchored SnO_2 for high-performance perovskite solar cells. *Energy Environ. Sci.* **2018**, *11*, 3463–3471. [CrossRef]
97. Zhang, F.; Zhu, K. Additive Engineering for Efficient and Stable Perovskite Solar Cells. *Adv. Energy Mater.* **2019**, *10*, 1902579. [CrossRef]
98. Xiao, M.; Huang, F.; Huang, W.; Dkhissi, Y.; Zhu, Y.; Etheridge, J.; Gray-Weale, A.; Bach, U.; Cheng, Y.B.; Spiccia, L. A fast deposition-crystallization procedure for highly efficient lead iodide perovskite thin-film solar cells. *Angew. Chem. Int. Ed.* **2014**, *53*, 9898–9903. [CrossRef]
99. Mei, Y.; Sun, M.; Liu, H.; Li, X.; Wang, S. Polymer additive assisted crystallization of perovskite films for high-performance solar cells. *Org. Electron.* **2021**, *96*, 106258. [CrossRef]
100. Wang, Z.; Chen, S.; Gao, X. PTB7 as additive in Anti-solvent to enhance perovskite film surface crystallinity for solar cells with efficiency over 21%. *Appl. Surf. Sci.* **2022**, *575*, 151737. [CrossRef]
101. Mei, D.; Qiu, L.; Chen, L.; Xie, F.; Song, L.; Wang, J.; Du, Y.; Xiong, J. Incorporating polyvinyl pyrrolidone in green anti-solvent isopropanol: A facile approach to obtain high efficient and stable perovskite solar cells. *Thin Solid Films* **2022**, *752*, 139196. [CrossRef]
102. Choi, H.; Liu, X.; Kim, H.I.; Kim, D.; Park, T.; Song, S. A Facile Surface Passivation Enables Thermally Stable and Efficient Planar Perovskite Solar Cells Using a Novel IDTT-Based Small Molecule Additive. *Adv. Energy Mater.* **2021**, *11*, 2003829. [CrossRef]
103. Guo, J.; Sun, J.; Hu, L.; Fang, S.; Ling, X.; Zhang, X.; Wang, Y.; Huang, H.; Han, C.; Cazorla, C.; et al. Indigo: A Natural Molecular Passivator for Efficient Perovskite Solar Cells. *Adv. Energy Mater.* **2022**, *12*, 2200537. [CrossRef]

104. Liu, Z.; Ono, L.K.; Qi, Y. Additives in metal halide perovskite films and their applications in solar cells. *J. Energy Chem.* **2020**, *46*, 215–228. [CrossRef]
105. Zhao, F.; Guo, Y.; Yang, P.; Tao, J.; Jiang, J.; Chu, J. Effect of Li-doped TiO$_2$ layer on the photoelectric performance of carbon-based CsPbIBr$_2$ perovskite solar cell. *J. Alloys Compd.* **2023**, *930*, 167377. [CrossRef]
106. Mo, H.; Wang, D.; Chen, Q.; Guo, W.; Maniyarasu, S.; Thomas, A.G.; Curry, R.J.; Li, L.; Liu, Z. Laser-Assisted Ultrafast Fabrication of Crystalline Ta-Doped TiO2 for High-Humidity-Processed Perovskite Solar Cells. *ACS Appl. Mater. Interfaces* **2022**, *14*, 15141–15153. [CrossRef] [PubMed]
107. Deng, J.; Zhang, H.; Wei, K.; Xiao, Y.; Zhang, C.; Yang, L.; Zhang, X.; Wu, D.; Yang, Y.; Zhang, J. Molecular Bridge Assisted Bifacial Defect Healing Enables Low Energy Loss for Efficient and Stable Perovskite Solar Cells. *Adv. Funct. Mater.* **2022**, *32*, 2209516. [CrossRef]
108. Luo, T.; Ye, G.; Chen, X.; Wu, H.; Zhang, W.; Chang, H. F-doping-Enhanced Carrier Transport in the SnO$_2$/Perovskite Interface for High-Performance Perovskite Solar Cells. *ACS Appl. Mater. Interfaces* **2022**, *14*, 42093–42101. [CrossRef]
109. Ma, H.; Wang, M.; Wang, Y.; Dong, Q.; Liu, J.; Yin, Y.; Zhang, J.; Pei, M.; Zhang, L.; Cai, W.; et al. Asymmetric organic diammonium salt buried in SnO$_2$ layer enables fast carrier transfer and interfacial defects passivation for efficient perovskite solar cells. *Chem. Eng. J.* **2022**, *442*, 136291. [CrossRef]
110. Chen, J.; Dong, H.; Li, J.; Zhu, X.; Xu, J.; Pan, F.; Xu, R.; Xi, J.; Jiao, B.; Hou, X.; et al. Solar Cell Efficiency Exceeding 25% through Rb-Based Perovskitoid Scaffold Stabilizing the Buried Perovskite Surface. *ACS Energy Lett.* **2022**, *7*, 3685–3694. [CrossRef]
111. Venu Rajendran, M.; Ganesan, S.; Sudhakaran Menon, V.; Raman, R.K.; Alagumalai, A.; Ashok Kumar, S.; Krishnamoorthy, A. Manganese Dopant-Induced Isoelectric Point Tuning of ZnO Electron Selective Layer Enable Improved Interface Stability in Cesium–Formamidinium-Based Planar Perovskite Solar Cells. *ACS Appl. Energy Mater.* **2022**, *5*, 6671–6686. [CrossRef]
112. Jiang, Q.; Zhang, L.; Wang, H.; Yang, X.; Meng, J.; Liu, H.; Yin, Z.; Wu, J.; Zhang, X.; You, J. Enhanced electron extraction using SnO$_2$ for high-efficiency planar-structure HC(NH$_2$)$_2$PbI$_3$-based perovskite solar cells. *Nat. Energy* **2016**, *2*, 16177. [CrossRef]
113. Che, Z.; Zhang, L.; Shang, J.; Wang, Q.; Zhou, Y.; Zhou, Y.; Liu, F. Organic ammonium chloride salt incorporated SnO$_2$ electron transport layers for improving the performance of perovskite solar cells. *Sustain. Energy Fuels* **2022**, *6*, 3416–3424. [CrossRef]
114. Arshad, Z.; Wageh, S.; Maiyalagan, T.; Ali, M.; Arshad, U.; Noor-ul, A.; Qadir, M.B.; Mateen, F.; Al-Sehemi, A.G. Enhanced charge transport characteristics in zinc oxide nanofibers via Mg^{2+} doping for electron transport layer in perovskite solar cells and antibacterial textiles. *Ceram. Int.* **2022**, *48*, 24363–24371. [CrossRef]
115. Adnan, M.; Usman, M.; Ali, S.; Javed, S.; Islam, M.; Akram, M.A. Aluminum Doping Effects on Interface Depletion Width of Low Temperature Processed ZnO Electron Transport Layer-Based Perovskite Solar Cells. *Front. Chem.* **2021**, *9*, 795291. [CrossRef]
116. Yang, W.S.; Park, B.-W.; Jung, E.H.; Jeon, N.J.; Kim, Y.C.; Lee, D.U.; Shin, S.S.; Seo, J.; Kim, E.K.; Noh, J.H.; et al. Iodide management in formamidinium-lead-halide-based perovskite layers for efficient solar cells. *Science* **2017**, *356*, 1376–1379. [CrossRef] [PubMed]
117. Min, H.; Ji, S.-G.; Seok, S.I. Relaxation of externally strained halide perovskite thin layers with neutral ligands. *Joule* **2022**, *6*, 2175–2185. [CrossRef]
118. Chen, Z.; Li, Y.; Liu, Z.; Shi, J.; Yu, B.; Tan, S.; Cui, Y.; Tan, C.; Tian, F.; Wu, H.; et al. Reconfiguration toward Self-Assembled Monolayer Passivation for High-Performance Perovskite Solar Cells. *Adv. Energy Mater.* **2023**, *13*, 2202799. [CrossRef]
119. Jeong, M.J.; Moon, C.S.; Lee, S.; Im, J.M.; Woo, M.Y.; Lee, J.H.; Cho, H.; Jeon, S.W.; Noh, J.H. Boosting radiation of stacked halide layer for perovskite solar cells with efficiency over 25%. *Joule* **2023**, *7*, 112–127. [CrossRef]
120. Rong, Y.; Jin, M.; Du, Q.; Shen, Z.; Feng, Y.; Wang, M.; Li, F.; Liu, R.; Li, H.; Chen, C. Simultaneous ambient long-term conductivity promotion, interfacial modification, ion migration inhibition and anti-deliquescence by MWCNT:NiO in spiro-OMeTAD for perovskite solar cells. *J. Mater. Chem. A* **2022**, *10*, 22592–22604. [CrossRef]
121. Yang, H.; Shen, Y.; Zhang, R.; Wu, Y.; Chen, W.; Yang, F.; Cheng, Q.; Chen, H.; Ou, X.; Yang, H.; et al. Composition-Conditioning Agent for Doped Spiro-OMeTAD to Realize Highly Efficient and Stable Perovskite Solar Cells. *Adv. Energy Mater.* **2022**, *12*, 2202207. [CrossRef]
122. Xu, D.; Gong, Z.; Jiang, Y.; Feng, Y.; Wang, Z.; Gao, X.; Lu, X.; Zhou, G.; Liu, J.M.; Gao, J. Constructing molecular bridge for high-efficiency and stable perovskite solar cells based on P3HT. *Nat. Commun.* **2022**, *13*, 7020. [CrossRef] [PubMed]
123. Sun, J.; Zhang, N.; Wu, J.; Yang, W.; He, H.; Huang, M.; Zeng, Y.; Yang, X.; Ying, Z.; Qin, G.; et al. Additive Engineering of the CuSCN Hole Transport Layer for High-Performance Perovskite Semitransparent Solar Cells. *ACS Appl. Mater. Interfaces* **2022**, *14*, 52223–52232. [CrossRef]
124. Wang, J.; Hu, Q.; Li, M.; Shan, H.; Feng, Y.; Xu, Z.-X. Poly(3-hexylthiophene)/Gold Nanorod Composites as Efficient Hole-Transporting Materials for Perovskite Solar Cells. *Sol. RRL* **2020**, *4*, 2000109. [CrossRef]
125. Park, S.; Kim, D.W.; Park, S.Y. Improved Stability and Efficiency of Inverted Perovskite Solar Cell by Employing Nickel Oxide Hole Transporting Material Containing Ammonium Salt Stabilizer. *Adv. Funct. Mater.* **2022**, *32*, 2200437. [CrossRef]
126. Liang, J.-W.; Firdaus, Y.; Azmi, R.; Faber, H.; Kaltsas, D.; Kang, C.H.; Nugraha, M.I.; Yengel, E.; Ng, T.K.; De Wolf, S.; et al. Cl$_2$-Doped CuSCN Hole Transport Layer for Organic and Perovskite Solar Cells with Improved Stability. *ACS Energy Lett.* **2022**, *7*, 3139–3148. [CrossRef]
127. Yue, X.; Zhao, X.; Fan, B.; Yang, Y.; Yan, L.; Qu, S.; Huang, H.; Zhang, Q.; Yan, H.; Cui, P.; et al. Surface Regulation through Dipolar Molecule Boosting the Efficiency of Mixed 2D/3D Perovskite Solar Cell to 24%. *Adv. Funct. Mater.* **2023**, *33*, 2209921. [CrossRef]
128. Guan, N.; Zhang, Y.; Chen, W.; Jiang, Z.; Gu, L.; Zhu, R.; Yadav, D.; Li, D.; Xu, B.; Cao, L.; et al. Deciphering the Morphology Change and Performance Enhancement for Perovskite Solar Cells Induced by Surface Modification. *Adv. Sci.* **2023**, *10*, 2205342. [CrossRef]

129. Zhang, C.; Shen, X.; Chen, M.; Zhao, Y.; Lin, X.; Qin, Z.; Wang, Y.; Han, L. Constructing a Stable and Efficient Buried Heterojunction via Halogen Bonding for Inverted Perovskite Solar Cells. *Adv. Energy Mater.* **2023**, *13*, 2203250. [CrossRef]
130. Wang, J.; Wang, Z.; Chen, S.; Jiang, N.; Yuan, L.; Zhang, J.; Duan, Y. Reduced Surface Hydroxyl and Released Interfacial Strain by Inserting CsF Anchor Interlayer for High-Performance Perovskite Solar Cells. *Sol. RRL* **2023**, *7*, 2200960. [CrossRef]
131. Zhang, K.; Zhao, G.; Ye, L.; Jia, N.; Liu, C.; Liu, S.; Ye, Q.; Wang, H. Bionic Levodopa-Modified TiO_2 for Preparation of Perovskite Solar Cells with Efficiency over 23%. *ACS Sustain. Chem. Eng.* **2022**, *10*, 16055–16063. [CrossRef]
132. Niu, Y.; He, D.; Zhang, X.; Hu, L.; Huang, Y. Enhanced Perovskite Solar Cell Stability and Efficiency via Multi-Functional Quaternary Ammonium Bromide Passivation. *Adv. Mater. Interfaces* **2023**, *10*, 2201497. [CrossRef]
133. Su, R.; Yang, X.; Ji, W.; Zhang, T.; Zhang, L.; Wang, A.; Jiang, Z.; Chen, Q.; Zhou, Y.; Song, B. Additive-associated antisolvent engineering of perovskite films for highly stable and efficient p–i–n perovskite solar cells. *J. Mater. Chem. C* **2022**, *10*, 18303–18311. [CrossRef]
134. Zhang, Y.; Chen, W.; Su, K.; Huang, Y.-Q.; Brooks, K.; Kinge, S.; Zhang, B.; Feng, Y.; Nazeeruddin, M.K. Bifunctional additive 2-amino-3-hydroxypyridine for stable and high-efficiency tin-lead perovskite solar cells. *J. Mater. Chem. C* **2023**, *11*, 151–160.
135. Xu, P.; Xie, L.; Yang, S.; Han, B.; Liu, J.; Chen, J.; Liu, C.; Jia, R.; Yang, M.; Ge, Z. Manipulating Halide Perovskite Passivation by Controlling Amino Acid Derivative Isoelectric Point for Stable and Efficient Inverted Perovskite Solar Cells. *Sol. RRL* **2023**, *7*, 2200858. [CrossRef]
136. Li, Y.; Shi, B.; Gao, F.; Wu, Y.; Lu, C.; Cai, X.; Li, J.; Zhang, C.; Liu, S.F. 2-Fluoro-4-iodoaniline passivates the surface of perovskite films to enhance photovoltaic properties. *Appl. Surf. Sci.* **2023**, *612*, 155787. [CrossRef]
137. Ni, M.; Qi, L. Orotic Acid as a Bifunctional Additive for Regulating Crystallization and Passivating Defects toward High-Performance Formamidinium-Cesium Perovskite Solar Cells. *ACS Appl. Mater. Interfaces* **2022**, *14*, 53808–53818. [CrossRef]
138. Li, Y.; Liu, L.; Zheng, C.; Liu, Z.; Chen, L.; Yuan, N.; Ding, J.; Wang, D.; Liu, S. Plant-Derived l-Theanine for Ultraviolet/Ozone Resistant Perovskite Photovoltaics. *Adv. Energy Mater.* **2023**, *13*, 2203190. [CrossRef]
139. Xu, Y.; Xiong, S.; Jiang, S.; Yang, J.; Li, D.; Wu, H.; You, X.; Zhang, Y.; Ma, Z.; Xu, J.; et al. Synchronous Modulation of Defects and Buried Interfaces for Highly Efficient Inverted Perovskite Solar Cells. *Adv. Energy Mater.* **2023**, *13*, 2203505. [CrossRef]
140. Huang, J.; Wang, H.; Chen, C.; Liu, S. Enhancement of $CsPbI_3$ perovskite solar cells with dual functional passivator 4-Fluoro-3-phenoxybenzaldehyde. *Surf. Interfaces* **2022**, *35*, 102477. [CrossRef]
141. Liu, B.; Wang, Y.; Wu, Y.; Zhang, Y.; Lyu, J.; Liu, Z.; Bian, S.; Bai, X.; Xu, L.; Zhou, D.; et al. Vitamin Natural Molecule Enabled Highly Efficient and Stable Planar n–p Homojunction Perovskite Solar Cells with Efficiency Exceeding 24.2%. *Adv. Energy Mater.* **2023**, *13*, 2203352. [CrossRef]
142. Park, H.; Heo, J.; Jeong, B.H.; Lee, J.; Park, H.J. Interface engineering of organic hole transport layer with facile molecular doping for highly efficient perovskite solar cells. *J. Power Sources* **2023**, *556*, 232428. [CrossRef]
143. Cheng, N.; Cao, Y.; Li, W.; Yu, Z.; Liu, Z.; Lei, B.; Zi, W.; Xiao, Z.; Tu, Y.; Rodríguez-Gallegos, C.D. SnO_2 electron transport layer modified with gentian violet for perovskite solar cells with enhanced performance. *Org. Electron.* **2022**, *108*, 106600. [CrossRef]
144. Shang, X.; Ma, X.; Meng, F.; Ma, J.; Yang, L.; Li, M.; Gao, D.; Chen, C. Zwitterionic ionic liquid synergistically induces interfacial dipole formation and traps state passivation for high-performance perovskite solar cells. *J. Colloid Interface Sci.* **2023**, *630*, 155–163. [CrossRef] [PubMed]
145. Zhuang, Q.; Wang, H.; Zhang, C.; Gong, C.; Li, H.; Chen, J.; Zang, Z. Ion diffusion-induced double layer doping toward stable and efficient perovskite solar cells. *Nano Res.* **2022**, *15*, 5114–5122. [CrossRef]
146. Yu, B.; Yu, H.; Sun, Y.; Zhang, J. Dual-layer synergetic optimization of high-efficiency planar perovskite solar cells using nitrogen-rich nitrogen carbide as an additive. *J. Mater. Chem. A* **2022**, *10*, 21390–21400. [CrossRef]

Disclaimer/Publisher's Note: The statements, opinions and data contained in all publications are solely those of the individual author(s) and contributor(s) and not of MDPI and/or the editor(s). MDPI and/or the editor(s) disclaim responsibility for any injury to people or property resulting from any ideas, methods, instructions or products referred to in the content.

Article

Rational Design of Novel Conjugated Terpolymers Based on Diketopyrrolopyrrole and Their Applications to Organic Thin-Film Transistors

Shiwei Ren [1,*,†], Yubing Ding [1,†], Wenqing Zhang [2], Zhuoer Wang [3], Sichun Wang [4,*] and Zhengran Yi [1,*]

1. Zhuhai-Fudan Innovation Research Institute, Hengqin 519000, China
2. Key Laboratory of Organic Solids, Institute of Chemistry, Chinese Academy of Sciences, Beijing 100190, China
3. Key Laboratory of Colloid and Interface Chemistry of Ministry of Education School of Chemistry and Chemical Engineering, Shandong University, Jinan 250100, China
4. Laboratory of Molecular Materials and Devices, Department of Materials Science, Fudan University, Shanghai 200438, China
* Correspondence: shiwei_ren@fudan.edu.cn (S.R.); scwang@ciac.ac.cn (S.W.); zhengranyi@fudan-zhuhai.org.cn (Z.Y.)
† These authors contributed equally to this work.

Citation: Ren, S.; Ding, Y.; Zhang, W.; Wang, Z.; Wang, S.; Yi, Z. Rational Design of Novel Conjugated Terpolymers Based on Diketopyrrolopyrrole and Their Applications to Organic Thin-Film Transistors. *Polymers* **2023**, *15*, 3803. https://doi.org/10.3390/polym15183803

Academic Editor: Tengling Ye

Received: 21 August 2023
Revised: 9 September 2023
Accepted: 16 September 2023
Published: 18 September 2023

Copyright: © 2023 by the authors. Licensee MDPI, Basel, Switzerland. This article is an open access article distributed under the terms and conditions of the Creative Commons Attribution (CC BY) license (https://creativecommons.org/licenses/by/4.0/).

Abstract: Organic polymer semiconductor materials, due to their good chemical modifiability, can be easily tuned by rational molecular structure design to modulate their material properties, which, in turn, affects the device performance. Here, we designed and synthesized a series of materials based on terpolymer structures and applied them to organic thin-film transistor (OTFT) device applications. The four polymers, obtained by polymerization of three monomers relying on the Stille coupling reaction, shared comparable molecular weights, with the main structural difference being the ratio of the thiazole component to the fluorinated thiophene (Tz/FS). The conjugated polymers exhibited similar energy levels and thermal stability; however, their photochemical and crystalline properties were distinctly different, leading to significantly varied mobility behavior. Materials with a Tz/FS ratio of 50:50 showed the highest electron mobility, up to 0.69 cm^2 V^{-1} s^{-1}. Our investigation reveals the fundamental relationship between the structure and properties of materials and provides a basis for the design of semiconductor materials with higher carrier mobility.

Keywords: polymer semiconductor materials; conjugated polymer; Stille coupling; OTFT device; terpolymer structures

1. Introduction

Research on polymers with conjugated architectures with alternating single and double bonds has attracted much attention because they are important candidate semiconductors for organic thin-film transistors (OTFTs) [1,2]. Device performance can be significantly enhanced by improving the molecular structure of organic semiconductors, and materials with carrier mobility in the range of 0.1–1 cm^2 V^{-1} s^{-1} have potential applications in the fabrication of lightweight and flexible optoelectronic devices such as radio-frequency electronic trademarks, smart cards, sensors, logic circuits, and e-paper [3–5]. Organic semiconductor materials are generally classified into three categories: N-type, P-type, and ambipolar materials, according to the carrier species. N-type materials are mainly responsible for electron transport, and their research progress lags far behind that of P-type materials, which are responsible for hole transport [6–8]. The research on N-type materials is mainly constrained by the lack of diversity of acceptor species, and most of the selective acceptor species are concentrated in the systems of lactam structures, such as diketopyrrolopyrrole (DPP), isoindigo (IID), naphthalene diimide (NDI), perylene diimide (PDI), and so on [9,10]. The acceptor units affect the lowest unoccupied molecular orbital (LUMO) energy level of the materials, which determines their ability to attract electrons. Based on the failure to effectively develop new potential acceptor structural units, it is extremely important to modify the existing structures. From the perspective of structure-determined properties, the introduction of functional groups with strong electron-withdrawing ability is the most direct way to lower the LUMO energy level of the material. Some typical strong electron-withdrawing groups include fluorine atoms (F), chlorine atoms, bromine atoms, carbonyl groups, cyano groups, dicyanomethylene groups, nitrogen atoms, and so on [5,11,12]. On the other hand, increasing the proportion of acceptors in the molecular structure is an effective way to increase the electron mobility. Chen et al. summarized the two-acceptor strategy and three-acceptor strategy, which involve further insertion of acceptor units within the polymer system based on the D-A structure in order to form a D-A-A or D-A-A-A structure, respectively [13]. Recently, we reported the significant effect of the all-acceptor strategy on the improvement of electron mobility, further enriching the development and structural design strategies of N-type materials [14]. In addition to this, the research on terpolymers is receiving increasing attention because they can achieve flexible adjustment of energy levels by modulating the ratio of the components, and because their synthesis does not require much effort [15–17].

Here, we chose to study a polymer based on DPP combined with thiophene, which was first reported in 2015 [18]. The polymer shows a moderate hole mobility of 0.09 cm^2 V^{-1} s^{-1} and is a P-type unipolar material that does not display electron transport properties (Figure 1). In order to improve its electron mobility, we envisaged the introduction of a strong electron-absorbing group F on thiophene, namely, fluorinated thiophene (FS). The F atom features a small atomic radius and a strong electron-accepting ability, which is an outstanding advantage for lowering the LUMO energy level of the material and not affecting the planarity of the whole molecule [19,20]. At the same time, the hydrophobicity of the F-containing polymer prevents moisture and oxygen from diffusing into the film, thereby improving the stability of the device in air. Furthermore, a strong electron-withdrawing unit, thiazole–thiazole (Tz), is introduced into the two-component polymers, and it is used as the third component to form the new ternary polymers [21,22]. In order to fully investigate the maximum electron mobility of this series of materials, four polymer materials with FS/Tz ratios of 100:0, 75:25, 50:50, and 25:75 were designed and prepared sequentially.

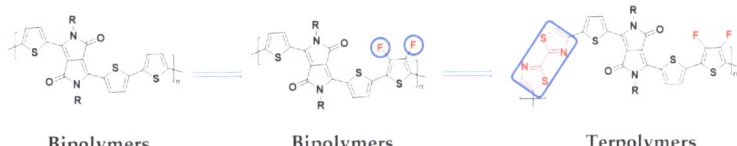

Bipolymers   Bipolymers   Terpolymers

Figure 1. Molecular structure design strategies and chemical structure schematics from bipolymer to terpolymer.

2. Materials and Methods

Materials: 3,6-di(thiophen-2-yl)-2,5-dihydropyrrolo[3,4-c]pyrrole-1,4-dione (DPP-S), (3,4-difluorothiophene-2,5-diyl)bis(trimethylstannane) (FS-Sn), N-bromosuccinimide (NBS), 5,5′-bis(trimethylstannyl)-2,2′-bithiazole (Tz-Sn), potassium carbonate (K$_2$CO$_3$), Pd catalysts, and organic solvents (e.g., methanol, chloroform (CHCl$_3$), dimethylformamide (DMF), petroleum ether (PE), dichloromethane (DCM), etc.) were purchased from Derton Optoelectronic Materials Science Technology Co., Ltd. (Shenzhen, China) and Sigma-Aldrich (St. Louis, MI, USA), and they were used as received. DPP-S was synthesized as described in the previous literature [23]. Synthesis of DPP-S-C$_8$C$_{10}$ monomer: Argon was passed into a solution of DPP-S (3.50 g, 11.67 mmol) in anhydrous DMF (50 mL), and K$_2$CO$_3$ (4.83 g, 35.00 mmol, 3.00 eq) was added in batches over ten minutes. The mixture was slowly heated to 80 °C, and 9-(bromomethyl)nonane (9.25 g, 25.67 mmol, 2.20 eq) was added dropwise to the flask using a syringe. The mixture was stirred at 100 °C for 12 h. The mixture was then extracted with DCM, washed with water and brine, and dried with Na$_2$SO$_4$. After removal of the solvent under reduced pressure, the residue was purified by silica gel chromatography by eluate (PE: DCM = 4:1) to afford an orange–red powder (6.13 g, 61.0%); δ ^{1}H NMR (400 MHz, chloroform-d) δ 8.88 (d, J = 3.9 Hz, 1H), 7.62 (d, J = 5.0 Hz, 1H), 7.27 (t, J = 5.7 Hz, 1H), 4.01 (d, J = 7.6 Hz, 2H), 1.95–1.86 (m, 1H), 1.35–1.12 (m, 32H), 0.92–0.83 (m, 6H); ^{13}C NMR (100 MHz, chloroform-d) δ 161.78, 140.46, 135.25, 130.48, 129.88, 128.42, 107.98, 46.25, 37.77, 31.95, 31.91, 31.22, 31.21, 30.05, 30.04, 29.68, 29.66, 29.59, 29.53, 29.39, 29.33, 26.24, 22.72, 22.70, 14.16, 14.15 (Figures S1 and S2 in the Supplementary Materials). Mass for $C_{54}H_{89}N_2O_2S_2{}^+$: 861.6360; found: 861.6359 (Figure S5). Synthesis of DPP-S-Br monomer: To a solution of DPP-S-C_8C_{10} (2.50 g, 2.90 mmol) in chloroform (25 mL), we passed argon and then added NBS (1.08 g, 6.09 mmol, 2.10 eq). The reaction was completed by stirring the mixture at 60 °C for 0.5 h. The mixture was then extracted with DCM. After removal of the solvent under reduced pressure, the residue was purified by silica gel chromatography with the eluent (PE:DCM = 5:1) to give an orange–red powder (2.80 g, 95.2%); ^{1}H NMR (400 MHz, CDCl$_3$) δ 8.62 (d, J = 4.2 Hz, 1H), 7.21 (d, J = 4.2 Hz, 1H), 3.92 (d, J = 7.7 Hz, 2H), 1.94–1.81 (m, J = 7.2, 6.7 Hz, 1H), 1.38–1.13 (m, 32H), 0.91–0.85 (m, 6H); ^{13}C NMR (75 MHz, CDCl$_3$) δ 161.41, 139.41, 135.31, 131.43, 131.19, 118.95, 108.04, 46.36, 37.77, 31.93, 31.89, 31.19, 29.99, 29.65, 29.56, 29.50, 29.37, 29.29, 26.19, 22.70, 22.67, 14.12 (Figures S3 and S4). Mass for $C_{54}H_{86}Br_2N_2O_2S_2$: 1016.4497; found: 1016.4487 (Figure S6). Synthesis of PDPP−S−FS$_n$−Tz$_m$ polymer: Tz 50%: DPP-S-Br (200 mg, 196.23 μmol), FS-Sn (43.7 mg, 98.11 μmol 0.5 eq), Tz-Sn (48.5 mg, 98.11 μmol 0.5 eq), and triethyl phosphite (P(o-ty)$_3$, 4.80 mg, 15.69 μmol) were dissolved in dry chlorobenzene (10 mL). The Schlenk tube was vented with argon over ten minutes, after which tris(dibenzylideneacetone)dipalladium ([Pd$_2$(dba)$_3$] catalyst, 3.6 mg, 3.92 μmol) was quickly added. The polymerization reaction was stirred at 130 °C for three days and gradually returned to room temperature. The purification was carried out by Soxhlet extraction with hexane (12 h), methanol (12 h), ethyl acetate (8 h), and acetone (10 h). Finally, the target polymer was obtained using chloroform phase. The fractions were evaporated and concentrated, precipitated into methanol (200 mL), and filtered to give the polymeric material in the form of a dark powder, which was dried under vacuum for 5 h (80 °C) to give Tz 50% (199.4 mg, 88.8%). The other three polymerizations were synthesized under similar reaction conditions and showed similar yields, except that the ratio of the two monomers varied.

Characterization: Nuclear magnetic resonance spectra of small molecules and polymers were measured in deuterated chloroform (NMR, Bruker AVANCE 400, Mannheim, Germany). Solid-state NMR: ^{13}C cross-polarization magic angle spinning nuclear magnetic resonance (^{13}C CP/MAS NMR) spectra were recorded on a Bruker Avance III 400MHz spectrometer (Germany). Samples were packed in 4 mm ZrO$_2$ rotors, which were spun at 8 kHz in a double-resonance MAS probe. Mass spectrometry analyses were performed in ESI mode on a magnetic resonance mass spectrometer (solariX, Memmingen, Germany). The molecular weight of the polymer was estimated by high-temperature gel permeation chromatography (Agilent PL-GPC 220, Santa Clara, CA, USA). The solvent flow phase chosen here was 1,3,5-trichlorobenzene to fully ensure the solubility of the material. Thermal stability measurements were performed on a thermal analysis system in a nitrogen atmosphere with a heating rate of 10 °C min^{-1} (HITACHI STA200, Tokyo, Japan). Elemental analysis was conducted using an organic elemental analyzer (UNICUBE-Elementar, Langenselbold, Germany) in CHNS mode. The Fourier-transform infrared (FT-IR) spectra were collected using an FT-IR spectrometer (VERTEX 70v, Bruker, Germany) in a vacuum. Photochemical analyses were conducted using a UV–vis spectrometer (UV3600i, Shimadzu, Japan) using the standard quartz cell and the quartz chip (15 mm × 15 mm × 1 mm). The concentration of the solution was approximately 0.5 mg/mL, and the solvent was anhydrous chlorobenzene. The polymer semiconductor solution was spin-coated onto pre-cleaned quartz plates, followed by thermal annealing at 150 °C for ten minutes. Electrochemical measurements were performed under an argon atmosphere using cyclic voltammetry in acetonitrile solutions containing tetrabutylammonium hexafluorophosphate (TBAPF$_6$) (CHI760E, CH Instruments, Bee Cave, TX, USA) at a scan rate of 50 mV/s. An Ag/AgCl electrode, platinum electrodes, and glassy carbon electrodes were used as reference, counter, and working electrodes, respectively. Then, 5 µL of a solution of the anhydrous chlorobenzene of the polymer, in the concentration range of 0.5 mg/mL to 1 mg/mL, was weighed sequentially with a pipette gun, dropped onto a glassy carbon electrode, and allowed to evaporate naturally at room temperature (the diameter of the circular hole of the electrode used was 4 mm).

OTFT device: Highly doped N-type silicon (Si) wafers with a silicon dioxide (SiO_2) content of 300 nm were used as substrates, which were ultrasonically cleaned in deionized water, acetone, and isopropanol for 5 min, sequentially, and then processed in the UV zone for 25 min. After that, the substrate was modified with octadecyltrichlorosilane (OTS). Specifically, 1.3 µL of OTS and 1 mL of trichloroethylene were thoroughly mixed and then spin-coated on the substrate at 3000 rpm for 30 s. The substrate was then placed in a glass desiccator and dried under an atmosphere of NH_3 for 8 h. For the preparation of semiconductor films, the conjugated polymers were pre-dissolved in chlorobenzene solvent (concentration 5 mg/mL) and heated overnight at 80 °C with stirring. Subsequently, the polymer solution was spin-coated onto OTS-treated Si wafers at 2000 rpm for 1 min, and then annealed at 200 °C for 30 min in a glove box (nitrogen atmosphere). Finally, source/drain electrodes of Au were deposited by evaporation in a vacuum (W/L = 5, W = 1000 µm, L = 200 µm) to finish the device preparation. The electrical performance of the OTFT device was assessed using a B1500A semiconductor parameter analyzer (Keithley, Cleveland, OH, USA) in a N_2 glovebox.

GIWAXS: The grazing-incidence wide-angle X-ray scattering images were acquired on beamline BL1W1A of the Beijing Synchrotron Radiation Facility. An Eiger detector was employed, and the pixel was 0.075 mm. The beam center and the sample-to-detector distance were calibrated with LaB_6. The monochromatic wavelength of the light source was 1.5406 Å, the photon energy was 8.05 keV, and the grazing-incidence angle was 0.2°. The exposure time of the samples was 30 s.

AFM: Atomic force microscopy images were obtained using an SPA300HV from Seiko Instruments Inc. (Chiba, Japan) in tapping mode (scanning range 5×5 µm). The scanning probe was performed using the OLTESPA-R3 silicon cantilever (Bruker, Germany). The elastic coefficient of the probe was 2 N/m.

3. Results

3.1. Synthesis and Structural Analysis

The synthesis route of the series of polymers $PDPP-S-FS_n-Tz_m$ obtained based on Stille coupling polymerization is shown in Figure 2. DPP-S was synthesized with good yields using commercially available 2-cyanothiophene as a precursor. In order to improve the solubility of DPP-S, its N position was chemically modified to introduce an alkyl chain. Then, 9-(bromomethyl)nonadecane was introduced under base reaction conditions relying on an alkylation reaction to produce $DPP-S-C_8-C_{10}$, which possessed excellent solubility in common organic solvents such as n-hexane, chloroform, and ethyl acetate. Br atoms can be easily introduced on thiophene using NBS as a reagent, and the resulting DPP-S-Br can be used as a monomer for Stille coupling. The other two Sn-containing monomers, Tz-Sn and FT-Sn, were polymerized with DPP-S-Br under palladium catalysis. The Stille coupling reaction was carried out over a period of 72 h to ensure that the target products were polymerized to a sufficient extent. Chlorobenzene was chosen as the solvent due to its high boiling point and its excellent solubility for the polymers. The molecular weight of the polymers increased as the polymerization process progressed, resulting in a corresponding decrease in solubility in other common organic solvents. In order to ensure that the molecular weights of the four polymers were within a certain range, we strictly controlled the reaction time and the chemical equivalent ratios, including the amounts of catalyst and ligand. The polymers were purified using the Soxhlet extraction technique, and the successive use of hexane, acetone, methanol, and ethyl acetate effectively removed oligomers and small molecules from the reaction system. Polymers or oligomers with

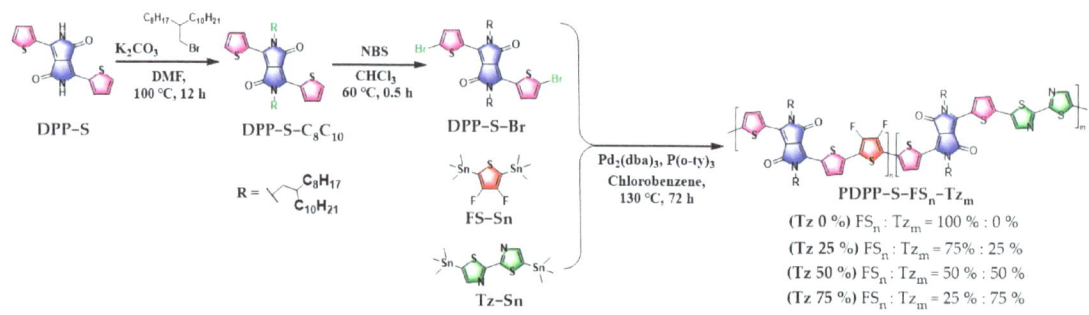

Figure 2. Synthesis of DPP-based monomer and PDPP−S−FS$_n$−Tz$_m$ polymers: Tz 0%, Tz 25%, Tz 50%, Tz 75%.

The weight-average molecular weight (Mw) and number-average molecular weight (Mn) of the four polymers were in the region of 100.0 kDa and 25.0 kDa, respectively, as measured by high-temperature gel permeation chromatography (Figure S7). The relatively uniform distribution of Mn confirmed the effect of a suitable catalyst-to-ligand ratio and reaction time on polymer purity. The polymer dispersion index (PDI) calculated from the ratio of Mw to Mn is shown in Table 1 below. Based on the Mn values, it is roughly estimated that 21–25 repeating units were polymerized for each main chain. Elemental analyses of the four elements C, H, N, and O contained in the polymer Tz 0%, with theoretical values of 71.20%, 9.10%, 2.65%, and 9.90%, respectively, show that the contents of all four elements deviated from the theoretical values by less than 1% (Table 1). The relative increase in the proportions of N and S with increasing Tz content further indicated that the materials were sufficiently pure to be free of catalyst and ligand components. Liquid NMR analyses of the four polymers were difficult due to the poor quality of their spectra. On the one hand, this was due to their poor solubility, and on the other hand, it was due to the inability to clearly observe microscopic variations in the proportions of the components (Figures S8–S11). The series of polymers were further characterized by NMR spectroscopy with solid-state NMR (Figures S12–S15). The ^{13}C NMR spectra were divided into an aromatic part and an alkyl chain part and corresponded well to the molecular composition of the polymers. The carbonyl group appeared near 160 ppm and, together with the typical peak of the central carbon peak of the thiazole, formed the leftmost broad single peak of low field in the spectrum. In addition, the four polymers were characterized by FT-IR spectroscopy, the results of which are shown in Figure S16. The typical carbonyl peaks of the polymers all appear near 1668 cm^{-1}.

Table 1. The results of molecular weight and elemental analyses of the four polymers.

	Mn	Mw	PDI	C [1]	H [1]	N [1]	S [1]
				(%)	(%)	(%)	(%)
Tz 0%	24,541	85,717	3.49	71.20	9.10	2.65	9.90
Tz 25%	21,933	96,864	4.41	70.85	9.01	3.41	10.69
Tz 50%	28,506	139,502	4.89	70.73	8.85	4.04	11.08
Tz 75%	24,390	107,112	4.39	70.19	8.82	4.57	11.83

[1] Based on the average of two tests.

To test the thermal stability of this series of polymers, thermogravimetric (TGA) tests were performed under nitrogen, as shown below. All four polymers underwent significant decomposition up to 383 °C, with a 10% weight loss at 409 °C (Figure S17). There were no significant differences in overall thermal stability, except for a linear relationship at 70% weight loss, where the higher the Tz component, the more difficult it was to decompose completely. The polymers exhibited very poor solubility in non-chlorinated solvents such as ethanol, hexane, and tetrahydrofuran. Good solubility was only found in chlorinated solvents such as dichloromethane, chloroform, chlorobenzene, and trichlorobenzene at room temperature, and the solubility was further improved by heating to 50 °C. The dissolubility of the polymers in chloroform and chlorobenzene at room temperature was approximately 8 mg/mL and 15 mg/mL, respectively, and chlorobenzene was chosen as the solvent for subsequent characterization measurements and device preparation.

3.2. Photochemical Properties

The UV–vis absorption spectra of the polymer system in solution and film are shown below in Figure 3, and the corresponding spectral data are summarized in Table 2. Two absorption bands in the range of 350–500 nm and 600–900 nm can be observed for the polymers in solution. The high-energy absorption band located in the short-wavelength band can be attributed to the π–π* transition (band I), while the low-energy absorption band belongs to the intramolecular charge-transfer transition (band II). With the increase in Tz content, the main absorption peak (λ_{max}) of absorption band I was significantly redshifted, from 406 nm to 444 nm (Figure 3a). The main absorption peak of absorption band II was blueshifted with increasing Tz content, and its corresponding λ_{max} was 906 nm, 899 nm, 895 nm, and 765 nm, respectively. The two polymers Tz 0% and Tz 75% exhibited distinct 0–0 and 0–1 peaks with shoulders at 767 nm and 701 nm, respectively. In contrast, the Tz 25% and Tz 50% polymers presented relatively broad absorption bands, which may be related to their reduced coplanarity due to better solubility. There was a redshift of close to 10 nm in the main absorption peak in band II from solution to film (Figure 3b). For example, the absorption of the Tz 75% polymer increased from 762 nm to 772 nm. These changes are related to the coplanar structure and better solid-state stacking with π–π stacking distances between the polymers. The optical bandgap of the polymer in solution (E_g^{soln}) was nearly 1.38 eV, and in the thin-film state the bandgap (E_g^{film}) was reduced to approximately 1.35 eV, as ascertained at the onset of the UV–visible absorption at 900 nm and 920 nm, respectively. The UV measurement diagrams using chloroform as the solvent also show dual absorption bands (Figure S18), and the data for the specific maximum absorption peaks are presented in Table S1. The UV absorption spectra of the dimers of the polymers calculated on the basis of the DFT again show two absorption bands, the results of which are shown in Figure S19.

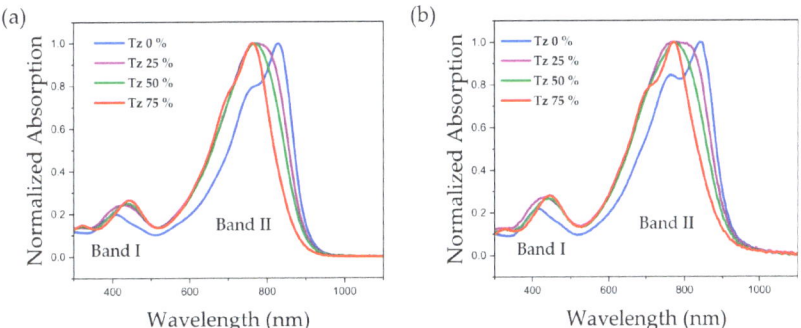

Figure 3. Normalized UV–vis spectra of the four polymers in (**a**) chlorobenzene solutions and (**b**) annealed thin films.

Table 2. UV–vis absorption of the four polymers.

	λ_{max} (nm) [1]	λ_{max} (nm) [1]	λ_{onset} (nm) [1]	λ_{max} (nm) [2]	λ_{max} (nm) [2]	λ_{onset} (nm) [2]	E_g^{soln} (eV) [3]	E_g^{film} (eV) [4]
Tz 0%	829	406	906	841	421	925	1.34	1.31
Tz 25%	775	427	899	782	433	921	1.35	1.32
Tz 50%	769	441	895	775	442	914	1.36	1.33
Tz 75%	762	444	878	772	446	900	1.38	1.36

[1] In solution. [2] In film. [3] Calculations come from $1240/\lambda_{onset}$ in solution. [4] Calculations come from $1240/\lambda_{onset}$ in film.

3.3. Electrochemical Properties

The redox potentials of the polymers of the PDPP−S−FS$_n$−Tz$_m$ series were investigated using cyclic voltammetry (CV). Their corresponding electrochemical data are summarized in Table 3. The differences in the LUMO energy levels calculated from the onset reduction potential for the four polymers were very small, in the vicinity of −3.68 eV (Figure 4). The extremely low-lying LUMO energy levels of the polymers are related to the presence of three strong electron-withdrawing components in the polymer backbone. On the other hand, it can be seen that by changing the proportion of F atoms or thiazoles in the molecular architecture, the effect on the LUMO energy level of the material is insignificant. In addition, the difference between their reduction peaks is also small, and the electrochemical reduction properties of the four polymers are similar. The polymers show obvious reversible cyclic reduction characteristics, and their reduction peaks are much more obvious than the oxidation peaks, indicating that their electron mobility is higher than their hole mobility. It is worth mentioning that the HOMO energy level of the polymer Tz 0% was significantly different from that of the other three polymers. Tz 0% possessed the highest HOMO energy level of −3.66 eV, which at the same time led to a minimum energy gap of merely 1.04 eV. The energy gap of the polymers was in the vicinity of 1.34 eV, which is close to their optical bandgaps.

Table 3. Electrochemical characteristics of the four polymers.

	E_{red} (V)	E_{red}^{onset} (V)	LUMO (eV) [5]	E_{ox} (V)	E_{ox}^{onset} (V)	HOMO (eV) [6]	E_g^{cv} (eV) [7]
Tz 0%	−0.96	−0.75	−3.66	0.49	0.29	−4.70	1.04
Tz 25%	−0.94	−0.72	−3.69	0.79	0.58	−4.99	1.30
Tz 50%	−0.91	−0.71	−3.70	0.79	0.60	−5.01	1.31
Tz 75%	−0.94	−0.76	−3.65	0.78	0.62	−5.03	1.38

[5] $E_{LUMO} = -4.80\ eV - [(E_{red}^{onset}) - E_{1/2}(ferrocene)]$. [6] $E_{HOMO} = -4.80\ eV - [(E_{ox}^{onset}) - E_{1/2}(ferrocene)]$. [7] $E_g^{cv} = E_{HOMO} - E_{LUMO}$.

3.4. OTFT Performance

In order to characterize the charge transport behavior of these materials, we fabricated bottom-gate top-contact (BGTC)-structured OTFT devices based on PDPP-S-FS$_n$-Tz$_m$, and the corresponding device configurations are shown in Figure 5a. Considering that the HOMO level of the polymer matched that of the gold electrode (Au, ~−5.00 eV), Au was selected as the contact electrode. The field-effect mobility (μ) was calculated from the following equation:

$$I_D = (W/2L)\ \mu C_i\ (V_{GS} - V_{TH})^2$$

where I_D is the drain current, C_i is the capacitance of 300 nm silicon dioxide ($C_i = 10\ nF\ cm^{-2}$), W and L are the width and length of the channel, respectively, and V_{GS} and V_{TH} are the gate voltage and the threshold voltage, respectively. The average mobility values were obtained based on measurements performed on eight devices.

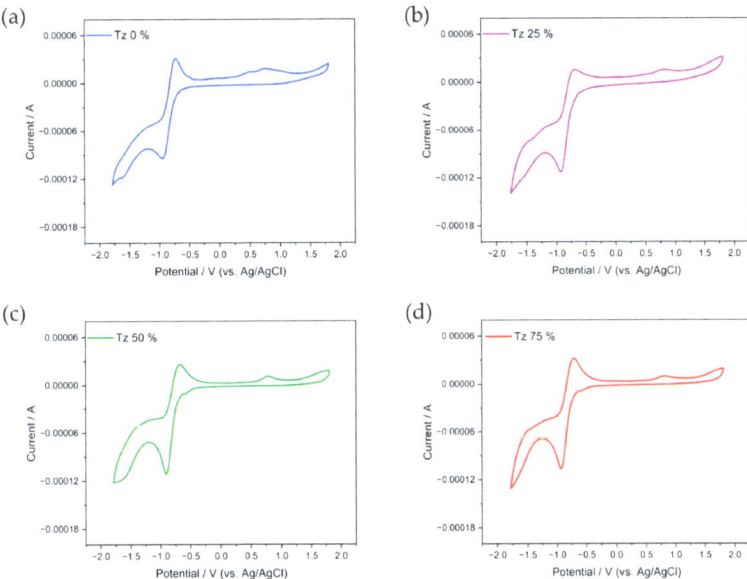

Figure 4. CV of (**a**) Tz 0%-, (**b**) Tz 25%-, (**c**) Tz 50%-, and (**d**) Tz 75%-based films in the acetonitrile solution with positive sweeps.

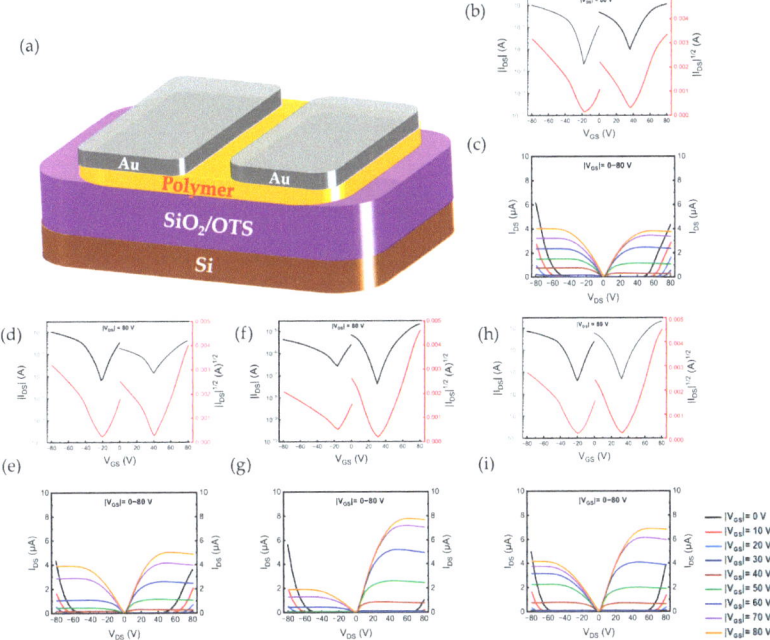

Figure 5. (**a**) OTFT devices with BGTC architectures. (**b,d,f,h**) Current–voltage characteristics of OTFTs measured under vacuum: Tz 0%, Tz 25%, Tz 50%, and Tz 75%, respectively. (**c,e,g,i**) Output characteristics of Tz 0%-, Tz 25%-, Tz 50%-, and Tz 75%-based devices, respectively (Different colored lines represent the gate voltage, with the left side showing a negative voltage and the right side a positive voltage).

Table 4 summarizes the electron and hole mobility extracted from the transfer characteristic curve. It is evident that the OTFT devices based on PDPP-S-FS$_n$-Tz$_m$ have N-type dominant ambipolar transport characteristics, with electron mobility one order of magnitude higher than hole mobility (Figure 5b,d,f,h). The maximum electron mobility value in the Tz 0% polymer containing a thiazole-free system was only 0.38 cm^2 V^{-1} s^{-1}. The electron mobility improved with the increase in the thiazole ratio. Carrier mobility as high as 0.69 cm^2 V^{-1} s^{-1} was shown at a thiazole–fluorothiophene ratio of 50:50. Continuing to increase the proportion of thiazole to 75% did not cause a further increase in mobility, which dropped to 0.57 cm^2 V^{-1} s^{-1}. On the other hand, the increase in electron mobility was inversely proportional to its hole mobility. The polymer Tz 50% exhibited the highest electron mobility while having the lowest hole mobility among the four materials. The Tz 0% showed the highest hole mobility of 0.07 cm^2 V^{-1} s^{-1}, and its average hole mobility was in the vicinity of 0.05 cm^2 V^{-1} s^{-1}. The output behaviors of the polymer-based devices are shown in Figure 5c,e,g,i, respectively.

Table 4. Electron and hole transport properties of PDPP−S−FS$_n$−Tz$_m$-based OTFT devices.

	Coating Speed (mm/s)	Annealing (°C)	Max Electron Mobilities (cm^2/(V s))	Electron Mobilities [1] (cm^2/(V s))	Max Hole Mobilities [1] (cm^2/(V s))	Hole Mobilities [1] (cm^2/(V s))
Tz 0%	2000	200	0.38	0.25 ± 0.11	0.070	0.053 ± 0.013
Tz 25%	2000	200	0.48	0.36 ± 0.08	0.064	0.050 ± 0.011
Tz 50%	2000	200	0.69	0.50 ± 0.08	0.036	0.022 ± 0.014
Tz 75%	2000	200	0.57	0.42 ± 0.10	0.051	0.035 ± 0.016

[1] Based on eight OTFT devices for each condition.

3.5. Crystallinity Analysis of Polymer Films

We investigated the relationship between the crystallinity of polymer films and the electrical performance of OTFT devices by grazing-incidence wide-angle X-ray scattering (GIWAXS). Figure 6a–d show 2D-GIWAXS images of four polymer films prepared by spin-coating, and Figure 6e shows the corresponding in-plane and out-of-plane 1D profiles based on 2D-GIWAXS of the annealed films. The polymers all showed distinct out-of-plane (100) and in-plane (010) peaks, indicating that their films were predominantly edge-on in orientation. The polymer Tz 0% showed only (100) in the out-of-plane direction, while the other three polymers showed enhanced (100) diffraction intensity in the out-of-plane direction, accompanied by the appearance of secondary diffraction (200), which fully explains their enhanced crystallinity. This also explains the fact that the electron mobility of Tz 0% was the worst among the four. On the other hand, the presence of in-plane (100) diffraction for Tz 0% and Tz 25% indicated the gradual appearance of a face-on orientation. The stacking pattern of face-on crossed with edge-on is unfavorable for electron transport. The layered side-chain distances of Tz 0% and Tz 25% were 22.43 Å and 21.65 Å, respectively, which are loose compared to the stacking of Tz 50% and Tz 75% (Table 5). Tz 50% showed the shortest π–π stacking distance, which was calculated based on the in-plane (010) diffraction peaks to be only 3.95 Å. Tight stacking facilitates carrier transport and helps to achieve high mobility in OTFT devices, which explains why Tz 50% was the best performer of the four.

3.6. Morphological Analysis of Polymer Films

The morphology of the polymer semiconductor films was investigated by atomic force microscopy (AFM), and Figure S20 shows the film height diagrams of the four polymer materials after annealing at 200 °C. Overall, the surface morphology of the polymer films shows a clear uniform distribution. The Tz 50% polymer film shows a dense fibrous structure, and the ordered interchain stacking in the fibers facilitates the charge transport. In contrast, the fibrillar shape of Tz 0% and Tz 25% gradually disappears into disordered aggregates. These results are consistent with the GIWAXS results, which are unfavorable for the electrical properties of the polymeric materials as the crystallinity becomes weaker.

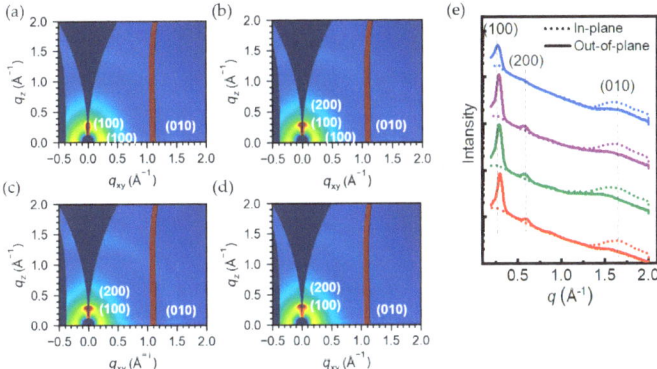

Figure 6. The 2D-GIWAXS patterns of (a) Tz 0%, (b) Tz 25%, (c) Tz 50%, and (d) Tz 75%, and (e) 1D-GIWAXS films annealed at 200 °C (blue: Tz 0%; purple: Tz 25%; green: Tz 50%; red: Tz 75%).

Table 5. Crystallographic information of the PDPP−S−FS$_n$−Tz$_m$ series of polymers.

	In-Plane (010) Peak Position (Å^{-1})	In-Plane (010) π-Spacing (Å)	In-Plane (100) Peak Position (Å^{-1})	In-Plane (100) d-Spacing (Å)	Out-of-Plane (100) Peak Position (Å^{-1})	Out-of-Plane (100) d-Spacing (Å)
Tz 0%	1.58	3.98	0.27	23.26	0.28	22.43
Tz 25%	1.58	3.98	0.26	24.15	0.29	21.65
Tz 50%	1.59	3.95	—	—	0.30	20.93
Tz 75%	1.58	3.97	—	—	0.30	20.93

4. Discussion

Compared to the hole mobility of the polymer PDPP-T based on DPP and thiophene, with moderate hole mobility, which was first reported in 2015, this work significantly improved the electron mobility of this type of material (Table 6). This was firstly related to the introduction of fluorinated thiophene, where the strongly electronegative F atom significantly reduced the frontline orbital energy level of the material. Bura et al. reported that DPP-based polymers prepared by introducing F atoms on the thiophene ring showed an electron mobility of 0.51 cm^2 V^{-1} s^{-1} [24]. By replacing the electron donor thiophene with a benzene ring and further introducing a fluorine atom, the polymer DPP-F$_1$Ph showed moderate electron mobility and a balanced bipolar character [25]. In addition to fluorine atoms, chlorine atoms are also promising electron-withdrawing moieties. Geng's group reported that the polymer DPPTh-4Cl, prepared by introducing four Cl atoms to a thiophene ring, showed a high mobility of 0.81 cm^2 V^{-1} s^{-1}, which was attributed by the authors to the structural coplanarity caused by a weak Cl...S interaction and very low LUMO energy levels [26]. Other strong electron-attracting groups, such as cyano groups, can also be introduced on top of the F atom. Zhang et al. reported last year that the polymer F2CNTVT-DPP, which introduces −F and −CN on (E)-1,2-di(thiophen-2-yl)ethene (TVT), showed a maximum mobility of 2.03 cm^2 V^{-1} s^{-1} [27]. On the other hand, this research is also related to simple terpolymerization to improve mobility. Similar work was reported by Chen et al. The polymer P2DPP-2FBTz, based on two DPPs and another receptor as a three-component composition, showed an average electron mobility of 0.61 cm^2 V^{-1} s^{-1} [28]. Yi et al. reported DPP-2T-DPP-TBT, a bipolar material with high electron mobility, which was also obtained using three-component copolymerization [29]. In addition to the classical acceptor units used to prepare N-type materials, more and more research is focusing on the creation of new structures to improve the electron mobility of polymers. The polymer PFIDTO-T, with a ladder-like aromatic diketone structure, demonstrates highly promising electron mobility [30]. The polymer PBN-27, constructed with a bis-B-N-bridged bipyridine unit and a benzobithiazole unit, was developed and showed moderate N-type electron mobility [31]. Recently, the polymer PNFFN, constructed from IID-based isomers developed by Yu et al., showed the highest mobility of 1.82 cm^2 V^{-1} s^{-1}, which was further increased to 10.74 cm^2 V^{-1} s^{-1} when poly-F atoms were introduced to the system [32]. Guo et al. reported an amide-based macro-heterocyclic polymer PDTzNTI, consisting of sulfur and nitrogen atoms, with unipolar N-type transport and significant electron mobility [33].

Table 6. Typical examples of polymer-based organic thin-film transistors in the last five years.

Materials	Electron Mobilities (cm^2/(V s)	Hole Mobilities (cm^2/(V s)	Device Structure	Average Electron Mobilities (cm^2/(V s)
DPP-T	–	0.094	BGBC	–
DPP-FDT	0.51	0.80	BGBC	0.50
DPP-F$_1$Ph	0.26	0.20	BGTC	0.30
DPPTh-4ClBT	0.82	0.73	TGBC	–
F2CNTVT-DPP	2.03	–	TGBC	1.50
P2DPP-2FBTz	0.68	–	TGBC	0.61
DPP-2T-DPP-TBT	3.84	3.01	TGBC	3.00
PFIDTO-T	0.27	–	TGBC	0.03
PBN-27	0.34	–	TGBC	0.32
PNFFN-DTE	1.82	–	TGBC	1.30
PNFFN-FDTE	10.74	–	TGBC	10.56
PDTzNTI	1.22	–	TGBC	0.87

5. Conclusions

To summarize, we report in this work the changes in carrier mobility caused by the continued introduction of new acceptor units into two-component acceptor materials. The synthesis of these four polymers was achieved via Stille coupling polymerization, and proper control of the reaction and purification conditions allowed for the preparation of a series of polymers with high molecular weights and a relatively homogeneous molecular weight distribution. The polymers showed N-dominant properties and the highest electron mobility of 0.69 cm^2 V^{-1} s^{-1} at a half–half ratio of thiazole to fluorothiophene, and the increase in mobility was mainly related to the close intramolecular Π-distance and good crystallinity, as shown by GIWAXS and film morphology analysis. Further increasing the proportion of thiazole failed to improve the electron mobility of the material in OTFT devices. Therefore, sophisticated molecular structure design strategies and component ratio modulation are key factors in controlling the material properties and device performance. Further development of N-type unipolar and ambipolar organic semiconductor polymer materials and enhancement of electron mobility are still in progress.

Supplementary Materials: The following supporting information can be downloaded at: https://www.mdpi.com/article/10.3390/polym15183803/s1, Figure S1: The ^{1}H NMR spectra of compound DPP-S-C$_8$C$_{10}$. Figure S2: The ^{13}C NMR spectra of compound DPP-S-C$_8$C$_{10}$. Figure S3: The ^{1}H NMR spectra of compound DPP-S-Br. Figure S4: The ^{13}C NMR spectra of compound DPP-S-Br. Figure S5: Mass spectra of compound DPP-S-C$_8$C$_{10}$. Figure S6: Mass spectra of compound DPP-S-Br. Figure S7: GPC data (cumulative percent curves and molecular weight distribution) of the four polymers. Figures S8–S11: The ^{1}H NMR spectra of polymers Tz 0% to Tz 75%, respectively. Figures S12–S15: The ^{13}C NMR spectra of polymers Tz 0% to Tz 75%, respectively. Figure S16: FT-IR spectra of the four polymers. Figure S17: TGA analysis of the four polymers. Figure S18: Normalized UV–vis spectrum of the four polymers in chloroform solutions and in annealed thin films. Figure S19: Theoretical simulations of (a) electrostatic potential surfaces and (b) non-covalent interaction scattering of the dimer; (c,d) UV–vis spectrum of the dimer of the polymer calculated by DFT at B3LYP/6-31G (d). Figure S20: AFM height images of polymers in annealed films of (a) Tz 0%, (b) Tz 25%, (c) Tz 50%, and (d) Tz 75%. Table S1: UV–vis absorption of the four polymers.

Author Contributions: Conceptualization, S.R. and Z.Y.; methodology, S.W.; software, S.W.; validation, S.R., Z.W. and W.Z.; formal analysis, S.W.; investigation, S.R.; resources, S.R.; data curation, Y.D.; writing—original draft preparation, Y.D.; writing—review and editing, S.R., S.W. and Z.Y; visualization, Z.Y.; supervision, Z.Y.; project administration, Z.Y.; funding acquisition, S.R. All authors have read and agreed to the published version of the manuscript.

Funding: This research was funded by the China Postdoctoral Science Foundation (2022TQ0399).

Institutional Review Board Statement: Not applicable.

Data Availability Statement: Not applicable.

Acknowledgments: We thank the Hengqin Postdoctoral Administrative Centre for supporting this project. We thank Mingling Zhu and Zhao Yang for photochemical measurements and mass testing at the Institute of Chemistry, Chinese Academy of Sciences (ICCAS). The measurements of solid-state NMR were performed at the Center for Physicochemical Analysis and Measurements in ICCAS. The help from Aijiao Guan and Junfeng Xiang was acknowledged. Thanks to Mingchao Shao for the discussion on solid-state NMR spectra.

Conflicts of Interest: The authors declare no conflict of interest

References

1. Kim, M.; Ryu, S.U.; Park, S.A.; Choi, K.; Kim, T.; Chung, D.; Park, T. Donor–Acceptor-Conjugated Polymer for High-Performance Organic Field-Effect Transistors: A Progress Report. *Adv. Funct. Mater.* **2019**, *30*, 1904545. [CrossRef]
2. Sirringhaus, H. 25th Anniversary Article: Organic Field-Effect Transistors: The Path Beyond Amorphous Silicon. *Adv. Mater.* **2014**, *26*, 1319–1335. [CrossRef]
3. Tang, Z.; Wei, X.; Zhang, W.; Zhou, Y.; Wei, C.; Huang, J.; Chen, Z.; Wang, L.; Yu, G. An A–D–A′–D′ strategy enables perylenediimide-based polymer dyes exhibiting enhanced electron transport characteristics. *Polymer* **2019**, *180*, 121712. [CrossRef]
4. Wu, W.; Liu, Y.; Zhu, D. π-Conjugated molecules with fused rings for organic field-effect transistors: Design, synthesis and applications. *Chem. Soc. Rev.* **2010**, *39*, 1489–1502. [CrossRef] [PubMed]
5. Yang, J.; Zhao, Z.; Wang, S.; Guo, Y.; Liu, Y. Insight into High-Performance Conjugated Polymers for Organic Field-Effect Transistors. *Chem* **2018**, *4*, 2748–2785. [CrossRef]
6. Anthony, J.E.; Facchetti, A.; Heeney, M.; Marder, S.R.; Zhan, X. n-Type organic semiconductors in organic electronics. *Adv. Mater.* **2010**, *22*, 3876–3892. [CrossRef] [PubMed]
7. Zhang, Y.; Wang, Y.; Gao, C.; Ni, Z.; Zhang, X.; Hu, W.; Dong, H. Recent advances in n-type and ambipolar organic semiconductors and their multi-functional applications. *Chem. Soc. Rev.* **2023**, *52*, 1331–1381. [CrossRef]
8. Zhao, Y.; Guo, Y.; Liu, Y. 25th Anniversary Article: Recent Advances in n-Type and Ambipolar Organic Field-Effect Transistors. *Adv. Mater.* **2013**, *25*, 5372–5391. [CrossRef]
9. Griggs, S.; Marks, A.; Bristow, H.; McCulloch, I. n-Type organic semiconducting polymers: Stability limitations, design considerations and applications. *J. Mater. Chem. C Mater.* **2021**, *9*, 8099–8128. [CrossRef]
10. Park, K.H.; Go, J.Y.; Lim, B.; Noh, Y.Y. Recent progress in lactam-based polymer semiconductors for organic electronic devices. *J. Polym. Sci.* **2021**, *60*, 429–485. [CrossRef]
11. Cheon, H.J.; An, T.K.; Kim, Y.-H. Diketopyrrolopyrrole (DPP)-Based Polymers and Their Organic Field-Effect Transistor Applications: A Review. *Macromol. Res.* **2022**, *30*, 71–84. [CrossRef]
12. Ren, S.; Yassar, A. Recent Research Progress in Indophenine-Based-Functional Materials: Design, Synthesis, and Optoelectronic Applications. *Materials* **2023**, *16*, 2474–2505. [CrossRef] [PubMed]
13. Chen, J.; Yang, J.; Guo, Y.; Liu, Y. Acceptor Modulation Strategies for Improving the Electron Transport in High-Performance Organic Field-Effect Transistors. *Adv. Mater.* **2021**, *34*, 2104325–2104355. [CrossRef] [PubMed]
14. Ren, S.; Zhang, W.; Wang, Z.; Yassar, A.; Liao, Z.; Yi, Z. Synergistic Use of All-Acceptor Strategies for the Preparation of an Organic Semiconductor and the Realization of High Electron Transport Properties in Organic Field-Effect Transistors. *Polymers* **2023**, *15*, 3392–3404. [CrossRef] [PubMed]
15. Kang, T.E.; Kim, K.-H.; Kim, B.J. Design of terpolymers as electron donors for highly efficient polymer solar cells. *J. Mater. Chem. A* **2014**, *2*, 15252–15267. [CrossRef]
16. Dang, D.; Yu, D.; Wang, E. Conjugated Donor–Acceptor Terpolymers Toward High-Efficiency Polymer Solar Cells. *Adv. Mater.* **2019**, *31*, 1807019. [CrossRef] [PubMed]
17. Luo, H.; Liu, Z.; Zhang, D. Conjugated D–A terpolymers for organic field-effect transistors and solar cells. *Polym. J.* **2017**, *50*, 21–31. [CrossRef]
18. Mueller, C.J.; Singh, C.R.; Fried, M.; Huettner, S.; Thelakkat, M. High Bulk Electron Mobility Diketopyrrolopyrrole Copolymers with Perfluorothiophene. *Adv. Funct. Mater.* **2015**, *25*, 2725–2736. [CrossRef]
19. Jiang, W.; Yu, X.; Li, C.; Zhang, X.; Zhang, G.; Liu, Z.; Zhang, D. Fluoro-substituted DPP-bisthiophene conjugated polymer with azides in the side chains as ambipolar semiconductor and photoresist. *Sci. China Chem.* **2022**, *65*, 1791–1797. [CrossRef]

20. Lee, M.; Kim, T.; Nguyen, H.V.T.; Cho, H.W.; Lee, K.-K.; Choi, J.-H.; Kim, B.; Kim, J.Y. Regio-regular alternating diketopyrrolopyrrole-based D1–A–D2–A terpolymers for the enhanced performance of polymer solar cells. *RSC Adv.* **2019**, *9*, 42096–42109. [CrossRef]
21. Ly, J.T.; Burnett, E.K.; Thomas, S.; Aljarb, A.; Liu, Y.; Park, S.; Rosa, S.; Yi, Y.; Lee, H.; Emrick, T.; et al. Efficient Electron Mobility in an All-Acceptor Napthalenediimide-Bithiazole Polymer Semiconductor with Large Backbone Torsion. *ACS Appl. Mater. Interfaces* **2018**, *10*, 40070–40077. [CrossRef] [PubMed]
22. Yuan, Z.; Buckley, C.; Thomas, S.; Zhang, G.; Bargigia, I.; Wang, G.; Fu, B.; Silva, C.; Brédas, J.-L.; Reichmanis, E. A Thiazole–Naphthalene Diimide Based n-Channel Donor–Acceptor Conjugated Polymer. *Macromolecules* **2018**, *51*, 7320–7328. [CrossRef]
23. Yi, Z.; Yan, Y.; Wang, H.; Li, W.; Liu, K.; Zhao, Y.; Gu, G.; Liu, Y. Chain-Extending Polymerization for Significant Improvement in Organic Thin-Film Transistor Performance. *ACS Appl. Mater. Interfaces* **2022**, *14*, 36918–36926. [CrossRef] [PubMed]
24. Bura, T.; Beaupré, S.; Ibraikulov, O.A.; Légaré, M.-A.; Quinn, J.; Lévêque, P.; Heiser, T.; Li, Y.; Leclerc, N.; Leclerc, M. New Fluorinated Dithienyldiketopyrrolopyrrole Monomers and Polymers for Organic Electronics. *Macromolecules* **2017**, *50*, 7080–7090. [CrossRef]
25. Park, J.H.; Jung, E.H.; Jung, J.W.; Jo, W.H. A fluorinated phenylene unit as a building block for high-performance n-type semiconducting polymer. *Adv. Mater.* **2013**, *25*, 2583–2588. [CrossRef] [PubMed]
26. Sui, Y.; Shi, Y.; Deng, Y.; Li, R.; Bai, J.; Wang, Z.; Dang, Y.; Han, Y.; Kirby, N.; Ye, L.; et al. Direct Arylation Polycondensation of Chlorinated Thiophene Derivatives to High-Mobility Conjugated Polymers. *Macromolecules* **2020**, *53*, 10147–10154. [CrossRef]
27. Zhang, C.; Tan, W.L.; Liu, Z.; He, Q.; Li, Y.; Ma, J.; Chesman, A.S.R.; Han, Y.; McNeill, C.R.; Heeney, M.; et al. High-Performance Unipolar n-Type Conjugated Polymers Enabled by Highly Electron-Deficient Building Blocks Containing F and CN Groups. *Macromolecules* **2022**, *55*, 4429–4440. [CrossRef]
28. Chen, J.; Jiang, Y.; Yang, J.; Sun, Y.; Shi, L.; Ran, Y.; Zhang, Q.; Yi, Y.; Wang, S.; Guo, Y.; et al. Copolymers of Bis-Diketopyrrolopyrrole and Benzothiadiazole Derivatives for High-Performance Ambipolar Field-Effect Transistors on Flexible Substrates. *ACS Appl. Mater. Interfaces* **2018**, *10*, 25858–25865. [CrossRef]
29. Yi, Z.; Jiang, Y.; Xu, L.; Zhong, C.; Yang, J.; Wang, Q.; Xiao, J.; Liao, X.; Wang, S.; Guo, Y.; et al. Triple Acceptors in a Polymeric Architecture for Balanced Ambipolar Transistors and High-Gain Inverters. *Adv. Mater.* **2018**, *30*, 1801951–1801958. [CrossRef]
30. Tao, X.; Li, W.; Wu, Q.; Wei, H.; Yan, Y.; Zhao, L.; Hu, Y.; Zhao, Y.; Chen, H.; Liu, Y. Ladder-Like Difluoroindacenodithiophene-4,9-dione Derivative: A New Acceptor System for High-Mobility n-Type Polymer Semiconductors. *Adv. Funct. Mater.* **2022**, *33*, 2210846–2210857. [CrossRef]
31. Cao, X.; Li, H.; Hu, J.; Tian, H.; Han, Y.; Meng, B.; Liu, J.; Wang, L. An Amorphous n-Type Conjugated Polymer with an Ultra-Rigid Planar Backbone. *Angew. Chem. Int. Ed. Engl.* **2023**, *62*, e202212979–e202212986. [CrossRef] [PubMed]
32. Zhang, W.; Shi, K.; Lai, J.; Zhou, Y.; Wei, X.; Che, Q.; Wei, J.; Wang, L.; Yu, G. Record-High Electron Mobility Exceeding 16 cm^2 V^{-1} s^{-1} in Bisisoindigo-Based Polymer Semiconductor with a Fully Locked Conjugated Backbone. *Adv. Mater.* **2023**, *35*, e2300145–e2300155. [CrossRef] [PubMed]
33. Shi, Y.; Guo, H.; Qin, M.; Zhao, J.; Wang, Y.; Wang, H.; Wang, Y.; Facchetti, A.; Lu, X.; Guo, X. Thiazole Imide-Based All-Acceptor Homopolymer: Achieving High-Performance Unipolar Electron Transport in Organic Thin-Film Transistors. *Adv. Mater.* **2018**, *30*, 1705745–1705753. [CrossRef] [PubMed]

Disclaimer/Publisher's Note: The statements, opinions and data contained in all publications are solely those of the individual author(s) and contributor(s) and not of MDPI and/or the editor(s). MDPI and/or the editor(s) disclaim responsibility for any injury to people or property resulting from any ideas, methods, instructions or products referred to in the content.

Article

Synergistic Use of All-Acceptor Strategies for the Preparation of an Organic Semiconductor and the Realization of High Electron Transport Properties in Organic Field-Effect Transistors

Shiwei Ren [1,2,*], Wenqing Zhang [3], Zhuoer Wang [4], Abderrahim Yassar [5], Zhiting Liao [1] and Zhengran Yi [1,*]

1. Zhuhai Fudan Innovation Institute, Guangdong-Macao Deep-Cooperation Zone of Hengqin, Zhuhai 519001, China; zhitingliao@126.com
2. Department of Materials Science, Fudan University, Shanghai 200433, China
3. Key Laboratory of Organic Solids, Institute of Chemistry, Chinese Academy of Sciences, Beijing 100190, China; zhangwq@iccas.ac.cn
4. Key Laboratory of Colloid and Interface Chemistry of Ministry of Education School of Chemistry and Chemical Engineering, Shandong University, Jinan 250100, China; wangzhuoer94@163.com
5. Laboratory of Physics of Interfaces and Thin Films-CNRS, Ecole Polytechnique, Institut Polytechnique de Paris, 91128 Palaiseau, France; abderrahim.yassar@polytechnique.edu
* Correspondence: shiwei_ren@fudan.edu.cn (S.R.); zhengranyi@fudan-zhuhai.org.cn (Z.Y.)

Citation: Ren, S.; Zhang, W.; Wang, Z.; Yassar, A.; Liao, Z.; Yi, Z. Synergistic Use of All-Acceptor Strategies for the Preparation of an Organic Semiconductor and the Realization of High Electron Transport Properties in Organic Field-Effect Transistors. *Polymers* 2023, *15*, 3392. https://doi.org/10.3390/polym15163392

Academic Editor: Tengling Ye

Received: 31 July 2023
Revised: 10 August 2023
Accepted: 11 August 2023
Published: 13 August 2023

Copyright: © 2023 by the authors. Licensee MDPI, Basel, Switzerland. This article is an open access article distributed under the terms and conditions of the Creative Commons Attribution (CC BY) license (https://creativecommons.org/licenses/by/4.0/).

Abstract: The development of n-type organic semiconductor materials for transporting electrons as part of logic circuits is equally important to the development of p-type materials for transporting holes. Currently, progress in research on n-type materials is relatively backward, and the number of polymers with high electron mobility is limited. As the core component of the organic field-effect transistor (OFET), the rational design and judicious selection of the structure of organic semiconductor materials are crucial to enhance the performance of devices. A novel conjugated copolymer with an all-acceptor structure was synthesized based on an effective chemical structure modification and design strategy. PDPPTT-2Tz was obtained by the Stille coupling of the DPPTT monomer with 2Tz-SnMe$_3$, which features high molecular weight and thermal stability. The low-lying lowest unoccupied molecular orbital (LUMO) energy level of the copolymer was attributed to the introduction of electron-deficient bithiazole. DFT calculations revealed that this material is highly planar. The effect of modulation from a donor–acceptor to acceptor–acceptor structure on the improvement of electron mobility was significant, which showed a maximum value of 1.29 cm^2 V^{-1} s^{-1} and an average value of 0.81 cm^2 V^{-1} s^{-1} for electron mobility in BGBC-based OFET devices. Our results demonstrate that DPP-based polymers can be used not only as excellent p-type materials but also as promising n-type materials.

Keywords: n-type materials; organic semiconductor; conjugated copolymer; Stille coupling; OFET; electron mobility

1. Introduction

Since the traditional concept that organic compounds can only be used as insulating materials has been broken, the field of organic electronics has been extensively studied and rapidly developed over the past few decades [1]. The most fundamental research object in this field is organic semiconductor materials, which refer to organic small molecules and oligomer and polymer materials with π-conjugated architecture [2–4]. Organic materials are available from a wide variety of sources, with tailorable chemical structures, modifiable functional groups, and high flexibility. Through rational chemical design and molecular structure strategy modulation, the organic materials prepared have been applied in a variety of functional devices, such as the organic field-effect transistor (OFET), organic light-emitting diode (OLED), organic photovoltaic (OPV), organic light-emitting transistor (OLET), organic electrochemical transistor (OECT), and organic thermoelectric

(TE) and so on [5–10]. Among them, OFETs are also organic thin-film transistors (OTFTs), which are regarded as the basic building blocks of organic electronic devices for carrying the function of transporting electrons or holes [11]. Electron-transporting materials are categorized as n-type materials, while p-type materials transport holes and materials that satisfy the need to transport both electrons and holes are referred to as ambipolar semiconductors. The current literature on organic semiconductors has focused on the high performance of p-type materials, while the development of electron transport materials has lagged behind. A variety of systems have been developed with hole mobilities above 10 cm^2 V^{-1} s^{-1}, while ambipolar or n-type polymer semiconductors with electron mobilities above 10 cm^2 V^{-1} s^{-1} are quite rare. The primary reason for such a problem is the limited diversity of materials due to the low variety of acceptors. Common organic semiconductors with high LUMO energy levels have large electron–injection barriers. The diversity of available acceptor choices is limited, with most n-type materials being based only on systems consisting of isoindgio (IID), naphthalene diimide (NDI), perylene diimide (PDI), bithiophene imide (BTI), benzothiadiazo (BT) and diketopyrrolopyrrole (DPP) as acceptor units [12–21]. Further, air sensitivity is a particular issue in the development of n-type OSCs. The presence of electron traps significantly inhibits the performance of n-type devices. The fact is that the advancement of n-type organic semiconductors can contribute to a wide variety of smart applications. Notably, some unique areas, such as organic logic circuits, new integrated display technologies, and organic thermoelectrics, require high mobility and high-performance n-type organic semiconductors.

It is generally believed that a LUMO energy level near −4.0 eV favors the stability of the material in the air. It is relatively easy and efficient to achieve this energy level requirement using chemical strategies. It is beneficial to address the effect of molecular structure design strategies on material properties exemplified by DPP. As a weak acceptor structural unit, it can be polymerized with donor units such as thiophene, di-thiophene, or tri-thiophene to obtain polymeric materials with alternating donor–acceptor (D–A) arrangements, which correspond to hole mobilities in the range of 0.04–3.46 cm^2 V^{-1} s^{-1} [13]. The chemical modification of the donor material to introduce strong electron-withdrawing groups such as –F atoms, –Cl atoms, or cyano contributes to lowering the LUMO energy level of the material, thereby increasing the electron mobility of the material. The copolymer of DPP combined with 3,4-difluorothiophe shows a 0.22 cm^2 V^{-1} s^{-1} hole mobility and 0.19 cm^2 V^{-1} s^{-1} electron mobility [16,22]. Lowering the proportion of donor motifs in the system, in other words increasing the proportion of acceptors in the polymer, means that the obtained P2DPP-based polymers exhibit ambipolar performance with maximum electron and hole mobilities of 3.01 cm^2 V^{-1} s^{-1} and 4.16 cm^2 V^{-1} s^{-1}, respectively [23]. Further, the three-acceptor strategy was validated, and the P2DPP-BT-based polymers exhibited hole and electron mobilities of 3.52 cm^2 V^{-1} s^{-1} and 2.83 cm^2 V^{-1} s^{-1}, respectively. Similarly, other acceptor-based building blocks are mostly used in a combined donor–acceptor mode, and the chemical structure of the donor unit is continuously modified to control LUMO energy levels and the performance of the material. Di–receptor and tri–acceptor strategies are currently effective strategies to enhance the device performance of n-type materials. Here, we propose a strategy for the structural design of all-acceptor (A–A) copolymers. We envision that the polymerization of electron-deficient bithiazole (2Tz) as the acceptor unit with the DPPTT acceptor block could significantly improve the electron transport properties of the material, prompting the conversion of the material from the p-type of PDPPTT-2T to the n-type of PDPPTT-2Tz (Scheme 1). This design principle is in line with previous reports that bithiazoles lead to considerably higher ionization potentials and electron affinities compared to electron-rich bithiophene (2T) [24].

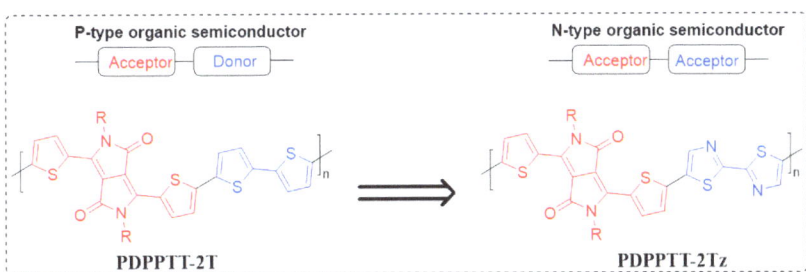

Scheme 1. From D—A to A—A design strategies.

2. Materials and Methods

Materials and synthesis: Chemical reagents, organic solvents, and catalysts were purchased from Aldrich and used as received. Specific synthetic pathways for monomer-based materials **1–3** are described in the Supplementary Information (SI). Compound **3** (200 mg, 0.18 mmol), 2Tz-SnMe$_3$ (87.88 mg, 0.18 mmol 1.0 eq), tris(dibenzylideneacetone)dipalladium ([Pd$_2$(dba)$_3$], 3.24 mg, 3.54 μmol), tri(o-tolyl) phosphine (P(o-tol)$_3$, 4.31 mg, 14.18 μmol), and anhydrous chlorobenzene (CB, 9 mL) was added to a Schlenk tube. This tube was charged with argon through a freeze-pump-thaw cycle three times. The reaction was polymerized at 130 °C for three days. After the polymerization was complete, sodium diethyldithiocarbamate trihydrate was added to remove [Pd$_2$(dba)$_3$], and the mixture was stirred for 30 min at 60 °C before being precipitated into methanol (350 mL). The precipitated product was filtered and purified via Soxhlet extraction with hexane (12 h), methanol (12 h), and acetone (12 h) to give fractions of blue, light red, and blue-violet colors. Finally, the target polymer was obtained using the chloroform phase, the fractions of which were dissolved in chloroform showing a dark green color. This fraction was concentrated by evaporation, precipitated into methanol (100 mL), and filtered to obtain the polymeric material PDPPTT-2Tz (184.4 mg, 91.7%) in the form of a dark powder.

Instrumentation: Nuclear magnetic resonance spectra recorded on a Bruker AVANCEIII (400 MHz, German) spectrometer with deuterated chloroform were used as the solvent. ^{1}HNMR chemical shifts were referenced relative to internal tetramethylsilane. The splitting patterns were designated as s (singlet); d (doublet); t (triplet); and m (multiplet). Molecular weight was determined by gel permeation chromatography (Agilent PL-GPC 50, United States of America), utilizing 1,2,4-trichlorobenzene as an eluent. Mass spectrometry was performed on an Autoflex III (Bruker Daltonics Inc, German) MALDI-TOF spectrometer. The elemental analysis was measured by an organic elemental analyzer (Thermo Scientific Flash 2000, USA). UV-Vis spectra were recorded on a Varian Cary model 500 UV-Vis-NIR spectrophotometer (Agilent, USA) using standard quartz cells of 1 cm width and solvents of a spectroscopic grade. Electrochemical measurements were carried out in an acetonitrile solution with tetra-n-butylammonium hexafluorophosphate using a Metrohm Autolab PGSTAT12 Potentiostat. An Ag/AgCl electrode and platinum wire were used as the reference electrode and counter electrode, respectively. Density Functional Theory (DFT) calculations were performed using the B3LYP-D3/def2tzvp basis set of the Gaussian 16 program to elucidate the highest occupied molecular orbital (HOMO) and LUMO levels after optimizing low-energy conformation using the same method. Fourier transform infrared measurements were performed on a Thermo Scientific iN10 spectrometer. Thermogravimetric and Differential Scanning Calorimetry measurements were performed on a Mettler TGA/DSC thermal analysis system with a heating rate of 10 °C min^{-1} in a nitrogen atmosphere.

Device: The substrates for polymer-based OFETs were subjected to cleaning using ultrasonication in deionized water, acetone, and isopropanol. The cleaned substrates were dried under nitrogen gas and then treated with plasma for 10 min. Before the deposition of polymer semiconductors, octadecyltrichlorosilane (OTS) treatment was performed on the SiO$_2$ gate dielectrics in a vacuum to form an OTS self-assembled monolayer. The field-effect characteristics of the devices were determined in the glove box by using a 4200 SCS semiconductor parameter analyzer (Keithley, China). Different channel lengths (L) of the FET devices (L = 5, 10, 20, 30, 40, and 50 μm) and the same channel widths (W) of 1400 μm were used to optimize device performance.

3. Results

3.1. Synthesis

PDPPTT-2Tz was synthesized following the procedure summarized in Scheme 2. The long side chain was introduced at the N-position of **1** through an alkylation reaction to generate intermediate **2** and enhance its solubility in organic solvents. Bromination under heating conditions was utilized to minimize the reaction time, whereby N-bromosuccinimide was employed to afford **3**, which could serve as a precursor for the Stille coupling reaction. Polymerization occurred via the dibromo-DPPTT monomer (**3**) and 5,5'-bis(trimethylstannyl)-2,2'-bithiazole (2Tz-SnMe$_3$) with palladium (0) as the catalyst. It is worth noting that polymerization can often be switched on within ten minutes using stoichiometric equivalents of catalysts and ligands. An obvious experimental phenomenon is that within ten minutes from room temperature to 130 °C, the color of the solution can change from red to purple and blue to dark green. The dark green color is a sign that the polymerization of DPP-type polymers has started. The purification of target polymers relies mainly on the Soxhlet extraction technique. Hexane, methanol, and acetone were sequentially used to remove oligomers and low molecular weight fractions from the polymer system, and high molecular weight macromolecular fractions were collected by the chloroform solvent, resulting in a 91.7% yield. As mentioned earlier, the choice of these two monomers is favorable for device performance. Firstly, Br-containing and Sn-containing monomers can easily undergo the Stille coupling polymerization reaction, which reduces the difficulty of material generation. At the same time, its ability to draw electrons as an electron acceptor can significantly reduce the LUMO energy level of the material. The material energy level matches more closely with the Au electrode energy level, enabling the easy injection of electrons. Finally, both acceptor units have a planar structure, which is favorable for electron mobility.

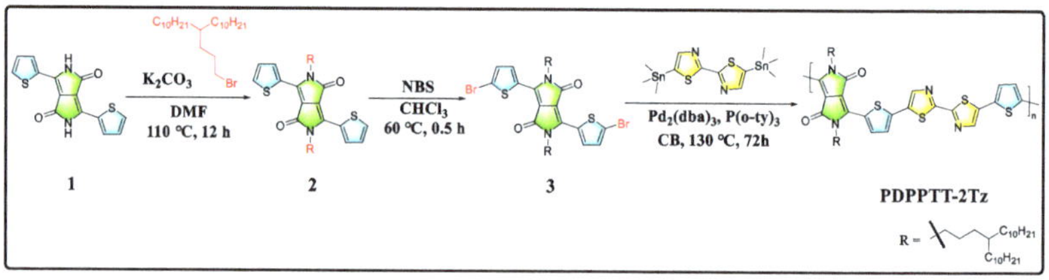

Scheme 2. Synthesis of DPP-based small molecular and PDPPTT-2Tz copolymer.

High-temperature GPC tests (150 °C) showed that the polymer has an extremely high molecular weight with a narrow polymer dispersibility index (PDI). The Mn, Mw, and PDI of **PDPPTT-2Tz** exhibited 171.2 kDa, 257.8 kDa, and 1.50, respectively (Figure S1 in SI). For the mostly DPP-based polymers whose molecular weights (Mn) were generally less than 100 kg mol^{-1}, we attributed this to the minimal use of catalysts with reasonable ligand ratios combined with long polymerization times [25–27]. In general, the molecular weight of the polymer is related to device performance [28,29]. High molecular weight components predominantly adopt a planar π-stacking conformation, which allows for efficient interchain charge carrier transport [30]. Along with this, the narrow distribution favors carrier transport compared to a high PDI [31]. On the other hand, the higher molecular weight decreased the solubility of the polymer, which was almost insoluble in non-chlorinated solvents such as tetrahydrofuran, ethyl acetate, ethanol, etc [32]. The solubility of the polymer in chloroform and chlorobenzene was 3.5 mg/mL and 7.5 mg/mL, respectively, when heated to 50 °C. The reason for using chlorobenzene as a solvent at the beginning of the reaction was designed to avoid the precipitation of high molecular weight polymers with reduced solubility during the reaction. Thermogravimetric analysis (TGA) indicated that the polymer had excellent thermal stability, with a thermal decomposition temperature (5% weight loss) of 416.5 °C in the nitrogen atmosphere (Figure 1). Differential scanning calorimetry analysis indicated that there was no significant phase transition in the temperature range of 300–350 °C (Figure S2), which could be attributed to the rigid skeleton and high molecular weight of **PDPPTT-2Tz**. Due to the poor solubility of polymers, the NMR test could not be used to characterize the copolymer structure. We, therefore, used infrared spectroscopy to compare the difference in absorption between the monomers and polymers (Figure S3). The two materials presented completely different peaks in the wavenumber range of 600–4000 cm^{-1}, with the typical carbonyl characteristic peak migrating finely from symmetry at 1672 cm^{-1} to 1665 cm^{-1} and showing asymmetry. The new peaks at around 850 cm^{-1} as well as at around 3500 cm^{-1} were attributed to the insertion of the thiazole unit. The elemental analysis of polymers is another means of testing their composition and purity. The data and average values of the two measurements for the C, H, N, and S content of these molecules are in good agreement with the theoretical values calculated from the composition of repeating units (Table S1). These indicate that the material is of high purity and that impurities, including catalysts and ligands, have been purified out.

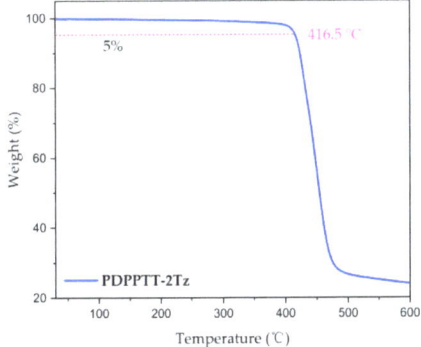

Figure 1. TGA of the PDPPTT-2Tz polymer in nitrogen.

3.2. Density Functional Theory Calculations

The optimized molecular geometries, highest occupied molecular orbital (HOMO) distribution, lowest occupied molecular orbital (LUMO), and the energy gap (ΔEc) were calculated using DFT at the B3LYP-D3/def2tzvp level [33,34] on the basis of the Gaussian 16 program [35,36]. The low-energy conformation of the methyl-substituted trimers was applied instead of the long-conjugated system of the polymer [37]. Notably, the effect of adding dispersion correction to the theoretical calculations was intended to further precise the intramolecular forces and energy levels of the material with only a minor additional computational overhead [38,39]. The smaller thiophene unit only had one α-hydrogen atom, which allowed for favorable intramolecular sulfur–oxygen interactions with a calculated S...O distance of 2.98 Å, which was less than the sum of the van der Waals radii of 3.32 Å [40]. The central backbone is oriented in a nearly coplanar manner with its neighboring thiophene units at a dihedral angle of 10.17° to each other (Figure 2a). The introduction of the 2Tz group did not significantly affect the planarity of the main chain structure. This is due, on the one hand, to the fact that 2Tz itself consists of two thiazole units linked in a trans conformation with a dihedral angle between the thiazole rings close to 180°. On the other hand, the dihedral angle between the thiophene unit and the neighboring thiazole in DPP is only 4.95° (Figure 2b). The calculated LUMO energy level was −3.79 eV, and the lower energy level coincided with the all-acceptor molecular structure design strategy (Table 1). Thereby, it allowed for the easier electron injection and operational stability of the resulting polymer. The energy gap (ΔE_g) obtained from theoretical calculations was 1.81 eV. Non-covalent interaction scattering and reduced density gradient analyses were performed to observe and differentiate non-covalent interactions within the molecule, as shown in Figure 2c [41]. Theoretical simulations of Electrostatic potential surfaces and the molecular orbital maps of the trimers are shown in Figures S4 and S5 of the SI, respectively.

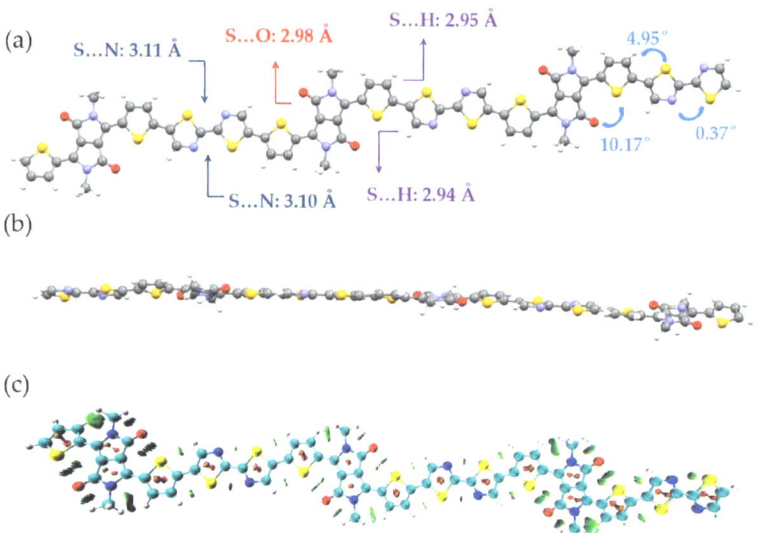

Figure 2. Top (**a**) and Side (**b**) views of the optimized conjugated backbone conformation of PDPPTT-2Tz (methyl-substituted trimers); (**c**) Non-covalent interactions within the molecule.

Table 1. Theoretically calculated values of frontier orbital energy levels of methyl-substituted trimers.

	HOMO (eV)	LUMO (eV)	ΔE_g (eV)
PDPPTT-2Tz	−5.16	−3.79	1.81

3.3. Photophysical Properties and Electrochemical Properties

The UV-visible absorption spectra of **PDPPTT-2Tz** in the solution and in thin films are illustrated in Figure 3 below, and the spectral characteristics are summarized in Table 2. Two absorption peaks could be seen in the solution, including a high-energy absorption peak in the range of 350–500 nm and a low-energy absorption peak in the range of 600–900 nm, which were attributed to the π–π* transition and the intramolecular charge transfer transition, respectively (Figure 3a). The maximum absorption peak (λ_{max}) of copolymer **PDPPTT-2Tz** in the chloroform solution was 765 nm. Compared to the solution absorption, the λ_{max} of the film exhibited a pronounced redshift of 10 nm. The changes observed in the absorption spectra from the solution to the solid state were the result of enhanced intermolecular interactions and increased molecular ordering levels. J aggregation could be determined by observing the difference in the intensity ratios of the first two vibrational peaks (often referred to as the 0–0 and 0–1 transitions) in the absorption spectrum. The films show an optical band gap (E_g^{opt}) of approximately 1.31 eV, as determined from the onset of UV-visible absorption at approximately 946 nm.

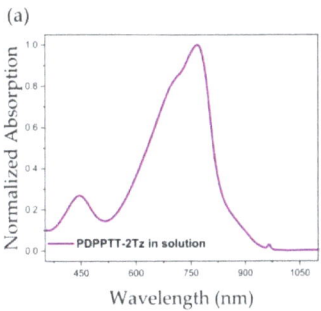 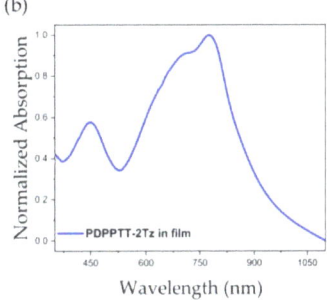

Figure 3. UV-vis spectra of **PDPPTT-2Tz** (**a**) In solution; (**b**) In thin film.

Table 2. Optical and electrochemical characteristics of copolymer **PDPPTT-2Tz**.

	λ_{max} solution (nm)	λ_{max}^{film} (nm)	E_g^{opt} (eV) [1]	E_{red} (eV)	HOMO (eV) [2]	LUMO (eV) [3]	E_g^{cv} (eV) [4]
PDPPTT-2Tz	765	775	1.31	−0.99	−5.08	−3.72	1.36

[1] $E_g^{opt} = 1240/\lambda_{onset}$; [2] $E_{LUMO} = -4.80 \text{ eV} - [(E_{red}^{onset}) - E_{1/2}(\text{ferrocene})]$; [3] $E_{HOMO} = -4.80 \text{ eV} - [(E_{ox}^{onset}) - E_{1/2}(\text{ferrocene})]$; [4] $E_g^{cv} = E_{HOMO} - E_{LUMO}$.

The redox potential of the **PDPPTT-2Tz** film was investigated using cyclic voltammetry (CV) in a 0.1 M TBAPF$_6$ solution in anhydrous acetonitrile. The polymer film was prepared as follows. Firstly, 1 mg of the sample was dissolved in 1 mL of CB, and 0.5 mL of the above solution was aspirated with a needle. In total, 1–2 drops were placed on the electrode while waiting for the reagent to evaporate. The test conditions were measured at room temperature in a nitrogen atmosphere. The reduction peaks were shown to be quasi-reversible, which was attributed to the carbonyl group within the skeleton and was progressively reduced. On the other hand, the oxidation process showed a significantly lower and less reversible current, suggesting that **PDPPTT-2Tz** could favor electron transport rather than hole conduction. The HOMO level was calculated using the onset of the oxidation potential measured by CV, and the LUMO of the polymers was calculated using the onset reduction potential with an offset of −4.80 eV for saturated Ag/AgCl (Figure 4). The ferrocene reference was obtained under the same test conditions as 0.39 V. We observed an onset oxidation potential of 0.67 V and an onset reduction potential of −0.69 V, corresponding to −5.08 eV for HOMO and −3.72 eV for LUMO (Table 2). The energy gap (E_g^{cv}) calculated from electrochemical measurements was 1.36 eV, which was close to the optical bandgap (E_g^{opt}). These low-lying LUMO energy levels are consistent with the DFT calculations, indicating that the all-acceptor strategy effectively decreased the energy levels of the material.

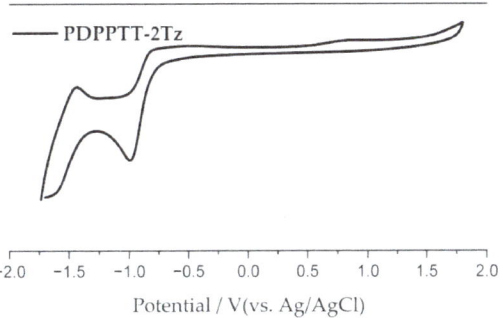

Figure 4. CV of **PDPPTT-2Tz** film in an acetonitrile solution with a scan rate of 50 mV s^{-1}.

3.4. OFET Performance

In order to investigate the charge transfer characteristics of **PDPPTT-2Tz**, we fabricated OFET devices with a bottom–gate bottom–contact (BGBC) structure, where the corresponding BGBC device configuration is shown in Figure 5a. Polymer-based OFETs were fabricated on a highly doped silicon wafer with a 300 nm SiO_2 insulator, which was used as the gate electrode. The source-drain gold electrodes were formed by photolithography. Then, a layer of the polymer semiconductor film was deposited on the OTS-treated substrates by spin-coating from the polymer solution in hot o-dichlorobenzene (5 mg/mL) at a speed of 2000 rpm for 60 s. For annealing the semiconductor film, the samples were further placed on a hotplate in a glove box at 220 °C for 20 min before cooling down to room temperature. The carrier mobility, threshold voltage, and on/off current ratio extracted from the transfer characteristics are summarized in Table 3. The field-effect mobility in saturation (μ) was calculated from the following equation:

$$I_{DS} = (W/2L)\, C_i\, \mu\, (V_{GS} - V_{th})^2 \qquad (1)$$

where W/L is the channel width/length, C_i is the gate dielectric layer capacitance per unit area, and V_{GS} and V_{th} are the gate voltage and threshold voltage, respectively.

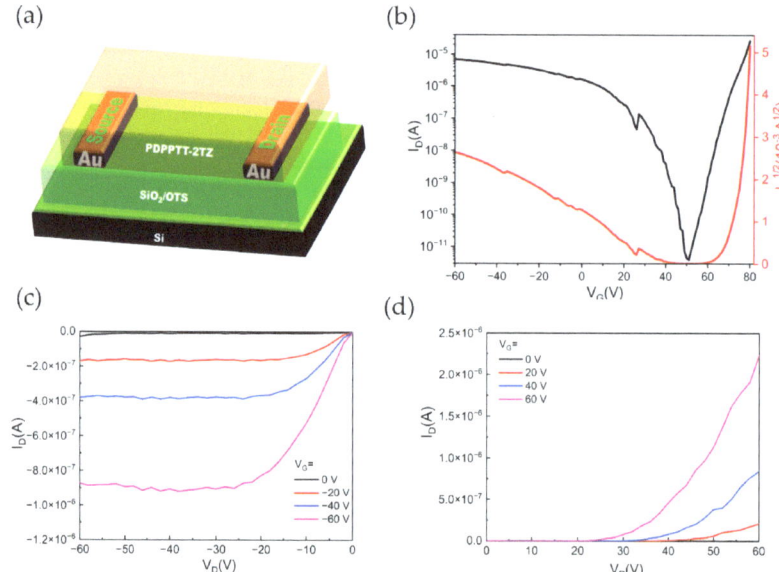

Figure 5. (a) OFET with BGBC architectures; (b) Current−voltage characteristics of OFET based on PDPPTT-2Tz measured under a vacuum at optimized annealing temperature (220 °C); (c,d) Output characteristics of **PDPPTT-2Tz** based device.

Table 3. Electron transport properties of **PDPPTT-2Tz** on OTS-modified BGTC OFET devices.

	Coating Speed (mm/s)	Annealing (°C)	Mobilities [1] (cm^2/(V s))	Max Mobilities (cm^2/(V s))	Threshold Voltages (V)	$I_{ON/OFF}$
PDPPTT-2Tz	2000	220	0.81	1.29	72	10^6

[1] Based on four devices for each condition.

As shown in Figure 5b, it is clear to see that the **PDPPTT-2Tz**-based OFET device showed N-type dominant carrier transfer characteristics with a maximum μ_e value of 1.29 $cm^2\ V^{-1}\ s^{-1}$ with a channel length of 40 μm. The maximum possible difference between the I_{on} and I_{off} states was > 10^6, which was calculated between the maximum gate bias (+80 V) and the minimum gate bias just below the switching voltage, showing that the material had an excellent switching property [42,43]. The average μ_e value of 0.81 cm^2 V^{-1} s^{-1} relatively outperformed N-type devices based on DPP-based polymers [44,45]. On the other hand, the hole mobility was extremely low in the range of 0.003–0.007 cm^2 V^{-1} s^{-1}, which could be related to the nature of the system as electron deficient. The output characteristics of the **PDPPTT-2Tz**-based device are shown in Figure 5c,d.

4. Discussion

Here, we believe that the significant improvement in electron mobility was due to multiple contributing factors. Firstly, the reduction in the LUMO energy level due to the all-acceptor strategy is crucial, as this was favorable for tuning the energy bands of the semiconductor as well as for matching with the working electrode (Au). Secondly, the effect of molecular weight on mobility cannot be ignored, especially since the materials polymerized in this paper had extremely high molecular weights and narrow dispersions. The increase in the effective conjugation length was the key to improving carrier mobility. At the same time, a reasonable choice of acceptors resulted in a more planar structure of the overall molecule. It favored intermolecular stacking and π–π interactions, thus facilitating electron transport. Although carriers in the range of 1.0 cm^2 V^{-1} s^{-1} are already sufficient for the functioning of most basic devices, such as e-paper, sensors, electronic labels, etc., there is still a gap with materials with electron mobility of more than 10 cm^2 V^{-1} s^{-1}, as reported in the literature. This is related to the structure and molecular design strategy of the material itself on the one hand and the preparation method of polymer semiconductor films on the other. The enhancement of carrier transport in polymer semiconductors by external forces conferred by rod coating, film scraping, spraying, and solution shearing was decisive for the performance of the devices. In addition, doping and blending might be other options for enhancing the device's performance.

Research in this field for n-type organic semiconductor materials and their applications in functional devices still has some limitations and is a major direction for future research. Firstly, most polymers are based on Stille coupling polymerization, which requires the use of toxic Sn-containing reagents. On the one hand, this increases the cost of synthesis, and on the other hand, it can easily contaminate human beings and the environment. The current alternative is to use C–H activation for polymerization, which reduces or even eliminates the use of Sn reagents. However, this method also has the disadvantages of low yield, low molecular weight polymerization, and poor generalizability. Second, the reproducibility of the polymers is inferior, as the molecular weight or dispersion of the polymers cannot be exactly the same from each batch, which is insufficient compared to the purity of the small molecules. The optimization of coupling polymerization reaction conditions, the development of efficient green polymerization methods, and the use of mild and environmentally friendly solvents and catalysts are essential points to explore. Finally, n-type semiconductor transistor materials are currently tested in glove boxes, and it is difficult to prepare materials that are stable in the air for long periods of time with long lifetimes.

Realizing the stretchable properties of the material while satisfying the mobility could be the next major extension of this project. The intrinsic stretchability of materials is important because many electronic devices need to be stretchable and bendable in order to achieve a wide range of functionalities. Compared to small molecule semiconductors, polymer semiconductors have an advantage in the field of stretchability due to their longer and softer chain segments. In general, it is more difficult to balance the mechanical and electrical properties of materials. Most of the materials reported exhibit good electron mobility under unstressed conditions, but their electrical properties tended to degrade rapidly under 50% or greater stress. The PDPPTT-2Tz-based material possesses good electrical properties and can withstand the loss of some of its electron mobility while accommodating stretchable performance.

5. Conclusions

A novel organic semiconductor polymer, **PDPPTT-2Tz**, was designed based on an all-acceptor (A–A) structural strategy and was prepared by the palladium-catalyzed Stille polycondensation of electron-deficient 2Tz and DPPTT monomers. A polymer with high molecular weight and narrow dispersion can be prepared by the design of suitable experimental conditions. The enhancement of electron mobility by the A–A structure is significant compared to D–A structure polymers. DFT calculated results show that PDPPTT-2Tz has a lower LUMO energy level, which contributes to an improvement in its electron attraction. The two selected monomers with planar surfaces can still maintain good planarity when forming the polymer, given that the main chain has a tendency to have a large planar. The high molecular weight and planarity further favor its electron transport properties, and the polymer has been used in high-performance OFET devices showing good electron mobility and on-off ratios. Subsequent work on the stability of the device and further enhancement of the carrier mobility and its stretchable properties is still in progress.

Supplementary Materials: The following supporting information can be downloaded at: https://www.mdpi.com/article/10.3390/polym15163392/s1, Figure S1: GPC data (molecular weight distribution and cumulative percent curves) for PDPPTT-2Tz polymer; Figure S2: DSC of the PDPPTT-2Tz polymer in nitrogen; Figure S3: Infrared spectra of the PDPPTT-2Tz polymer; Figure S4: Electrostatic potential surfaces map; Figure S5: Molecular orbital maps; Figure S6: ^{1}H NMR spectra of Compound 2; Figure S7: ^{13}C NMR spectra of Compound 2; Figure S8: ^{1}H NMR spectra of Compound 3; Figure S9: ^{13}C NMR spectra of Compound 3; Table S1: Elemental analysis for C68H106N4O2S4.

Author Contributions: Conceptualization, Z.Y.; methodology, S.R. and W.Z.; software, Z.W.; validation, Z.W.; formal analysis, W.Z.; investigation, W.Z.; resources, Z.W. and Z.Y.; data curation, S.R. and W.Z.; writing—original draft preparation, S.R. and Z.L.; writing—review and editing, Z.W. and W.Z.; visualization, A.Y.; supervision, S.R. and Z.W.; project administration, Z.Y.; funding acquisition, S.R. All authors have read and agreed to the published version of the manuscript.

Funding: This research was funded by the China Scholarship Council (CSC) for the PhD funding (201808070090) and the fellowship of China Post-doctoral Science Foundation (2022TQ0399).

Institutional Review Board Statement: Not applicable.

Informed Consent Statement: Not applicable.

Data Availability Statement: Not applicable.

Acknowledgments: S.R. would like to thank the Hengqin Administrative Committee for supporting grants for the research projects. I would like to thank Jinyang Chen for her academic discussions and guidance on this article.

Conflicts of Interest: The authors declare no conflict of interest.

References

1. Zhang, Y.; Wang, Y.; Gao, C.; Ni, Z.; Zhang, X.; Hu, W.; Dong, H. Recent advances in n-type and ambipolar organic semiconductors and their multi-functional applications. *Chem. Soc. Rev.* **2023**, *52*, 1331–1381. [CrossRef] [PubMed]
2. Anthony, J.E.; Facchetti, A.; Heeney, M.; Marder, S.R.; Zhan, X. n-Type organic semiconductors in organic electronics. *Adv. Mater.* **2010**, *22*, 3876–3892. [CrossRef] [PubMed]
3. Mishra, A.; Bauerle, P. Small molecule organic semiconductors on the move: Promises for future solar energy technology. *Angew. Chem. Int. Ed. Engl.* **2012**, *51*, 2020–2067. [CrossRef] [PubMed]
4. Chen, J.; Zhang, W.; Wang, L.; Yu, G. Recent Research Progress of Organic Small-Molecule Semiconductors with High Electron Mobilities. *Adv. Mater.* **2023**, *35*, e2210772. [CrossRef]
5. Lee, W.; Park, Y. Organic Semiconductor/Insulator Polymer Blends for High-Performance Organic Transistors. *Polymers* **2014**, *6*, 1057–1073. [CrossRef]
6. Liu, Y.; Guo, Y.; Liu, Y. High-Mobility Organic Light-Emitting Semiconductors and Its Optoelectronic Devices. *Small Struct.* **2020**, *2*, 2000083. [CrossRef]
7. Bronstein, H.; Nielsen, C.B.; Schroeder, B.C.; McCulloch, I. The role of chemical design in the performance of organic semiconductors. *Nat. Rev. Chem.* **2020**, *4*, 66–77. [CrossRef]

8. Ren, S.; Yassar, A. Recent Research Progress in Indophenine-Based-Functional Materials: Design, Synthesis, and Optoelectronic Applications. *Materials* **2023**, *16*, 2474. [CrossRef] [PubMed]
9. Duan, J.; Zhu, G.; Chen, J.; Zhang, C.; Zhu, X.; Liao, H.; Li, Z.; Hu, H.; McCulloch, I.; Nielsen, C.B.; et al. Highly Efficient Mixed Conduction in a Fused Oligomer n-Type Organic Semiconductor Enabled by 3D Transport Pathways. *Adv. Mater.* **2023**, *35*, e2300252. [CrossRef]
10. Ren, S.; Habibi, A.; Wang, Y.; Yassar, A. Investigating the Effect of Cross-Conjugation Patterns on the Optoelectronic Properties of 7,7′Isoindigo-Based Materials. *Electronics* **2023**, *12*, 3313. [CrossRef]
11. Liu, K.; Ouyang, B.; Guo, X.; Guo, Y.; Liu, Y. Advances in flexible organic field-effect transistors and their applications for flexible electronics. *NPJ Flex. Electron.* **2022**, *6*, 1. [CrossRef]
12. Griggs, S.; Marks, A.; Bristow, H.; McCulloch, I. n-Type organic semiconducting polymers: Stability limitations, design considerations and applications. *J. Mater. Chem. C Mater.* **2021**, *9*, 8099–8128. [CrossRef] [PubMed]
13. Yi, Z.; Wang, S.; Liu, Y. Design of High-Mobility Diketopyrrolopyrrole-Based pi-Conjugated Copolymers for Organic Thin-Film Transistors. *Adv. Mater.* **2015**, *27*, 3589–3606. [CrossRef]
14. Sun, Y.; Di, C.A.; Xu, W.; Zhu, D. Advances in n-Type Organic Thermoelectric Materials and Devices. *Adv. Electron. Mater.* **2019**, *5*, 1800825. [CrossRef]
15. Wei, X.; Zhang, W.; Yu, G. Semiconducting Polymers Based on Isoindigo and Its Derivatives: Synthetic Tactics, Structural Modifications, and Applications. *Adv. Funct. Mater.* **2021**, *31*, 2010979. [CrossRef]
16. Liu, Q.; Bottle, S.E.; Sonar, P. Developments of Diketopyrrolopyrrole-Dye-Based Organic Semiconductors for a Wide Range of Applications in Electronics. *Adv. Mater.* **2020**, *32*, e1903882. [CrossRef]
17. Ren, S.; Habibi, A.; Ni, P.; Nahdi, H.; Bouanis, F.Z.; Bourcier, S.; Clavier, G.; Frigoli, M.; Yassar, A. Synthesis and characterization of solution-processed indophenine derivatives for function as a hole transport layer for perovskite solar cells. *Dyes Pigments* **2023**, *213*, 111136. [CrossRef]
18. Ye, T.; Jin, S.; Singh, R.; Kumar, M.; Chen, W.; Wang, D.; Zhang, X.; Li, W.; He, D. Effects of solvent additives on the morphology and transport property of a perylene diimide dimer film in perovskite solar cells for improved performance. *Sol. Energy* **2020**, *201*, 927–934. [CrossRef]
19. Lei, T.; Wang, J.Y.; Pei, J. Design, synthesis, and structure-property relationships of isoindigo-based conjugated polymers. *Acc. Chem. Res.* **2014**, *47*, 1117–1126. [CrossRef]
20. Lin, Y.C.; Chen, C.H.; Tsai, B.S.; Hsueh, T.F.; Tsao, C.S.; Tan, S.; Chang, B.; Chang, Y.N.; Chu, T.Y.; Tsai, C.E.; et al. Alkoxy- and Alkyl-Side-Chain-Functionalized Terpolymer Acceptors for All-Polymer Photovoltaics Delivering High Open-Circuit Voltages and Efficiencies. *Adv. Funct. Mater.* **2023**, *33*, 2215095. [CrossRef]
21. Wang, J.; Feng, K.; Jeong, S.Y.; Liu, B.; Wang, Y.; Wu, W.; Hou, Y.; Woo, H.Y.; Guo, X. Acceptor-acceptor type polymers based on cyano-substituted benzochalcogenadiazole and diketopyrrolopyrrole for high-efficiency n-type organic thermoelectrics. *Polym. J.* **2022**, *55*, 507–515. [CrossRef]
22. Mueller, C.J.; Singh, C.R.; Fried, M.; Huettner, S.; Thelakkat, M. High Bulk Electron Mobility Diketopyrrolopyrrole Copolymers with Perfluorothiophene. *Adv. Funct. Mater.* **2015**, *25*, 2725–2736. [CrossRef]
23. Yang, J.; Wang, H.; Chen, J.; Huang, J.; Jiang, Y.; Zhang, J.; Shi, L.; Sun, Y.; Wei, Z.; Yu, G.; et al. Bis-Diketopyrrolopyrrole Moiety as a Promising Building Block to Enable Balanced Ambipolar Polymers for Flexible Transistors. *Adv. Mater.* **2017**, *29*, 1606162–1606169. [CrossRef] [PubMed]
24. Su, H.L.; Sredojevic, D.N.; Bronstein, H.; Marks, T.J.; Schroeder, B.C.; Al-Hashimi, M. Bithiazole: An Intriguing Electron-Deficient Building for Plastic Electronic Applications. *Macromol. Rapid Commun.* **2017**, *38*, 1600610–1600634. [CrossRef]
25. Zhang, A.; Xiao, C.; Wu, Y.; Li, C.; Ji, Y.; Li, L.; Hu, W.; Wang, Z.; Ma, W.; Li, W. Effect of Fluorination on Molecular Orientation of Conjugated Polymers in High Performance Field-Effect Transistors. *Macromolecules* **2016**, *49*, 6431–6438. [CrossRef]
26. Sun, B.; Hong, W.; Aziz, H.; Abukhdeir, N.M.; Li, Y. Dramatically enhanced molecular ordering and charge transport of a DPP-based polymer assisted by oligomers through antiplasticization. *J. Mater. Chem. C* **2013**, *1*, 4423–4426. [CrossRef]
27. Di Pietro, R.; Erdmann, T.; Carpenter, J.H.; Wang, N.; Shivhare, R.R.; Formanek, P.; Heintze, C.; Voit, B.; Neher, D.; Ade, H.; et al. Synthesis of High-Crystallinity DPP Polymers with Balanced Electron and Hole Mobility. *Chem. Mater.* **2017**, *29*, 10220–10232. [CrossRef]
28. Peña-Alcántara, A.; Nikzad, S.; Michalek, L.; Prine, N.; Wang, Y.; Gong, H.; Ponte, E.; Schneider, S.; Wu, Y.; Root, S.E.; et al. Effect of Molecular Weight on the Morphology of a Polymer Semiconductor–Thermoplastic Elastomer Blend. *Adv. Electron. Mater.* **2023**, 2201055–2201068. [CrossRef]
29. Chu, T.-Y.; Lu, J.; Beaupré, S.; Zhang, Y.; Pouliot, J.-R.; Zhou, J.; Najari, A.; Leclerc, M.; Tao, Y. Effects of the Molecular Weight and the Side-Chain Length on the Photovoltaic Performance of Dithienosilole/Thienopyrrolodione Copolymers. *Adv. Funct. Mater.* **2012**, *22*, 2345–2351. [CrossRef]
30. Zen, A.; Pflaum, J.; Hirschmann, S.; Zhuang, W.; Jaiser, F.; Asawapirom, U.; Rabe, J.P.; Scherf, U.; Neher, D. Effect of Molecular Weight and Annealing of Poly(3-hexylthiophene)s on the Performance of Organic Field-Effect Transistors. *Adv. Funct. Mater.* **2004**, *14*, 757–764. [CrossRef]
31. Tripathi, A.S.M.; Sadakata, S.; Gupta, R.K.; Nagamatsu, S.; Ando, Y.; Pandey, S.S. Implication of Molecular Weight on Optical and Charge Transport Anisotropy in PQT-C12 Films Fabricated by Dynamic FTM. *ACS Appl. Mater. Interfaces* **2019**, *11*, 28088–28095. [CrossRef]

32. Kanimozhi, C.; Yaacobi-Gross, N.; Chou, K.W.; Amassian, A.; Anthopoulos, T.D.; Patil, S. Diketopyrrolopyrrole-diketopyrrolopyrrole-based conjugated copolymer for high-mobility organic field-effect transistors. *J. Am. Chem. Soc.* **2012**, *134*, 16532–16535. [CrossRef] [PubMed]
33. Lee, C.; Yang, W.; Parr, R.G. Development of the Colle-Salvetti correlation-energy formula into a functional of the electron density. *Phys. Rev. B* **1988**, *37*, 785–789. [CrossRef] [PubMed]
34. Becke, A.D. Density-functional thermochemistry. III. The role of exact exchange. *J. Chem. Phys.* **1993**, *98*, 5648–5652. [CrossRef]
35. Frisch, M.J.; Trucks, G.W.; Schlegel, H.B.; Scuseria, G.E.; Robb, M.A.; Cheeseman, J.R.; Scalmani, G.; Barone, V.; Petersson, G.A.; Nakatsuji, H.; et al. *Gaussian 16 Rev. C.01*; Gaussian Inc.: Wallingford, CT, USA, 2016.
36. Miehlich, B.; Savin, A.; Stoll, H.; Preuss, H. Results obtained with the correlation energy density functionals of becke and Lee, Yang and Parr. *Chem. Phys. Lett.* **1989**, *157*, 200–206. [CrossRef]
37. Zhang, W.; Shi, K.; Lai, J.; Zhou, Y.; Wei, X.; Che, Q.; Wei, J.; Wang, L.; Yu, G. Record-High Electron Mobility Exceeding 16 cm(2) V(-) (1) s(-) (1) in Bisisoindigo-Based Polymer Semiconductor with a Fully Locked Conjugated Backbone. *Adv. Mater.* **2023**, *35*, e2300145. [CrossRef]
38. Goerigk, L. A Comprehensive Overview of the DFT-D3 London-Dispersion Correction. In *Non-Covalent Interactions in Quantum Chemistry and Physics*; Elsevier: Amsterdam, The Netherlands, 2017; pp. 195–219.
39. Tunega, D.; Bucko, T.; Zaoui, A. Assessment of ten DFT methods in predicting structures of sheet silicates: Importance of dispersion corrections. *J. Chem. Phys.* **2012**, *137*, 114105. [CrossRef]
40. Nielsen, C.B.; Turbiez, M.; McCulloch, I. Recent advances in the development of semiconducting DPP-containing polymers for transistor applications. *Adv. Mater.* **2013**, *25*, 1859–1880. [CrossRef]
41. Lu, T.; Chen, F. Multiwfn: A multifunctional wavefunction analyzer. *J. Comput. Chem.* **2012**, *33*, 580–592. [CrossRef]
42. Saragi, T.P.I.; Fuhrmann-Lieker, T.; Salbeck, J. High ON/OFF ratio and stability of amorphous organic field-effect transistors based on spiro-linked compounds. *Synth. Metals* **2005**, *148*, 267–270. [CrossRef]
43. Dheepika, R.; Mohamed Imran, P.; Bhuvanesh, N.S.P.; Nagarajan, S. Solution-Processable Unsymmetrical Triarylamines: Towards High Mobility and ON/OFF Ratio in Bottom-Gated OFETs. *Chemistry* **2019**, *25*, 15155–15163. [CrossRef] [PubMed]
44. Liao, M.; Duan, J.; Peng, P.; Zhang, J.; Zhou, M. Progress in the synthesis of imide-based N-type polymer semiconductor materials. *RSC Adv.* **2020**, *10*, 41764–41779. [CrossRef] [PubMed]
45. Cheon, H.J.; An, T.K.; Kim, Y.-H. Diketopyrrolopyrrole (DPP)-Based Polymers and Their Organic Field-Effect Transistor Applications: A Review. *Macromol. Res.* **2022**, *30*, 71–84. [CrossRef]

Disclaimer/Publisher's Note: The statements, opinions and data contained in all publications are solely those of the individual author(s) and contributor(s) and not of MDPI and/or the editor(s). MDPI and/or the editor(s) disclaim responsibility for any injury to people or property resulting from any ideas, methods, instructions or products referred to in the content.

Article

Enhancing Blue Polymer Light-Emitting Diode Performance by Optimizing the Layer Thickness and the Insertion of a Hole-Transporting Layer

A. Saad [1], N. Hamad [1], Rasul Al Foysal Redoy [2], Suling Zhao [3,*] and S. Wageh [2,4,*]

1. Department of Physics, College of Science and Humanities in Al-Kharj, Prince Sattam bin Abdulaziz University, Al-Kharj 11942, Saudi Arabia; n.hamad@psau.edu.sa (N.H.)
2. Department of Physics, Faculty of Science, King Abdulaziz University, Jeddah 21589, Saudi Arabia; rredoy@stu.kau.edu.sa
3. Key Laboratory of Luminescence and Optical Information, Institute of Optoelectronics Technology, Beijing Jiaotong University, Beijing 100044, China
4. Physics and Engineering Mathematics Department, Faculty of Electronic Engineering, Menoufia University, Menouf 32952, Egypt
* Correspondence: slzhao@bjtu.edu.cn (S.Z.); wswelm@kau.edu.sa (S.W.)

Citation: Saad, A.; Hamad, N.; Redoy, R.A.F.; Zhao, S.; Wageh, S. Enhancing Blue Polymer Light-Emitting Diode Performance by Optimizing the Layer Thickness and the Insertion of a Hole-Transporting Layer. *Polymers* **2024**, *16*, 2347. https://doi.org/10.3390/polym16162347

Academic Editors: Tengling Ye and Bixin Li

Received: 8 June 2024
Revised: 7 August 2024
Accepted: 12 August 2024
Published: 20 August 2024

Copyright: © 2024 by the authors. Licensee MDPI, Basel, Switzerland. This article is an open access article distributed under the terms and conditions of the Creative Commons Attribution (CC BY) license (https://creativecommons.org/licenses/by/4.0/).

Abstract: Polymer light-emitting diodes (PLEDs) hold immense promise for energy-efficient lighting and full-color display technologies. In particular, blue PLEDs play a pivotal role in achieving color balance and reducing energy consumption. The optimization of layer thickness in these devices is critical for enhancing their efficiency. PLED layer thickness control impacts exciton recombination probability, charge transport efficiency, and optical resonance, influencing light emission properties. However, experimental variations in layer thickness are complex and costly. This study employed simulations to explore the impact of layer thickness variations on the optical and electrical properties of blue light-emitting diodes. Comparing the simulation results with experimental data achieves valuable insights for optimizing the device's performance. Our findings revealed that controlling the insertion of a layer that works as a hole-transporting and electron-blocking layer (EBL) could greatly enhance the performance of PLEDs. In addition, changing the active layer thickness could optimize device performance. The obtained results in this work contribute to the development of advanced PLED technology and organic light-emitting diodes (OLEDs).

Keywords: blue light-emitting diode; BP105; TAPC

1. Introduction

PLEDs, or conjugated polymer light-emitting diodes, are quite popular because of their simple solution procedure and possible high emission efficiency. As such, it is thought that PLEDs will serve as the foundation for the upcoming generation of flat-panel displays [1]. These specialized light-emitting diodes are essential parts of bright, energy-saving lighting and display systems. Blue PLEDs are a crucial invention in the future generation of lighting and visual solutions because they are essential for achieving full-color displays, maintaining color balance, and helping to minimize energy usage. Scientists and engineers have been working hard over the years to increase the efficiency of blue PLEDs, which will further qualify them for various uses. Controlling the layer thickness in PLEDs is an essential area of research for improving LED performance. The thickness of the organic layers in a PLED influences the probability of excitons (electron–hole pairs) recombining and the emission of light [2]. An optimal layer thickness ensures that a higher percentage of excitons recombine to produce light rather than being lost through nonradiative processes.

Consequently, controlling layer thickness could enhance light emission efficiency [3]. A well-tuned thickness can reduce resistive losses and increase the device's overall efficiency by facilitating more efficient charge transport. It is possible to generate optical resonance by varying the thickness of some layers, such as the emissive layer [4,5]. More effective light emission and a decrease in optical losses can be achieved by optimizing light confinement and extraction through this resonance [6]. Light emission in full-color displays can be controlled by varying the emissive layer thickness. Achieving a wide color spectrum and accurate color representation requires this fine control. However, studying the impact of different thicknesses presents challenges, as achieving such variations necessitates complex and costly deposition methods. Therefore, conducting simulations emerges as the preferable choice prior to experimental endeavors. Simulations offer a more cost-effective and efficient means of exploring the effects of layer thickness variations.

Generally, light-emitting devices consist of organic structures mainly based on the dual injection of holes and electrons from two electrodes with opposite charges. The injected charges should finally enter the emissive layer to form an exciton followed by light emission. Consequently, to enhance the device performance, multilayer structures should be preferred, including a separate hole-transport layer (HTL), electron-transport layer, and emissive layer (EML). The role of the HTL layer is to enhance the hole injection into the active layer, block the electron from overflowing, and confine the formed exciton in the active layer. Accordingly, the insertion of the HTL is a preferable choice for increasing the OLED efficiency.

In polymer light-emitting diode (PLED) research, running simulations in addition to conventional experimental work provides an economical, effective, and adaptable way to investigate a broad range of layer thickness scenarios. Through the exploration of unfeasible or extreme scenarios and the quick refinement of hypotheses, simulations maximize resource use and reduce risk. They help identify suitable experimental candidates by offering predictive insight into the behavior of layer thickness. Furthermore, simulations facilitate parametric analysis and expedite the experimentation procedure. Simulations supplement experimental research by addressing safety and environmental concerns, eliminating material waste, and testing hypotheses. This helps to advance the development of cutting-edge PLED technology and improves our understanding of PLED behavior.

Besides the importance of blue light-emitting devices (OLEDs) for application in flat panel displays and solid-state lighting [7], blue light has important therapeutic applications, including treating acne breakouts, psoriasis, seasonal affective disorders, and newborn jaundice [8]. Also, blue LEDs have applications in food safety, such as tomato treatment; blue light can delay ripening and, thereby, extend the storage life of tomatoes [9,10].

According to the abovementioned requirements, this paper introduces a simulation and the optimization of a blue light-emitting diode based on BP105 as an emission layer. The analysis of optical and electrical properties concerning changes in the active and electron-blocking layers' thickness was investigated. To conduct our simulations, we employed the commercially available software, Setfos v5.4, developed by Fluxim [11]. Utilizing this software, we successfully simulated experimental data for the blue multilayer polymer light-emitting diode, consisting of ITO, PEDOT: PSS, BP105, and LiF/Ca/Al [12,13]. After confirming the validity of the simulation method, we optimized the performance of the device by varying the thickness of the active layer. In addition, another design with the insertion of a layer that works simultaneously as HTL and electron blocking was investigated and optimized.

The selection of the PLED architecture as ITO, PEDOT: PSS, BP105, and LiF/Ca/Al arising from this arrangement has a good opportunity for further advancement and innovation in the optoelectronics industry. This arrangement combines the advantages of solution-processable materials such as PEDOT:PSS and BP105, creating a promising platform for cost-effective and scalable manufacturing. In this arrangement, the combination of LiF/Ca/Al was selected as one of the two electrodes. The LiF/Ca/Al electrode has excellent electrical performance and exhibits a low turn-on voltage. Furthermore, ITO, PEDOT: PSS, BP105, and LiF/Ca/Al designs have low heat generation and power consumption, thereby extending their lifespan. Moreover, the device produces high brightness through effective charge injection and transport systems.

2. Numerical Simulation of the Experimental Model

A drift-diffusion technique, which has been published in numerous other publications [14–16], serves as the foundation for the electrical simulations used in this study. The measured curves from the electrical experiments can be replicated by solving the equations in the steady-state, transient, and frequency domains. The computed exciton densities are fed into the optical solver for electro-optical simulations based on the transfer matrix and dipole emission models [17,18]. The required inputs for the simulation are the layer sequence with respective thicknesses, refractive indices, and each layer's electrical and excitonic material parameters. The results obtained in references [12,13] were simulated to confirm our computation's validity. The structure and the layer thicknesses of the blue light emitting device are as follows: ITO(100 nm), PEDOT: PSS(100 nm), BP105(80 nm), LiF(1 nm), Ca(10 nm), and Al(200 nm) [12,13]. The schematic construction of the device and band energy diagram are shown in Figure 1a,b. In this device, the active layer consists of a blue-emitting polymer known as BP105, which was developed by The Dow Chemical Company (Midland, MI, USA). The emission of BP105 is displayed in Figure 1c.

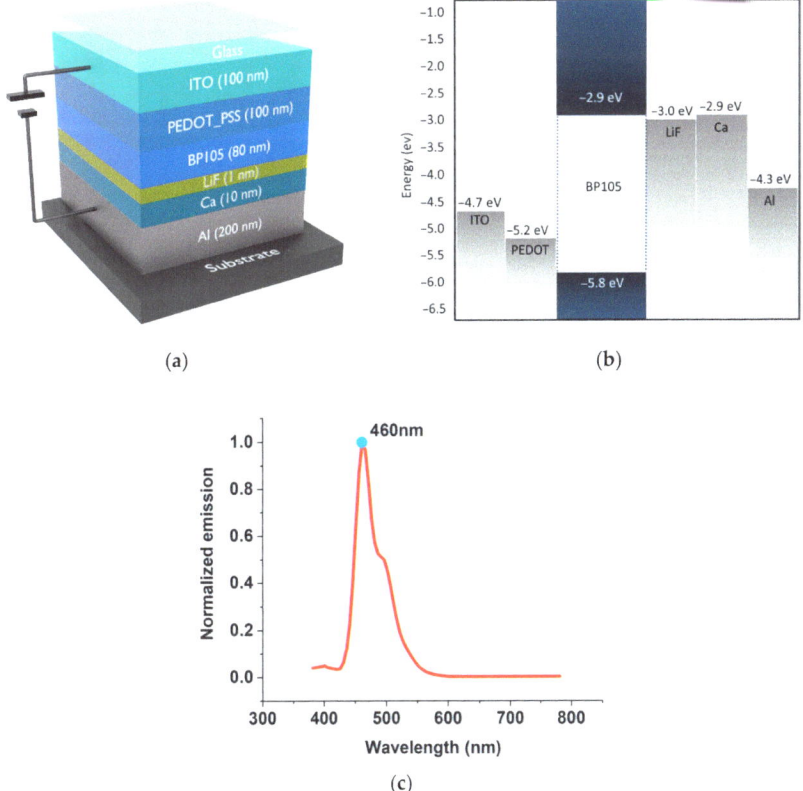

Figure 1. (a) Layer structure. (b) Band energy diagram. (c) Emission of BP105.

BP105, along with the top and bottom electrodes, acts as the main emission layer. ITO/PEDOT:PSS works as the top electrode (anode). ITO is a well-known transparent electrode with a work function of −4.7 eV. However, the potential barrier is large between the work function of ITO and the HOMO level of BP105, causing inefficient carrier injection. A hybrid electrode of ITO/PEDOT:PSS was introduced to lower this barrier. Comparatively to ITO alone (−4.7 eV), PEDOT's intrinsic work function (−5.2 eV) is closer to the HOMO level of BP105 (−5.8 eV). PEDOT:PSS molecules have intrinsic dipole moments due to their chemical structure. Variations in their work functions and Fermi levels may cause electrons to migrate between ITO and PEDOT:PSS when PEDOT:PSS is deposited on ITO. This electron movement builds both positive and negative charges at the contact, creating an electric dipole layer. The vacuum level is normally shifted upward by the development of this dipole layer at the ITO/PEDOT:PSS interface, raising the combined system's work function. This lowers the energy barrier for hole injection into the active layer by bringing the anode's effective work function closer to the HOMO level of BP105. PEDOT:PSS also provides a smoother and more consistent surface than the rough bare surface of ITO. Reducing the number of interface defects helps improve surface quality, thereby facilitating better electrical and physical contact with BP105.

For better performance, LED requires an electrode for the cathode that has a lower work function and is in good alignment with the LUMO level of BP105. For that, a combined electrode of LiF/Ca/Al was introduced to the structure. LiF has a very wide bandgap (~14 eV), which makes it an excellent insulator. Despite being an insulator, a very thin layer of LiF (1 nm) was used as a buffer layer to reduce any interface recombination and improve carrier confinement. However, the work function of LiF (−3.0 eV) and the LUMO level of BP105 still create a potential barrier. Ca was included to lower this barrier. The LUMO level of BP105 is well-aligned with the comparatively low work function of calcium, which is approximately −2.9 eV. The work function of the integrated system decreases when calcium deposits on LiF. This work function modification is crucial for aligning the energy levels between the cathode and the active layer, reducing the barrier for electron injection. However, calcium is a highly reactive metal that can trigger device degradation very quickly. A layer of aluminum on top of calcium can improve both the device's stability and carrier injections. Aluminum was chosen for its excellent electrical conductivity and its stability. Aluminum has a work function of about −4.3 eV. Although this is higher than that of calcium, the combination of LiF and Ca layers ensures that the overall injection barrier is minimized.

The optical and electrical parameters, including refractive index, HOMO-LUMO energy levels, charge mobility, and radiative and nonradiative decay rates of the materials, were applied from the built-in material database integrated into Setfos. The carrier mobility of BP105 was determined using the Extended Gaussian Disorder Model (EGDM). This model is widely employed to describe charge transport in organic materials, providing valuable insights into the mobility of charge carriers within the BP105 active layer of the device [19]. An Exponential Density of States (DOS) model was utilized to analyze the trap state. Within this model, the trapping density of electrons was set to 2×10^{24} m^{-3} [19]. This model allowed for a comprehensive investigation of electron trapping phenomena, providing a better understanding of the device's behavior. Recombination processes in the device were modeled using the Langevin model, where the recombination efficiency was specifically set to 1. The Langevin model is commonly used to describe the recombination of charge carriers within semiconductor materials and organic devices. In this case, the efficiency value of 1 suggests that all potential recombination events were considered without any reduction or losses, resulting in a comprehensive analysis of recombination processes. The simulations were conducted at a temperature of 293 Kelvin (293 K). Furthermore, in the simulations, the device was surrounded by glass, a substrate, and air, as illustrated in Figure 1a. These surrounding materials and conditions were integrated into the simulation setup to closely replicate the real-world environment in which the optoelectronic device operates.

The simulation of current efficiency and luminance of the device consisting of ITO(100 nm), PEDOT: PSS(100 nm), BP105(80 nm), LiF(1 nm), Ca(10 nm), Al(200 nm) are shown in Figure 2. In addition, the experimental data of similar construction published in references [12,13] are displayed to confirm the validity of our method. Clearly, the simulation closely matched the experimental results, affirming its accuracy and suitability in scientific contexts.

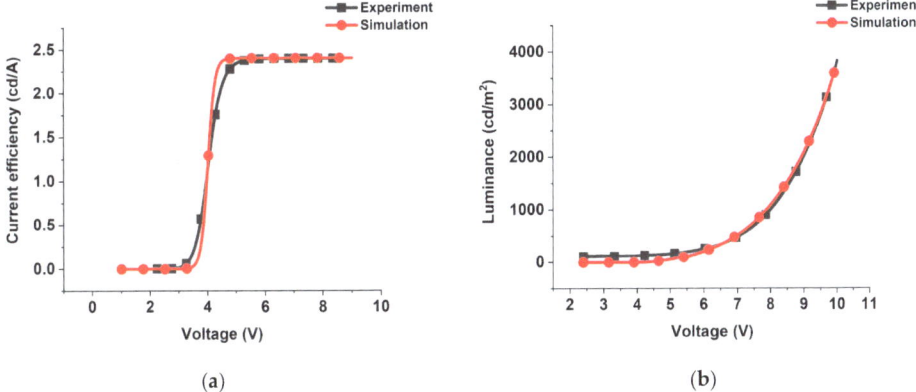

Figure 2. Validation of simulation results against experimental data for blue polymer light-emitting diode performance. (**a**) Current efficiency; (**b**) luminance.

3. Device Optimization

3.1. Varying Thickness of Active Layer

In the previous section, we demonstrated the ability of our simulations to accurately reproduce the experimental characterizations of the BP105 blue light-emitting polymer. Our modeling approach enables the simultaneous variation of parameters within the PLED stack and the scattering layer. Therefore, in this section, we present how the device can be further optimized.

The performance of the light-emitting diode was first optimized by varying the thickness of the emission layer. The effect of varying active layer thickness was investigated while keeping the other layer thicknesses constant. Figure 3a,b show the effect of changing BP105 thickness on the current efficiency and luminous efficacy with an applied voltage of 5 V. Increasing the thickness from 10 nm to 80 nm continuously causes an increase in current efficiency, and then the current efficiency has the first maximum value in the range of the thickness from 80 to 120 nm. Further, increasing thickness beyond 120 nm causes an increase in current efficiency, reaching the highest maximum at 180 nm thickness. The behavior for luminous efficacy with varying the thickness of the active layer showed two maximums, one at the thickness of the active layer of 60 nm and the second at the thickness of the active layer of 200 nm.

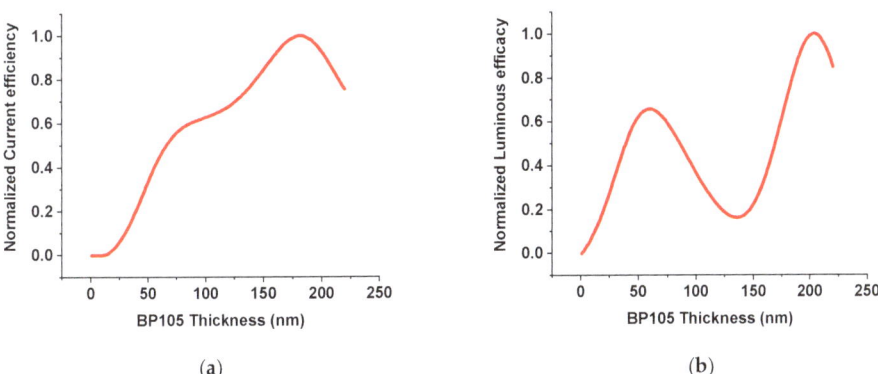

Figure 3. The effect of varying the active layer of BP105 thickness on (**a**) current efficiency; and (**b**) luminous efficacy.

According to the investigation of the device with various thicknesses of BP105, we found that the thicknesses of 60 nm, 80 nm, 180 nm, and 200 nm are important thicknesses of the active layer to study. Consequently, we investigated the current efficiency and luminous efficacy against voltage variation for the devices of BP105 with thicknesses of 60 nm, 80 nm, 180 nm, and 200 nm. Figure 4 shows the effect of variation of applied voltage on the current efficiency for different thicknesses of the emission layer.

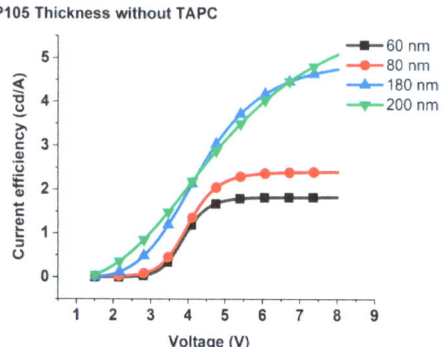

Figure 4. Current efficiency for different thicknesses of the emission layer of the device consists of ITO(100 nm), PEDOT: PSS(100 nm), BP105(60, 80, 180, and 200 nm), LiF(1 nm), Ca(10 nm), and Al(200 nm).

The increase in applied voltage leads to an increase in current efficiency continuously up to 5 V; beyond 5 V, the efficiency tends to saturate. As the voltage is increased with low values, the electric field becomes stronger, which facilitates the injection of more charge carriers into the device and, in consequence, causes an increase in current efficiency. While the saturation of current at higher voltage values can be ascribed to one of the following reasons: (1) there can be a mismatch between the densities of electrons and holes injected into the active layer. This charge imbalance can result in inefficient recombination processes, where excess carriers remain unpaired and contribute to nonradiative recombination channels. Consequently, the current efficiency may be saturate at higher voltage [20]. (2) At higher voltages, the density of charge carriers (electrons and holes) within the OLED increases. This elevated carrier density can lead to increased exciton-quenching effects, where excitons (electron-hole pairs) are more likely to undergo nonradiative decay processes, such as collisional quenching or exciton–exciton annihilation, rather than radiative recombination to emit light leads to current saturation [21]. To decide which mechanism is responsible for current saturation at higher voltage, we investigated the effect of applied voltage on the luminous efficacy. Figure 5 shows luminous efficacy against voltage variation for various devices that possess different thicknesses of active layer. Clearly, the luminous efficacy started to decrease at higher voltage, which indicates that the second mechanism is responsible for the saturation of current efficiency. At higher voltages, the process of collisional quenching or exciton–exciton annihilation and nonradiative recombination causes a decrease in the efficiency of converting injected charges into emitted photons and consequently, luminous efficacy decreases.

Increasing the thickness of active layers from 60 nm to 180 nm and 200 nm causes an enhancement of the current efficiency and luminous efficacy; these results can be attributed to the increasing thickness of Bp105 facilitating a higher density of polymer chains and a greater volume for exciton formation. This can increase the probability of exciton generation per unit area, leading to more efficient light emission and improved luminous efficacy. Furthermore, large thicknesses can help optimize the distribution and density of charge carriers within the device, leading to improved charge balance [22]. This can minimize nonradiative recombination processes and enhance light emission efficiency, thereby increasing luminous efficacy. Conversely, with the increasing thickness from 60 nm to 80 nm, the current efficiency increases, but luminous efficacy decreases. These results can be explained as follows. The thickness of 80 nm may exacerbate optical losses within the device due to weak microcavity effects (i.e., non-resonance optical length) [23]. This loss can reduce the amount of light extracted from the device and diminish luminous efficacy, even if the current efficiency improves.

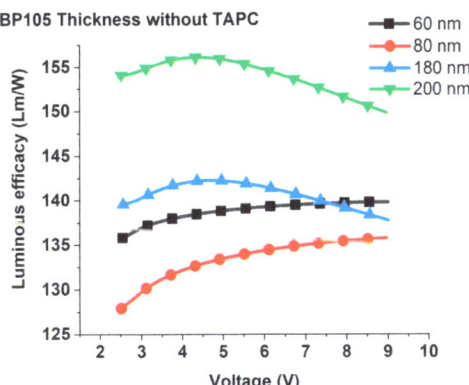

Figure 5. Luminous efficacy for different thicknesses of the emission layer.

3.2. Effect of Insertion of HTL Layer on the Performance of the OLED

Another strategy applied to improve the performance of the device is the introduction of TAPC (1,1-bis[(di-4-tolylamino)phenyl]cyclohexane) as an intermediary layer between PEDOT: PSS and the active layer. Due to the high triplet energy (2.9 eV) and LUMO levels (−2.0 eV) of TAPC, it works as a hole-transporting and electron-blocking layer. The schematic construction of the device with the insertion of the TAPC layer along with the band energy diagram is shown in Figure 6.

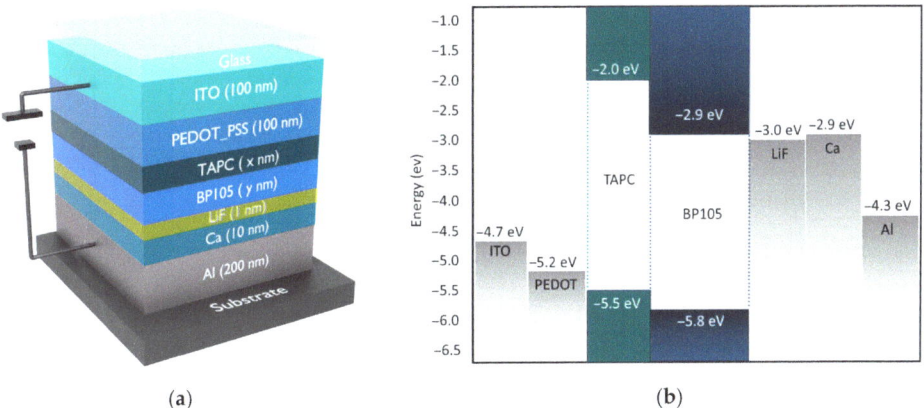

Figure 6. Layer structure and band energy diagram after adding TAPC. (**a**) Layer structure; (**b**) band energy diagram.

The changes in current efficiency and luminous efficacy with varying thicknesses of TAPC from 1 nm to 350 nm for the thicknesses of 60, 80, 180, and 200 nm of the BP105 layer are shown in Figure 7a,b. Obviously, three maximums appeared that achieve optimum current efficiency, which varied for different thicknesses of the active layer. Similar behavior was obtained for luminous efficacy. According to the appeared maximums with varying

TAPC thicknesses, we investigated the performance of various devices with thicknesses of 60 nm, 80 nm, 180 nm, and 200 nm for the BP105 layer.

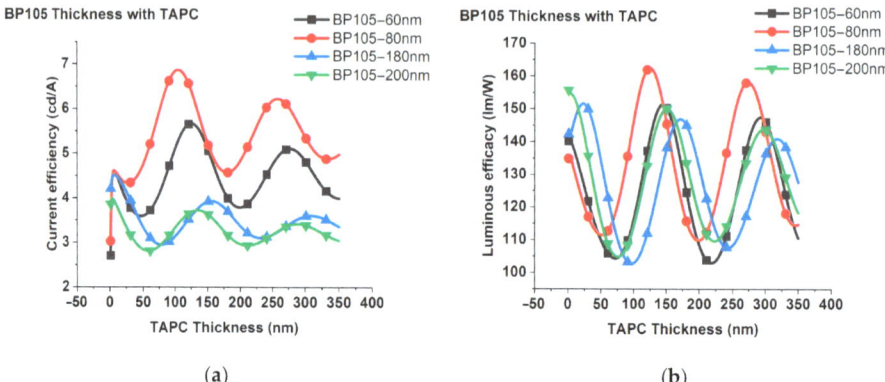

Figure 7. Varying the thicknesses of TAPC from 1 nm to 350 nm for the thicknesses of 60, 80, 180, and 200 nm of the BP105 layer. (**a**) Current efficiency; (**b**) luminous efficacy.

The effects of varying the thicknesses of TAPC on current efficiency and luminous efficacy at different applied voltages for the four devices with BP105 of thicknesses of 60 nm, 80 nm, 180 nm, and 200 nm were investigated. Figure 8 shows the current efficiency against the applied voltage for the device consisting of ITO(100 nm), PEDOT: PSS(100 nm), TAPC (x), BP105(60 nm), LiF(1 nm), Ca(10 nm), and Al(200 nm), with various thicknesses of TAPC hole-transporting layer thicknesses (6 nm, 125 nm, and 277 nm). At low voltage, the current efficiency was very small due to inefficient charge injection. As the voltage increased beyond the threshold (~3.5 V) up to nearly 6 V, the current efficiency rose as more charge carriers were injected and contributed to radiative recombination, and consequently, the current efficiency continuously increased. At higher voltages, the current efficiency was saturated, i.e., limiting the efficiency of converting injected charge carriers into emitted photons, leading to diminished current efficiency despite increasing voltage. The effect of variation of thickness at low voltages below 5 V on the current efficiency was insignificant, but for higher voltages and in the saturation region, applying the thickness of 125 nm, which is compatible with the second maximum, caused an increase in the current efficiency. Conversely, applying the thickness compatible with the third maximum of 277 nm, the current efficiency increased relative to the thickness of 6 nm and decreased relative to the thickness of 125 nm.

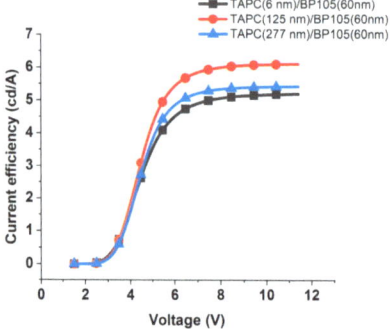

Figure 8. Current efficiency against the applied voltage for the device with 60 nm of BP105, with various thicknesses of TAPC hole-transporting layer thicknesses (6 nm, 125 nm, and 277 nm).

The behavior for the device with ITO(100 nm), PEDOT: PSS(100 nm), TAPC (x), BP105(80 nm), LiF(1 nm), Ca(10 nm), and Al(200 nm) and various TAPC hole-transporting layer thicknesses was similar to the device with BP105 of 60 nm but with an increase in current efficiency, as shown in Figure 9. We should keep in mind that the thickness that was given the highest values of luminous efficacy does not equal the thickness that has given higher current efficiency.

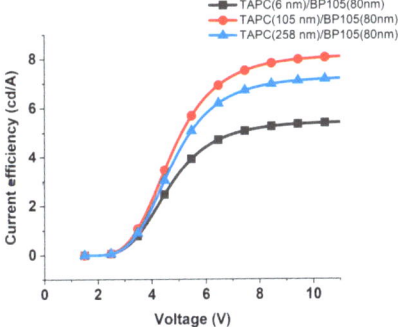

Figure 9. Current efficiency against the applied voltage for the device with 80 nm of BP105, with various thicknesses of TAPC hole-transporting layer thicknesses (6 nm, 105 nm, and 258 nm).

The behavior for devices with higher thicknesses of 180 nm and 200 nm of the active layer with the variation of TAPC thicknesses are different, as shown in Figure 10a,b. The current efficiency continuously increased and did not reach complete saturation. In addition, the current efficiency maximum values were less than those obtained for the devices with thinner active layers. These could be explained by the fact that the thinner layers lead to more efficient charge transport and reduced exciton quenching at the interfaces, resulting in higher current efficiency.

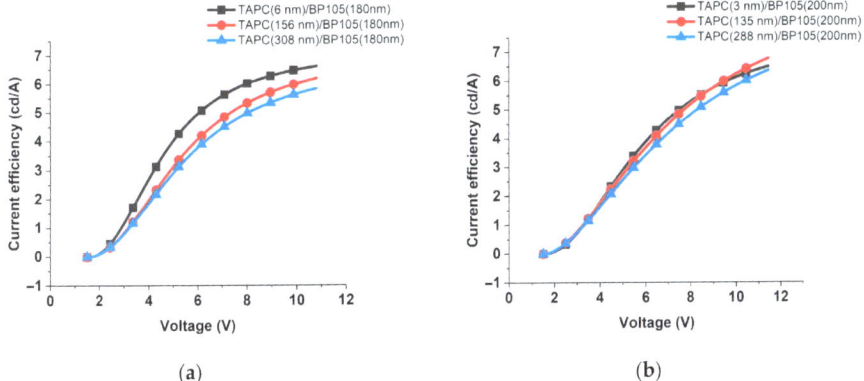

Figure 10. Current efficiency against the applied voltage for the device with (**a**) 180 nm and (**b**) 200 nm of BP105, with various thicknesses of TAPC hole-transporting layer thicknesses.

The effect of variation of TAPC thicknesses on the luminous efficacy of the devices with different thicknesses of active layer BP105 was investigated. The luminous efficacy for the devices with thicknesses of the active layer of 60 and 80 nm and various thicknesses of TAPC are shown in Figure 11a,b. The variation of TAPC layer thickness for the two devices showed similar behaviors but with the enhancement of efficacy for 80 nm of BP105. For the devices with a large active layer thickness of 180 nm and 200 nm, the trend of evolution of luminous efficacy in comparison with thinner devices was different. At low voltage, the luminous efficacy increased to a maximum, followed by a pronounced decrease at higher voltages, as shown in Figure 11c,d. These results could be attributed to the fact that there is an exciton quenching or non-radiative recombination process for the thicker active layer due to an imbalance in charge transportation within the device, leading to non-uniform current distribution and localized regions of high current density [24].

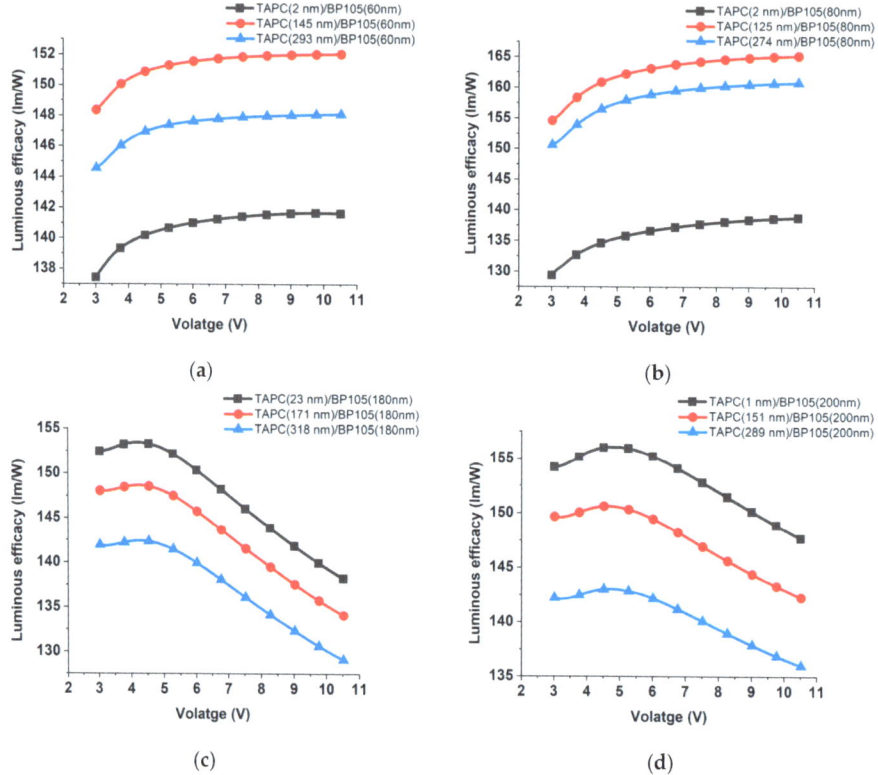

Figure 11. Luminous efficacy against the applied voltage for the device with (**a**) 60 nm, (**b**) 80 nm, (**c**) 180 nm, and (**d**) 200 nm of BP105, with various thicknesses of TAPC hole-transporting layer thicknesses.

The luminance of four devices with a thickness of the active layer of 60 nm, 80 nm, 180 nm, and 200 nm with varying thicknesses of the TAPC hole-transporting layer was investigated, as shown in Figure 12a–d. The luminance slowly increased at a lower voltage, reaching the built-in voltage value followed by an abrupt increase. Herein, we have a two-dimensional investigation. The first is the effect of changing hole-transporting thickness, and the second is the variation of the thickness of the active layer. For the diodes with thicknesses of active layers 60, 180, and 200 nm, the increase in hole-transporting layer thickness showed a pronounced decrease in luminance. However, the behavior of the device with an active layer thickness of 80 nm was different, and the maximum luminance obtained for the device with the thickness of TAPC was 100 nm. It is worth mentioning that the devices with the sum of the thicknesses of the hole-transporting layer and the active layer were close to 60, or multiple of them (i.e., the sum of HTL + EML = 180 nm) yielded the optimum luminance values. In addition, the luminance of devices with the sum of the small thicknesses of HTL + EML achieved higher luminance in comparison with the devices of large values of the sum of the thicknesses of HTL + EML. The decrease in luminance with thicker HTL can be produced due to the following factors: (1) hindered injection efficiency of holes into the emissive layer, leading to an imbalance in charge carriers; (2) exciton quenching at the interface between the HTL and the emissive layer; (3) the weak microcavity effect; and (4) thicker HTL layers resulting in more absorption of emitted light within the device, reducing the amount of light that escapes the device. The fourth reason can be excluded due to the absorption of TAPC located at higher energy relative to the emission wavelength of the device. As the devices with multiple thicknesses of HTL + EML achieve the optimum luminance, we can conclude that the main reasons that lead to a decrease in the luminance are the combination of the non-resonance thickness of microcavity and imbalance in charge carriers.

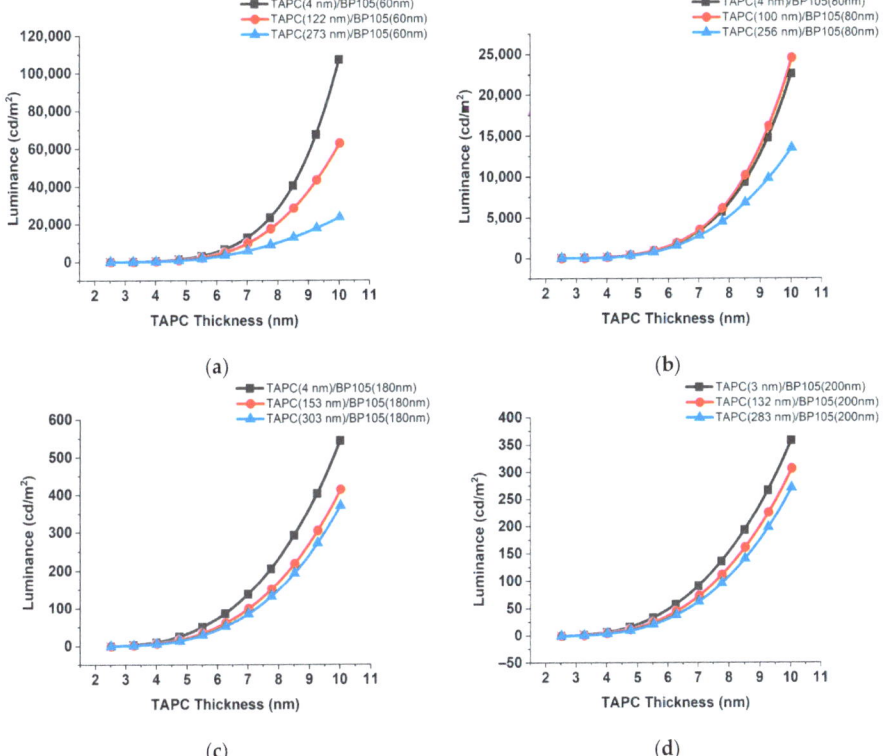

Figure 12. Luminance versus voltage for devices with different TAPC thicknesses for devices with a thickness of BP105 of (**a**) 60 nm, (**b**) 80 nm, (**c**) 180 nm, and (**d**) 200 nm.

3.3. Comparison of the Performance of the PLED with and without TAPC

In the previous sections, we detected the effective thickness of the active layer and the hole-transporting layer and their influence on the performance of the solar cell (Table 1). In this section, we will compare their performance. Figure 13 shows the current efficiency of devices with different combinations of TAPC and BP105. Comparing the current efficiency for the devices with and without the TAPC hole-transporting layer showed that the devices with the sum of the thickness HTL + EML or single EML close to 180 nm achieved the optimum current efficiency. The devices with the TAPC layer achieved the highest current efficiency relative to the devices without HTL. These results show that the improvement of hole injection, blocking electrons, and balance of carriers are the main factors that affect the current efficiency.

Table 1. Overview of every peak after thickness variation of BP105 from 1 nm to 225 nm and TAPC from 1 nm to 350 nm.

BP105 Thickness (Peaks)		TAPC Thickness (Peaks)		
		Current Efficiency	Luminous Efficacy	Luminance
Current efficiency	80 nm	6 nm 105 nm * 258 nm	2 nm 125 nm * 274 nm	4 nm 100 nm * 256 nm
	180 nm *	6 nm * 156 nm 308 nm	23 nm * 171 nm 318 nm	4 nm * 153 nm 303 nm
Luminous efficacy	60 nm	6 nm 125 nm * 277 nm	2 nm 145 nm * 293 nm	4 nm * 122 nm 273 nm
	200 nm *	3 nm * 135 nm 288 nm	1 nm * 151 nm 289 nm	3 nm * 132 nm 283 nm

Here, * represents the highest peak.

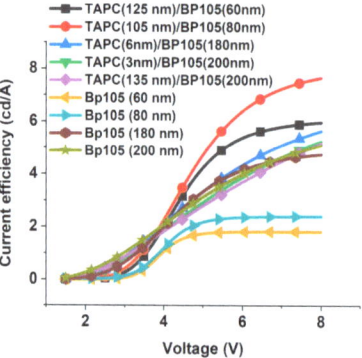

Figure 13. Comparison of the current efficiency of devices with different combinations of TAPC and BP105.

Figure 14 shows the comparison of the luminous efficacy of the devices with and without TAPC. Unlike current efficiency, the device with the sum of the thicknesses of BP105 and TAPC close to any specific thickness does not achieve the optimum luminous efficacy. Instead, the active layer thickness of 80 nm achieved the highest luminous efficacy, exceeding the similar device without TAPC at 4.5 V by about 21.66%. Although the active layer thickness less than or equal to 80 nm showed an increasing trend, the devices that have a thickness greater than or equal to 180 nm showed different behavior at a higher voltage. The luminous efficacy decreased at higher voltages for thicker devices.

A comparison of the luminance of the devices that included HTL showed a pronounced improvement relative to the devices without HTL (Figure 15). The luminance increased nearly two-fold for the PLED included TAPC relative to that one without HTL.

Astonishing results for the devices with higher thicknesses of TAPC of 273 nm and BP105 (60 nm) show that the change in luminance is negligible. These results mean that the factors that lead to the enhancement of luminance are comparable with factors that lead to decreasing the luminance at this thickness of TAPC.

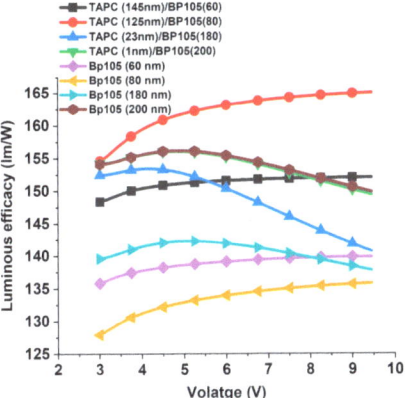

Figure 14. Comparison of the luminous efficacy of devices with different combinations of TAPC and BP105.

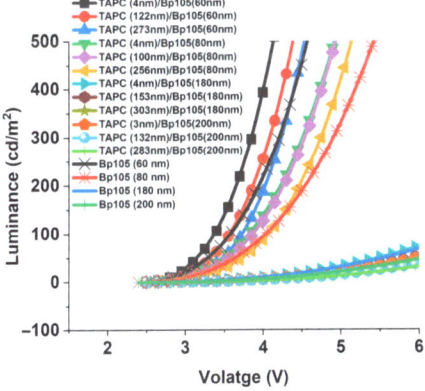

Figure 15. Comparison of the luminance of devices with different combinations of TAPC and BP105.

4. Conclusions

In this study, we conducted a comprehensive simulation of a blue light-emitting polymer device, focusing on the impact of an electron-blocking layer (EBL), the thickness variation in both the EBL and active layer (Bp105), on the device's optical and electrical performance. Our findings provide valuable insights into the enhancement of the device's efficiency and demonstrate the critical role of EBL thickness in optimizing the performance of blue light-emitting polymers.

Our investigation revealed that the incorporation of hole transporting and an electron-blocking layer significantly improved the overall performance of the device. This improvement can be attributed to the EBL's ability to confine electrons within the active layer, preventing them from escaping the device and reducing recombination losses.

Furthermore, we explored the impact of varying the active layer thickness (Bp105) combined with hole-transporting layer thickness and found that an active layer thickness of approximately 80 nanometers (nm) and a TAPC of 125 nm led to the best performance of luminance efficacy. This thickness optimization likely arises from the balance between efficient charge carrier transport and recombination probability within the active layer.

In summary, our study underscores the importance of the electron-blocking layer in enhancing the optical and electrical performance of blue light-emitting polymer devices. By fine-tuning the thickness of the EBL and active layer, we can achieve the best possible device performance. These findings have significant implications for the design and development of next-generation optoelectronic devices and hold promise for the advancement of organic light-emitting diodes (OLEDs) and other related technologies. Future research could further investigate the underlying mechanisms behind these enhancements and explore additional strategies to improve device efficiency and stability.

Author Contributions: Conceptualization, A.S., N.H., R.A.F.R., S.Z. and S.W.; methodology, A.S., N.H., R.A.F.R., S.Z. and S.W.; software, R.A.F.R. and S.W.; validation, R.A.F.R., S.Z. and S.W.; formal analysis, A.S., N.H., R.A.F.R. and S.W.; investigation, R.A.F.R., S.Z. and S.W.; data curation, A.S., N.H., R.A.F.R., S.Z. and S.W.; writing—original draft preparation, A.S. and R.A.F.R.; writing—review and editing, S.Z. and S.W.;; supervision, S.W.; project administration, S.W.; funding acquisition, A.S. All authors have read and agreed to the published version of the manuscript.

Funding: This research was funded by Prince Sattam bin Abdulaziz University through the project number (PSAU/2023/01/27742).

Institutional Review Board Statement: Not applicable.

Data Availability Statement: All data generated or analyzed during this study are included in this article.

Acknowledgments: The Authors extend their appreciation to Prince Sattam bin Abdulaziz University for funding this research work through the project number (PSAU/2023/01/27742).

Conflicts of Interest: The authors declare no conflicts of interest.

List of Abbreviations

OLED	Organic Light-Emitting Diode
HTL	Hole-transport Layer
ETL	Electron Transport Layer
EML	Emission Layer
EGDM	Extended Gaussian Disorder Model
DOS	Density of States
ITO	Indium Tin Oxide
PEDOT:PSS	Poly(3,4-ethylenedioxythiophene) polystyrene sulfonate
TAPC	1,1-bis[(di-4-tolylamino)phenyl]cyclohexane (a common HTL material)
BP105	A specific blue-emitting polymer
LiF	Lithium Fluoride
Ca	Calcium
Al	Aluminum

References

1. Friend, R.H.; Gymer, R.W.; Holmes, A.B.; Burroughes, J.H.; Marks, R.N.; Taliani, C.; Bradley, D.D.C.; Dos Santos, D.A.; Brédas, J.L.; Lögdlund, M.; et al. Electroluminescence in conjugated polymers. *Nature* **1999**, *397*, 121–128. [CrossRef]
2. Méhes, G.; Sandanayaka, A.S.D.; Ribierre, J.-C.; Goushi, K. Physics and Design Principles of OLED Devices. In *Handbook of Organic Light-Emitting Diodes*; Springer: Tokyo, Japan, 2020; pp. 1–73. [CrossRef]
3. Movla, H.; Ghaffari, S.; Ahmadinasab, H. Effects of the active region thickness on the exciton generation mechanisms in polymer light emitting diodes. *Optik* **2016**, *127*, 1814–1816. [CrossRef]
4. Dahal, E.; Allemeier, D.; Isenhart, B.; Cianciulli, K.; White, M.S. Characterization of higher harmonic modes in Fabry–Pérot microcavity organic light emitting diodes. *Sci. Rep.* **2021**, *11*, 8456. [CrossRef] [PubMed]
5. Papatryfonos, K.; de Oliveira, E.R.C.; Lanzillotti-Kimura, N.D. Effects of surface roughness and top layer thickness on the performance of Fabry-Perot cavities and responsive open resonators based on distributed Bragg reflectors. *arXiv* **2024**, arXiv:2309.13649. [CrossRef]
6. Hong, K.; Lee, J.-L. Review paper: Recent developments in light extraction technologies of organic light emitting diodes. *Electron. Mater. Lett.* **2011**, *7*, 77–91. [CrossRef]
7. Siddiqui, I.; Kumar, S.; Tsai, Y.-F.; Gautam, P.; Shahnawaz; Kesavan, K.; Lin, J.-T.; Khai, L.; Chou, K.-H.; Choudhury, A.; et al. Status and Challenges of Blue OLEDs: A Review. *Nanomaterials* **2023**, *13*, 2521. [CrossRef] [PubMed]
8. ZGarza, C.F.; Born, M.; Hilbers, P.A.J.; van Riel, N.A.W.; Liebmann, J. Visible Blue Light Therapy: Molecular Mechanisms and Therapeutic Opportunities. *Curr. Med. Chem.* **2019**, *25*, 5564–5577. [CrossRef]
9. Prasad, A.; Du, L.; Zubair, M.; Subedi, S.; Ullah, A.; Roopesh, M.S. Applications of Light-Emitting Diodes (LEDs) in Food Processing and Water Treatment. *Food Eng. Rev.* **2020**, *12*, 268–289. [CrossRef]

10. Lee, S.H.; Won, H.J.; Ban, S.; Choi, H.; Jung, J.H. Tomato Fruit Growth and Nutrient Accumulation in Response to Blue and Red Light Treatments during the Reproductive Growth Stage. *Horticulturae* **2023**, *9*, 1113. [CrossRef]
11. Fluxim, A.G. Simulation Software Setfos (Version 5.4). Available online: www.fluxim.com (accessed on 28 May 2024).
12. Tseng, S.-R.; Meng, H.-F.; Yeh, C.-H.; Lai, H.-C.; Horng, S.-F.; Liao, H.-H.; Hsu, C.-S.; Lin, L.-C. High-efficiency blue multilayer polymer light-emitting diode fabricated by a general liquid buffer method. *Synth. Met.* **2008**, *158*, 130–134. [CrossRef]
13. Huang, P.; Chou, C.; Chang, M.; Huang, W.; Lee, C.; Han, Y.; Liu, S. The Effect of Controlled Dopant Concentration on the Performance of Blue Polymer Light-emitting Diodes. *J. Chin. Chem. Soc.* **2011**, *58*, 326–331. [CrossRef]
14. Neukom, M.; Züfle, S.; Jenatsch, S.; Ruhstaller, B. Opto-electronic characterization of third-generation solar cells. *Sci. Technol. Adv. Mater.* **2018**, *19*, 291–316. [CrossRef] [PubMed]
15. Knapp, E.; Ruhstaller, B. The role of shallow traps in dynamic characterization of organic semiconductor devices. *J. Appl. Phys.* **2012**, *112*, 024519. [CrossRef]
16. Ruhstaller, B.; Carter, S.A.; Barth, S.; Riel, H.; Riess, W.; Scott, J.C. Transient and steady-state behavior of space charges in multilayer organic light-emitting diodes. *J. Appl. Phys.* **2001**, *89*, 4575–4586. [CrossRef]
17. Ruhstaller, B.; Knapp, E.; Perucco, B.; Reinke, N.; Rezzonico, D.; Mller, F. Advanced Numerical Simulation of Organic Light-emitting Devices. In *Optoelectronic Devices and Properties*; InTech: Houston, TX, USA, 2011. [CrossRef]
18. Altazin, S.; Penninck, L.; Ruhstaller, B. Outcoupling Technologies: Concepts, Simulation, and Implementation. In *Handbook of Organic Light-Emitting Diodes*; Springer: Tokyo, Japan, 2018; pp. 1–22. [CrossRef]
19. van Mensfoort, S.L.M.; Billen, J.; Carvelli, M.; Vulto, S.I.E.; Janssen, R.A.J.; Coehoorn, R. Predictive modelling of the current density and radiative recombination in blue polymer-based light-emitting diodes. *J. Appl. Phys.* **2011**, *109*, 064502. [CrossRef]
20. Cheng, H.; Feng, Y.; Fu, Y.; Zheng, Y.; Shao, Y.; Bai, Y. Understanding and minimizing non-radiative recombination losses in perovskite light-emitting diodes. *J. Mater. Chem. C Mater.* **2022**, *10*, 13590–13610. [CrossRef]
21. Ahmad, V.; Sobus, J.; Greenberg, M.; Shukla, A.; Philippa, B.; Pivrikas, A.; Vamvounis, G.; White, R.; Lo, S.-C.; Namdas, E.B. Charge and exciton dynamics of OLEDs under high voltage nanosecond pulse: Towards injection lasing. *Nat. Commun.* **2020**, *11*, 4310. [CrossRef] [PubMed]
22. Hart, K.; Hart, S.; Selvaggi, J.P. Modified charge carrier density for organic semiconductors modeled by an exponential density of states. *J. Comput. Electron.* **2021**, *20*, 259–266. [CrossRef]
23. Tian, Y.; Gan, Z.; Zhou, Z.; Lynch, D.W.; Shinar, J.; Kang, J.-H.; Park, Q.-H. Spectrally narrowed edge emission from organic light-emitting diodes. *Appl. Phys. Lett.* **2007**, *91*, 143504. [CrossRef]
24. Jiang, N.; Ma, G.; Song, D.; Qiao, B.; Liang, Z.; Xu, Z.; Wageh, S.; Al Ghamdi, A.A.; Zhao, S. Defects in lead halide perovskite light-emitting diodes under electric field: From behavior to passivation strategies. *Nanoscale* **2024**, *16*, 3838–3880. [CrossRef] [PubMed]

Disclaimer/Publisher's Note: The statements, opinions and data contained in all publications are solely those of the individual author(s) and contributor(s) and not of MDPI and/or the editor(s). MDPI and/or the editor(s) disclaim responsibility for any injury to people or property resulting from any ideas, methods, instructions or products referred to in the content.

Article

The Synthesis and Optical Property of a Ternary Hybrid Composed of Aggregation-Induced Luminescent Polyfluorene, Polydimethylsiloxane, and Silica

Nurul Amira Shazwani Zainuddin, Yusuke Suizu, Takahiro Uno and Masataka Kubo *

Division of Applied Chemistry, Graduate School of Engineering, Mie University, 1577 Kurimamachiya-cho, Tsu 514-8507, Japan; 422de02@m.mie-u.ac.jp (N.A.S.Z.); 423m327@m.mie-u.ac.jp (Y.S.); uno@chem.mie-u.ac.jp (T.U.)
* Correspondence: kubo@chem.mie-u.ac.jp; Tel.: +81-59-231-9410

Abstract: Tetraphenylethene (TPE) is known as a molecule that exhibits aggregation-induced emission (AIE). In this study, pendant hydroxyl groups were introduced onto polyfluorene with a TPE moiety. Sol-gel reactions of polydiethoxysiloxane (PDEOS) were carried out in the presence of hydroxyl-functionalized AIE polyfluorene (TPE-PF-OH) and polydimethylsiloxane carrying pendant hydroxyl groups (PDMS-OH) to immobilize AIE polyfluorene into a PDMS/SiO$_2$ hybrid in an isolated dispersion state. The luminescence intensity from this three-component hybrid increased with the increase in silica content. The luminescence intensity decreased with increasing external temperature. For the control experiment, sol-gel reactions of PDEOS were carried out in the presence of hydroxyl group-free polyfluorene (TPE-PF) and PDMS to obtain ternary composites. We found that the luminescence from this composite was not significantly affected by the silica content or external temperature. We synthesized temperature-responsive AIE materials without changing the concentration or aggregation state of the AIE molecules.

Keywords: aggregation-induced emission; polyfluorene; organic/inorganic hybrid; sol-gel method

Citation: Zainuddin, N.A.S.; Suizu, Y.; Uno, T.; Kubo, M. The Synthesis and Optical Property of a Ternary Hybrid Composed of Aggregation Induced Luminescent Polyfluorene, Polydimethylsiloxane, and Silica. *Polymers* **2024**, *16*, 3331. https://doi.org/10.3390/polym16233331

Academic Editors: Tengling Ye and Bixin Li

Received: 31 October 2024
Revised: 22 November 2024
Accepted: 25 November 2024
Published: 27 November 2024

Copyright: © 2024 by the authors. Licensee MDPI, Basel, Switzerland. This article is an open access article distributed under the terms and conditions of the Creative Commons Attribution (CC BY) license (https://creativecommons.org/licenses/by/4.0/).

1. Introduction

It is well known that conjugated polymers exhibit interesting electronic and optical properties [1]. Conjugated polymers such as polythiophenes and polyfluorenes have attracted much attention as luminescent polymers mainly for organic light-emitting diodes (OLEDs) and biomarkers [2]. Many conjugated luminescent polymers show strong luminescence in a solution but are partially or completely quenched in an aggregated state. This effect is known as aggregation-caused quenching (ACQ) [3]. Such concentration quenching leads to a decrease in the luminous efficiency of the light-emitting device. In the case of biomarkers, aggregation with biomolecules and analytical agents reduces sensitivity. On the other hand, some molecules such as tetraphenylethene (TPE) emit light in a solid state but not in a solution. This phenomenon is called aggregation-induced emission (AIE) [4–7]. The widely accepted reason for the AIE effect is the restriction of intramolecular motion (RIM), which is composed of both the restriction of intramolecular rotation (RIR) and the restriction of intramolecular vibration (RIV). In a solution state, exited energy is consumed by molecular motion, and the release of energy from the excited state to the ground state is non-radiative deactivation. In an aggregated state, on the other hand, spatial constraints and interactions with surrounding molecules significantly suppress these molecular motions, inhibiting the non-radiative deactivation pathway and resulting in luminescence [8].

AIE polymers may be superior to lower-molecular-weight AIE molecules in terms of prosessability, further functionalization, and thermal stability. Polymeric AIE molecules have many useful applications as new optical materials and have been widely investigated [9–11]. A typical AIE polymer is one in which a segment with an AIE functional group is introduced into the main chain or side chain moiety of the polymer. In most cases, an AIE–functional group has a propeller-shaped structure with rotatable periphery phenyl rings such as TPE and hexaphenylsilole. And such a polymer can be prepared by ether the polymerization of the AIE–functional monomer or post-polymerization modification. The main chain of the polymer may be a conjugated structure or an unconjugated structure. Compared to unconjugated AIE polymers, conjugated AIE polymers are expected to have higher luminescence efficiency. For example, Wu et al. synthesized a new series of TPE-containing conjugated polyfluorene copolymers through a palladium-catalyzed Suzuki polycondensation reaction and found that all polymers exhibited AIE properties thanks to the TPE moieties [12].

We have reported the incorporation of emitting polymers into silica as new emitting materials [13–15]. For example, we introduced pendant hydroxyl groups into polyfluorene and carried out a sol-gel reaction of tetraethoxysilane (TEOS) in the presence of hydroxyl-functionalized polyfluorene to obtain an organic/inorganic hybrid retaining the optical properties of the embedded polyfluorene [16]. Since we observed polymer aggregation in the resulting silica when we carried out the sol gel reaction of TEOS in the presence of a conjugated polymer without hydroxyl functionalities, hydrogen bonding between pendant hydroxyl groups and silanol groups in silica played an important role for homogeneous hybrid formation. Further, we successfully immobilized emitting polyfluorenes into silicone resin by the sol-gel reaction of TEOS in the presence of polyfluorene with pendant ethoxysilyl groups and silanol-terminated polydimethylsiloxane. Homogeneous hybridization proceeded because of the covalent bond formation between ethoxysilyl and silanol groups [17]. By immobilizing the luminescent polymer in a transparent solid matrix, energy transfer did not occur because, unlike in the solution state, the molecules lose their mobility and no intermolecular contact occurs. In other words, it corresponds to an isolated dispersion of molecules in a frozen state. On the other hand, if the solid matrix is soft like silicone, molecular motion is possible to some extent.

So far, the external environment affecting the aggregation-induced effect of polymers has been exclusively investigated in terms of solution concentration and solvent composition. We were interested in immobilizing AIE polyfluorene carrying TPE moieties into a solid matrix in an isolated dispersion. Our idea was to control the degree of the rotation of phenyl rings by changing the softness of the solid matrix. We hypothesized that by changing the softness of the external environment in which the aggregation-inducing molecules reside, the ease of intramolecular rotation can be changed and, as a result, the aggregation-inducing effect can be controlled. The solid matrix we focused on was a hybrid composed of PDMS and silica. Recently, hybrid materials composed of PDMS and SiO_2 have attracted much attention as biomaterials [18–20], photonic materials [21], and coating materials [22–24]. This diversity of applications is related to the flexibility of the materials ranging from a hard solid to a rubber-like substance.

The chemical structures of the compounds used in this study are shown in Figure 1. We synthesized AIE polyfluorene (TPE-PF-OH) which carries both TPE as an AIE active moiety and pendant hydroxyl groups which are capable of interacting with silica. Our preliminary experiment showed that the sol-gel reaction of polydiethoxysiloxane (PDEOS) in the presence of TPE-PF-OH and PDMS gave a translucent solid, indicating that the aggregation of TPE-PF-OH molecules took place in the solid matrix. Therefore, we prepared PDMS containing pendant hydroxyl groups (PDMS-OH), which can interact with TPE-PF-OH through hydrogen bonding. The sol-gel reactions of PDEOS were carried out in the presence of TPE-PF-OH and PDMS-OH to obtain TPE-PF-OH/PDMS-OH/SiO_2 hybrids with different amounts of silica. We measured the emission spectra of the resulting ternary TPE-PF-OH/PDMS-OH/SiO_2 hybrids to examine the effect of silica content and temperature on the luminescence intensity from the hybrid. For comparison, the sol-gel reactions of PDEOS were carried out in the presence of hydroxyl-free polyfluorene and PDMS to obtain ternary composites, in which AIE polyfluorene molecules were embedded in the solid matrix in the aggregated state. The optical properties of TPE-PF-OH/PDMS-OH/SiO$_2$ hybrids were compared with those of TPE-PF/PDMS/SiO$_2$ composites.

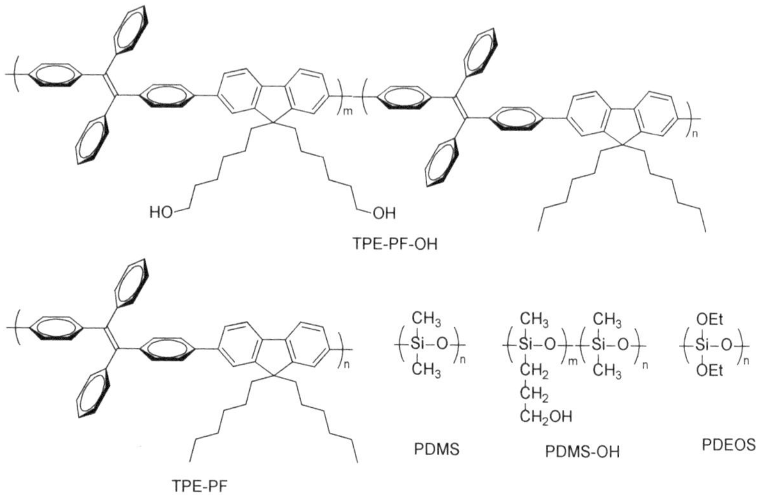

Figure 1. Chemical structure of the compounds used in this study.

2. Materials

2.1. Reagents

1,2-Bis(4-bromophenyl)-1,2-diphenylethene [25], 2,7-dibromo-9,9-bis(6-(2-tetrahydropyranyloxy) hexyl)fluorene [16], and 2,7-dibromo-9,9-dihexylfluorene [26] were prepared according to the reported procedures. The platinum–divinyltetramethyldisiloxane complex (3.0% Pt in vinyl-terminated PDMS) (SIP 6830), trimethylsilyl-terminated poly(dimethylsiloxane-co-methylhydrosiloxane) (PDMS-H) (molecular weight, 20,000–25,000; methylhydrosiloxane, 4–6 mol%), and trimethylsilyl-terminated polydimethylsiloxane (PDMS) (molecular weight, 26,000) were purchased from Gelest, Inc. (Morrisville, PA, USA) Tris(tris [3,5-bis(trifluoromethyl)phenyl]phosphine)palladium (0) was purchased from Wako Pure Chemical Industries, Ltd. (Osaka, Japan) Aliquat 336 was purchased from Aldrich (St. Louis, MO, USA). All other reagents were obtained from commercial sources and used as received.

2.2. Compounds

2.2.1. 1,2-Diphenyl-1,2-bis(4-(4,4,5,5-tetramethyl-1,3,2-dioxaborolan-2-yl)phenyl)ethene (**1**)

To a mixture of 1,2-bis(4-bromophenyl)-1,2-diphenylethene (2.8 g, 5.7 mmol), bis (pinacolato)diboron (3.8 g, 15 mmol) and KOAc (4.4 g, 45 mmol), degassed dioxane (50 mL) and [1,1'-bis(diphenylphosphino)ferrocene]palladium(II) dichloride (Pd(dppf)Cl$_2$, 60 mg) were added and the reaction mixture was stirred at 80 °C for 24 h. After cooling to room temperature, the reaction mixture was diluted with chloroform and washed with water. The organic layer was dried over anhydrous magnesium sulfate and placed under reduced pressure to remove the solvent. The residue was purified by column chromatography using a mixture of dichloromethane and hexane (2:1 v/v) as an eluent to give 2.2 g (67%) of compound **1** as a white solid; ^{1}H NMR (500 MHz, CDCl$_3$, δ): 7.53 (d, J = 7.2 Hz, 4H), 7.1–7.0 (m, 14H), 1.32 (s, 24H); ^{13}C NMR (125 MHz, CDCl$_3$, δ):146.9, 143.5, 141.4, 134.1, 131.4, 130.8, 127.8, 127.7, 126.6, 83.7, 24.9; IR (KBr, cm^{-1}): 2977, 1607; Anal. calcd. for C$_{38}$H$_{42}$B$_2$O$_4$: C 78.10, H 7.24; found: C 78.19, H 7.16.

2.2.2. TPE-PF-OTHP

A mixture of **1** (584 mg, 1.00 mmol), 2,7-dibromo-9,9-bis(6-(2-tetrahydropyranyloxy)hexyl)fluorene (**2**) (139 mg, 0.20 mmol), 2,7-dibromo-9,9-dihexylfluorene (**3**) (394 mg, 0.80 mmol), toluene (20 mL), 2 mol/L aqueous Na_2CO_3 (5 mL), and tris(tris [3,5-bis(trifluoromethyl)phenyl]phosphine)palladium (0) (10 mg) was deaerated by bubbling argon at least 10 min. The reaction mixture was heated at 90 °C for 72 h under argon and then poured into methanol. The precipitated polymer was purified by washing for 2 days in a Soxhlet apparatus with acetone to remove oligomers and catalyst residues to obtain 570 mg (81%) of TPE-PF-OTHP as a yellow powder; UV-vis (in THF): λ_{max} = 356 nm; IR (KBr, cm^{-1}): 2923, 2836, 1455, 693; GPC: M_w = 42,000, M_w/M_n = 2.5.

2.2.3. TPE-PF-OH

A mixture of TPE-PF-OTHP (0.29 g), 30 mL of THF, and 5 mL of 10% hydrochloric acid was stirred at 40 °C for 20 h. The reaction mixture was diluted with chloroform and washed with water. The organic layer was dried over anhydrous magnesium sulfate and placed under reduced pressure to remove the solvents. The residue was dissolved in a small amount of chloroform and then re-precipitated into methanol to obtain 0.18 g (63%) of TPE-PF-OH as a yellow powder; UV-vis (in THF): λ_{max} = 359 nm; IR (KBr, cm^{-1}): 2929, 2845, 1460, 722; GPC: M_w = 60,300, M_w/M_n = 1.9.

2.2.4. PDMS-OH

To a mixture of PDMS-SiH (5.0 g, SiH = 3.5 mmol) and allyl alcohol (0.81 g, 14 mmol) in 20 mL of toluene was added platinum catalyst (SIP 6830, 10 mg), and the reaction mixture was heated at 70 °C for 24 h. The rection mixture was pored into methanol to precipitate 3.8 g (71%) of PDMS-OH as colorless viscous oil; IR (NaCl, cm^{-1}): 2967, 1266, 1100, 1031, 797.

2.2.5. TPE-PF

A mixture of **1** (0.45 g, 0.77 mmol), **3** (0.38 g, 0.77 mmol), THF (20 mL), 2 mol/L aqueous Na_2CO_3 (4 mL), and tris(tris [3,5-bis(trifluoromethyl)phenyl]phosphine)palladium (0) (10 mg) was deaerated by bubbling argon at least 10 min. The reaction mixture was heated at 90 °C for 72 h under argon and then precipitated into methanol. The polymer was filtered and purified by washing for 2 days in a Soxhlet apparatus with acetone to remove oligomers and catalyst residues to obtain 0.40 g (78%) of TPE-PF as a yellow powder; UV-vis (in THF): λ_{max} = 369 nm; IR (KBr, cm^{-1}): 2922, 2864, 1465, 703; GPC: M_w = 13,200, M_w/M_n = 2.8.

2.2.6. TPE-PF-OH/PDMS-OH/SiO$_2$ Hybrid

In a typical example, to a mixture of PDMS-OH (0.60 g) and PDEOS (0.37 g), a solution of TPE-PF-OH (0.1 mg) in 10 mL of THF and 5 µL of dibutyltin dilaurate was added. The reaction mixture was stirred for 1 h and then allowed to stand at room temperature for one week to obtain 0.75 g of a pale-yellow transparent solid; IR (KBr, cm^{-1}): 2966, 1252, 1068, 1025, 708.

2.2.7. TPE-PF/PDMS/SiO$_2$ Composite

In a typical example, to a mixture of PDMS (0.60 g) and PDEOS (0.37 g), a solution of TPE-PF (0.1 mg) in 10 mL of THF and 5 µL of dibutyltin dilaurate was added. The reaction mixture was stirred for 1 h and then allowed to stand at room temperature for one week to obtain 0.75 g of a pale-yellow translucent solid; IR (KBr, cm^{-1}): 2967, 1251, 1068, 1021, 711.

2.3. Measurements

Nuclear magnetic resonance spectra (NMR) were recorded on 500 MHz for ^{1}H spectra and 125 MHz for ^{13}C spectra (ECZ500R, JEOL, Tokyo, Japan). The analysis was conducted at room temperature. The samples were dissolved in CDCl$_3$, with tetramethylsilane (TMS) serving as the internal standard. Photoluminescence spectra were recorded on a HAMA-MATSU Multi Channel Analyzer PMA-11 (Hamamatsu, Japan). The measurement was conducted at the exciting wavelength of 365 nm. Fourier transform infrared (FTIR) spectra and UV-vis spectra were recorded on JASCO FT/IR-4100 (Tokyo, Japan) and SHIMADZU UV-2550 (Kyoto, Japan), respectively. Elemental analysis was carried out using YANACO CHN-corder MT-5 (Kyoto, Japan). Gel permeation chromatography (GPC) was carried out on a Tosoh HLC-8020 chromatograph equipped with polystyrene gel columns (Tosoh Multipore HXL-M, Tokyo, Japan; exclusion limit = 2×10^6, 300×7.8 mm) and refractive/ultraviolet dual mode detectors. Tetrahydrofuran (THF) was used as the eluent at a flow rate of 1.0 mL/min. The calibration curves for GPC analysis were obtained using polystyrene standards.

3. Results and Discussion

3.1. Preparation of TPE-PF-OH

The synthetic pathway for PTE-PF-OH is shown in Figure 2. The key compound is a fluorene derivative **2** in which hydroxyl groups are protected by tetrahydropyranyl (THP) groups because the THP group, like other acetals and ketals, is inert under basic conditions during the Suzuki coupling reaction. The ternary copolymerization of **1**, **2** and **3** was carried out in the presence of a palladium catalyst. Since we already found that the introduction of 20 mol% hydroxyl group-containing monomer allowed for homogeneous mixing with silica [16], we added 20 mol% of monomer **2**. The ^{1}H NMR spectrum of the resulting TPE-PF-OTHP is shown in Figure 3. The peak at 0.6–0.8 ppm is due to the CH_3 protons of the hexyl group. The peaks at 4.5 and 3.8–3.2 ppm are assigned as CH and OCH_2 protons, respectively. The polymer composition of TPE-PF-OTHP was determined by ^{1}H NMR through the peak area ratio between the signals coming from CH_3 protons and those belonging to OCH_2 protons to be m:n = 19:81 which corresponded well with the expected value on the basis of the monomer feed ratio (m:n = 1:4). The THP group was then removed by conventional acid treatment in THF. Figure 4 shows ^{1}H NMR of TPE-PF-OH. The peaks due to THP groups disappeared completely, indicating the complete conversion of TPE-PF-OTHP to TPE-PF-OH.

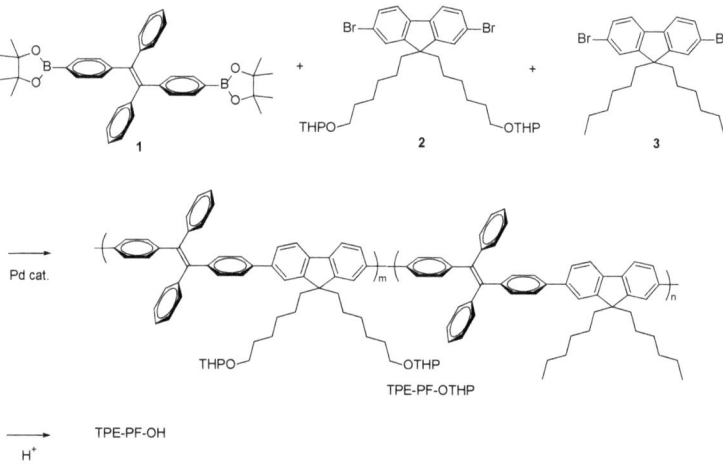

Figure 2. Synthetic pathway for TPE-PF-OH.

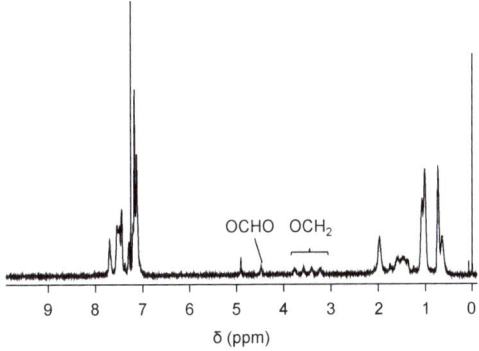

Figure 3. ^{1}H NMR spectrum of TPE-PF-OTHP.

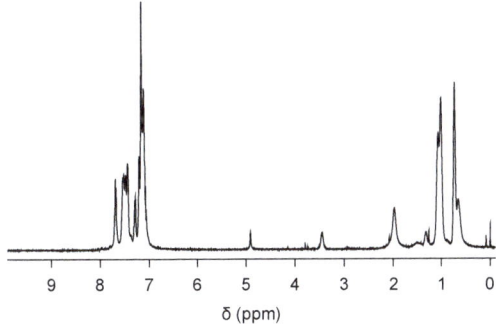

Figure 4. ^{1}H NMR spectrum of TPE-PF-OH.

Figure 5 shows the emission spectra of TPE-PF-OH in a diluted THF solution (0.1 mg/mL) and from thin film. Weak emission was observed in the THF solution (Figure 5a). This is probably because the intramolecular rotation process of the four phenyl rings in the TPE moieties in conjugated polymer may be limited in some degree, and weak emission still present in the solution state. On the other hand, much stronger emission was observed from the thin film of TPE-PF-OH (Figure 5b), indicating that TPE-PF-OH exhibits a typical AIE property [3].

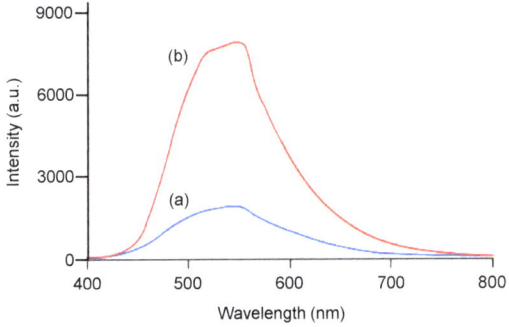

Figure 5. Emission spectra of TPE-PF-OH (a) in a THF solution (0.1 mg/mL) and (b) from thin film.

3.2. Preparation of PDMS-OH

In order to immobilize TPE-PF-OH in a PDMS/SiO$_2$ matrix in an isolated and dispersed state, we introduced pendant hydroxyl groups onto PDMS. The introduced hydroxyl groups on PDMS should help homogeneous mixing with emissive polymer with hydroxyl functionalities. We synthesized poly[dimethylsiloxane-*co*-methyl(3-hydroxypropyl)siloxane] (PDMS-OH) via the hydrosilylation reaction of trimethylsilyl-terminated poly (dimethylsiloxane-*co*-methylhydrosiloxane)] (PDMS-H) with an allyl alcohol in the presence of a platinum catalyst. Figure 6 shows the ^{1}H NMR spectra of PDMS-OH with that of the starting PDMS-H. PDMS-H shows an absorption peak at 4.6 ppm due to Si-H groups. After the hydrosilylation reaction, the peak at 4.6 ppm disappeared completely, while new peaks emerged at 3.3, 1.6, and 0.3 ppm coming from SiCH$_2$CH$_2$CH$_2$OH groups, indicating that Si-H groups were successfully converted to Si-CH$_2$CH$_2$CH$_2$OH groups. Figure 7 shows the IR spectra of PDMS before and after the hydrosilylation reaction. After hydrosilylation, the peak at 2155 cm^{-1} due to Si-H groups disappeared completely.

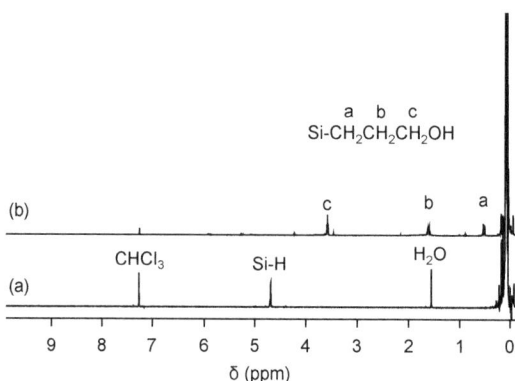

Figure 6. ^{1}H NMR spectra of (a) PDMS-H and (b) PDMS-OH.

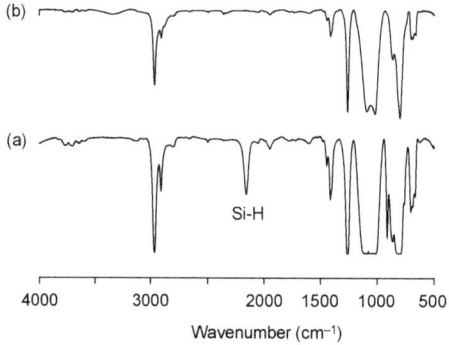

Figure 7. IR spectra of (a) PDMS-H and (b) PDMS-OH.

3.3. Preparation of TPE-PF-OH/PDMS-OH/SiO$_2$ and TPE/PDMS/SiO$_2$

We synthesized three TPE-PF-OH/PDMS-OH/SiO$_2$ hybrids with different silica contents. Although the preparation of hybrid elastomers composed of PDMS and SiO$_2$ was reported by the sol-gel reaction of tetraethoxysilane (TEOS) in the presence of hydroxy-terminated poly(dimethylsiloxane) [27], our preliminary experiments revealed that it was difficult to control the silica content in the hybrid due to volatilization of TEOS during the sol-gel reaction. We utilized poly(diethoxysilane) (PDEOS) as a silica precursor for a non-aqueous sol-gel reaction [28]. PDEOS is known as a non-volatile oligomeric form of TEOS. The sol-gel reaction conditions are summarized in Table 1. Since 1 g of PDEOS changes to about 0.41 g of silica after elimination of ethanol, silica content (wt%) in the hybrid can be easily calculated. The ternary hybrids obtained were transparent solid without polymer aggregation. The hybrid with 60 wt% silica was a hard solid, while the hybrid with 20 wt% silica was a rather rubbery solid.

Table 1. So-gel reaction conditions [1] for PTE-PF-OH/PDMS-OH/SiO$_2$ hybrids.

Entry	PDMS-OH, g	PDEOS, g	SiO$_2$ Content, wt%
1	0.60	0.37	20
2	0.45	0.73	40
3	0.30	1.1	60

[1] TPE-PF-OH = 0.1 mg, $(C_4H_9)_2Sn(OCOC_{11}H_{23})_2$ = 5 µL, THF = 10 mL, temp., = rt, time = 1 week.

For the control experiment, we synthesized AIE polyfluorene without hydroxyl functionalities by the Suzuki polycondesation reaction between **1** and **3** to obtain TPE-PF which is an alternating copolymer of TPE and 9,9-dihexylfluorene. The ^{1}H NMR of TPE-PF is shown in Figure 8. Then, we carried out the sol-gel reaction of PDEOS in the presence of TPE-PF and PDMS to obtain TPE-PF/PDMS/SiO$_2$ composites with different SiO$_2$ content. Table 2 summarizes the sol-gel reaction conditions. Since there is no interaction such as hydrogen bonding, electrostatic interaction, or π–π interaction among TPE-PF, PDMS, and silica, the resulting solid was an opaque solid, indicating that TPE-PF molecules aggregate in the solid matrix. This indicates that phenyl rings at a TPE moiety cannot rotate any more.

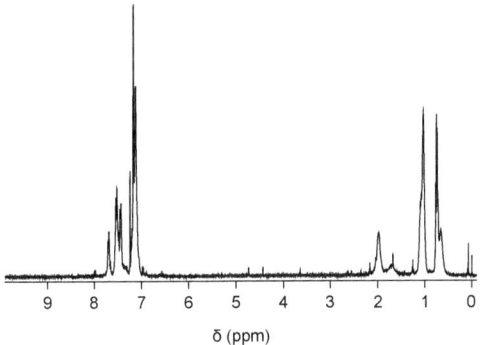

Figure 8. ^{1}H NMR spectrum of TPE-PF.

Table 2. So-gel reaction conditions [1] for PTE-PF/PDMS/SiO$_2$ composites.

Entry	PDMS, g	PDEOS, g	SiO$_2$ Content, wt%
1	0.60	0.37	20
2	0.45	0.73	40
3	0.30	1.1	60

[1] TPE-PF = 0.1 mg, $(C_4H_9)_2Sn(OCOC_{11}H_{23})_2$ = 5 µL, THF = 10 mL, temp., = rt, time = 1 week.

3.4. Effect of Silica Content on Emission

Figure 9a shows the emission spectra of TPE-PF-OF/PDMS-OH/SiO$_2$ hybrids with different silica contents at room temperature. It was obvious that emission intensity increased with the increase in silica content. This is reasonably explained by considering the restriction of intramolecular rotation of phenyl rings of TPE groups. It becomes difficult for phenyl rings to rotate when the solid matrix becomes harder [27].

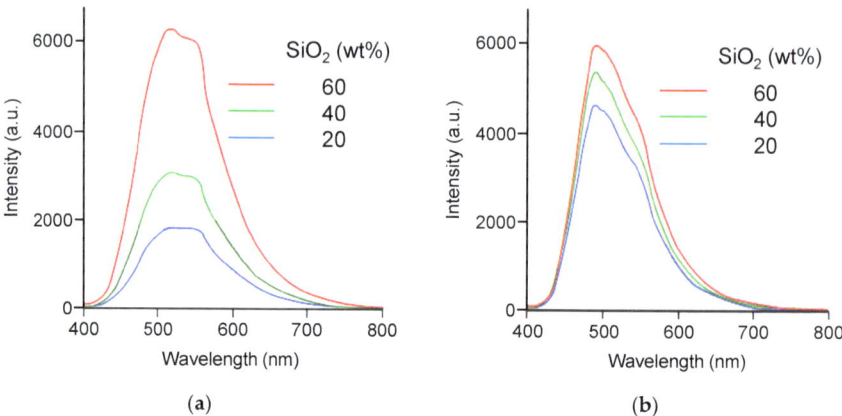

Figure 9. Emission spectra from (**a**) the TPE-PF-OH/PDMS-OH/silica hybrid and (**b**) TPE-PF/PDMS/silica composite at various silica contents.

Figure 9b shows the emission spectra of TPE-PF/PDMS/SiO$_2$ composites with different silica contents at room temperature. Although the emission intensity decreased with the decrease in silica content, the effect was less remarkable when compared with that for the TPE-PF-OH/PDMS-OH/SiO$_2$ system. This is because phenyl rings cannot rotate even if the matrix becomes soft.

3.5. Effect of Temperature on Emission

Next, we examined the effect of temperature on emission property. Figure 10a shows the emission spectra of the TPE-PF-OH/PDMS-OH/SiO$_2$ hybrid with 40 wt% silica at various temperatures. The fluorescence intensity decreased with the increase in temperature. The observed decrease in emission intensity can be explained by considering an easier rotation of pheny rings at a higher temperature.

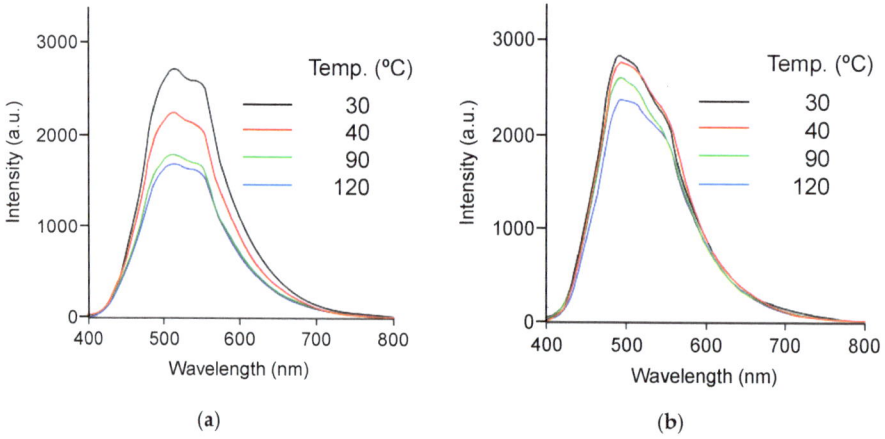

Figure 10. Emission spectra from (**a**) the TPE-PF-OH/PDMS-OH/SiO$_2$ hybrid and (**b**) TPE-PF/PDMS/SiO$_2$ composite at various temperatures.

Figure 10b shows the emission spectra of TPE-PF/PDMS/SiO$_2$ composites at various temperatures. Similar to the TPE-PF-OH/PDMS-OH/SiO$_2$ system, a decrease in emission intensity was observed as the external temperature increased. However, the decrease in emission intensity with increasing temperature is less pronounced than the TPE-PF-OH/PDMS-OH/SiO$_2$ system.

These experimental results show that the fluorescence intensity of TPE-PF/PDMS/SiO$_2$ composites is not significantly affected by the silica content or temperature. This is because TPE-PF molecules exist in an aggregated state due to the lack of interaction with the solid matrix, making it difficult for phenyl rings to rotate, supporting that the rotation of phenyl rings plays an important role in bringing about the AIE effect. There are not many reported papers of thermo-responsive AIE molecules. Liu et al. synthesized polyurethanes with soft and hard segments using AIE-active tetra-aniline derivative as the hard segment, and they found that these polyurethanes exhibited temperature-dependent fluorescent characteristics [29]. Ma et al. synthesized poly(N-isopropylacrylamide) (PNIPAAm) with an AIE moiety to observe thermo-induced emission due to the aggregation of PNIPAAm chains [30]. Our ternary hybrid containing AIE polyfluorene is another example of a thermo-responsive AIE material. The mechanism of the thermo-responsibility is coming from the change of softness of the transparent solid matrix in which AIE molecules are imbedded.

4. Conclusions

We synthesized two AIE polyfluorenes with a TPE moiety. One (TPE-PF-OH) had pendant hydroxyl groups while the other (TPE-PF) did not. And we converted PDMS-H to PDMS-OH with hydroxypropyl groups. We carried out the sol-gel reactions of PDEOS in the presence of TPE-PF-OH/PDMS-OH or TPE-PF/PDMS. The former gave transparent solids while the latter gave translucent solids. We examined emission properties by changing the silica content and temperature for these ternary systems. The effects of the silica content and temperature on fluorescence intensity differed significantly between the TPE-PF-OH/PDMS-OH/SiO$_2$ and TPE-PF/PDMS/SiO$_2$ systems. The luminescence intensity from TPE-PF-OH/PDMS-OH/SiO$_2$ was greatly affected by the silica content and temperature. On the other hand, the luminescence from TPE-PF/PDMS/SiO$_2$ was not significantly affected by the silica content or temperature. These results can be reasonably explained by considering the intramolecular motion of phenyl rings at a TPE moiety. We demonstrated that the luminescence properties of an AIE polymer can be altered by isolating and dispersing it in a transparent solid matrix of which the flexibility can be changed by external stimuli. Such materials are expected to be new temperature-responsive optical materials based on AIE phenomena.

Author Contributions: Conceptualization, M.K.; methodology, T.U.; preparation, N.A.S.Z. and Y.S.; investigation, N.A.S.Z. and Y.S.; writing—original draft preparation, N.A.S.Z.; writing—review and editing, M.K.; supervision, M.K.; project administration, M.K. All authors have read and agreed to the published version of the manuscript.

Funding: This research received no external funding.

Institutional Review Board Statement: Not applicable.

Data Availability Statement: The original contributions presented in this study are included in the article. Further inquiries can be directed to the corresponding author.

Acknowledgments: The authors acknowledge Masashi Tamura, the technical staff of the Faculty of Engineering, for measurement of the photoluminescence spectrum.

Conflicts of Interest: The authors declare no conflicts of interest.

References

1. Barford, W. *Electronic and Optical Properties of Conjugated Polymers*, 2nd ed.; Oxford University Press: Oxford, UK, 2013.
2. Müllen, K.; Scherf, U. *Organic Light Emitting Devices*; Wiley-VCH: Weinheim, Germany, 2006.
3. Hong, Y.; Lam, J.W.Y.; Tang, B.Z. Aggregation-Induced Emission. *Chem. Soc. Rev.* **2011**, *40*, 5361–5388. [CrossRef]
4. Ding, D.; Li, K.; Liu, B.; Tang, B.Z. Bioprobes Based on AIE Fluorogens. *Acc. Chem. Res.* **2012**, *46*, 2441–2453. [CrossRef] [PubMed]
5. Mei, J.; Hong, Y.; Lam, J.W.Y.; Qin, A.; Tang, Y.; Tang, B.Z. Aggregation-Induced Emission: The Whole Is More Brilliant than the Parts. *Adv. Mater.* **2014**, *26*, 5429–5479. [CrossRef]

6. Mei, J.; Leung, N.L.C.; Kwok, R.T.K.; Lam, J.W.Y.; Tang, B.Z. Aggregation-Induced Emission: Together We Shine, United We Soar! *Chem. Rev.* **2015**, *115*, 11718–11940. [CrossRef]
7. Wang, H.; Zhao, E.; Lam, J.W.Y.; Tang, B.Z. AIE Luminogens: Emission Brightened by Aggregation. *Mater. Today* **2015**, *18*, 365–377. [CrossRef]
8. Luo, J.; Xie, Z.; Lam, J.W.Y.; Cheng, L.; Tang, B.Z.; Chen, H.; Qiu, C.; Kwok, H.S.; Zhan, X.; Liu, Y.; et al. Aggregation-induced emission of 1-methyl-1,2,3,4,5-pentaphenylsilole. *Chem. Commun.* **2001**, *18*, 1740–1741. [CrossRef]
9. Qin, A.; Lam, J.W.Y.; Tang, B.Z. Luminogenic polymers with aggregation-induced emission characteristics. *Prog. Polym. Sci.* **2012**, *37*, 182–209. [CrossRef]
10. Hu, R.; Leung, N.L.C.; Tang, B.Z. AIE macromolecules: Syntheses, structures and functionalities. *Chem. Soc. Rev.* **2014**, *43*, 4494–4562. [CrossRef]
11. Hu, R.; Kang, Y.; Tang, B.Z. Recent advances in AIE polymers. *Polym. J.* **2016**, *48*, 359–370. [CrossRef]
12. Wu, W.; Ye, S.; Tang, R.; Huang, L.; Li, Q.; Yu, G.; Liu, Y.; Qin, J.; Li, Z. New tetraphenylethylene-containing conjugated polymers: Facile synthesis, aggregation-induced emission enhanced characteristics and application as explosive chemsensors and PLEDs. *Polymer* **2012**, *53*, 3163–3171. [CrossRef]
13. Kubo, M.; Takimoto, C.; Minami, T.; Uno, T.; Itoh, T.; Shoyama, M. Incorporation of π-Conjugated Polymer into Silica: Preparation of Poly[2-methoxy-5-(2-ethylhexyloxy)-1,4-phenylenevinylene]/Silica and Poly(3-hexylthiophene)/Silica Composites. *Macromolecules* **2005**, *38*, 7314–7320. [CrossRef]
14. Sugiura, Y.; Shoyama, M.; Inoue, K.; Uno, T.; Itoh, T.; Kubo, M. Emission from Silica Hybrid Containing RGB Fluorescent Conjugated Polymers. *Polym. Bull.* **2006**, *57*, 865–871. [CrossRef]
15. Kumazawa, N.; Towatari, M.; Uno, T.; Itoh, T.; Kubo, M. Preparation of Self-Doped Conducting Polycyclopentadithiophene and Its Incorporation into Silica. *J. Polym. Sci. Part A Polym. Chem.* **2014**, *52*, 1376–1380. [CrossRef]
16. Miyao, A.; Mori, Y.; Uno, T.; Itoh, T.; Yamasaki, T.; Koshio, A.; Kubo, M. Incorporation of Fluorene-Based Emitting Polymers into Silica. *J. Polym. Sci. Part A Polym. Chem.* **2010**, *48*, 5322–5328. [CrossRef]
17. Nishikawa, S.; Kami, S.; Nurul, A.; Badrul, H.; Uno, T.; Itoh, T.; Kubo, M. Hybridization of Emitting Polyfluorene with Silicone. *J. Polym. Sci. Part A Polym. Chem.* **2016**, *53*, 622–628. [CrossRef]
18. Yamashita, T.; Deguchi, K.; Nagotani, S.; Abe, K. Vascular Protection and Restoractive Therapy in Ischemic Stroke. *Cell Transpl.* **2011**, *20*, 95–97. [CrossRef]
19. Czarnobaj, K. Sol-gel-processed silica/polydimethylsiloxane/calcium xerogels aspolymeric matrices for Metronidazole delivery system. *Polym. Bull.* **2011**, *66*, 223–237. [CrossRef]
20. Prokopowicz, M.; Zeglinski, J.; Gandhi, A.; Sawicki, W.; Tofail, S.A. Bioactive silica-based drug delivery systems containing doxorubicin hydrochloride: In vitro studies. *Colloids Surf. B Biointerfaces* **2012**, *93*, 249–259. [CrossRef]
21. Zhang, X.-X.; Xia, B.-B.; Ye, H.-P.; Zhang, Y.-L.; Xiao, B.; Yan, L.-H.; Lv, H.-B.; Jiang, B. One-step sol-gel preparation of PDMS-silica ORMOSILs as environment-resistant and crack-free thick antireflective coatings. *J. Mater. Chem.* **2012**, *22*, 13132–13140. [CrossRef]
22. Deng, Z.-Y.; Wang, W.; Mao, L.-H.; Wang, C.-F.; Chen, S. Versatile superhydrophobic and photocatalytic films generated from TiO_2-SiO_2@PDMS and their applications on fabrics. *J. Mater. Chem. A* **2014**, *2*, 4178–4184. [CrossRef]
23. Kapridaki, C.; Pinho, L.; Mosquera, M.J.; Maravelaki-Kalaitzaki, P. Producing photoactive, transparent and hydrophobic SiO_2-crystalline TiO_2 nanocomposites at ambient conditions with application as self-cleaning coatings. *Appl. Catal. B Environ.* **2014**, *156–157*, 416–427. [CrossRef]
24. Elvira, M.R.; Mazo, M.A.; Tamayo, A.; Rubio, F.; Rubio, J. Study and Characterization of Organically Modified Silica-Zirconia Anti-Graffiti Coatings Obtained by Sol-Gel. *J. Chem. Chem. Eng.* **2013**, *7*, 120–131.
25. Kim, N.H.; Kim, D. Identification of the donor-substitution effect of tetraphenylethylene AIEgen: Synthesis, photophysical property analysis, and bioimaging applications. *Dyes Pigments* **2022**, *199*, 110698. [CrossRef]
26. Ranger, M.; Rondeau, D.; Leclerc, M. New Well-Defined Poly(2,7-fluorene) Derivatives: Photoluminescence and Base Doping. *Macromolecules* **1997**, *30*, 7686–7691. [CrossRef]
27. Mackenzie, J.D.; Huang, Q.; Iwamoto, T. Mechanical Properties of Ormosils. *J. Sol-Gel Sci. Technol.* **1996**, *7*, 151–161. [CrossRef]
28. Chemtob, A.; Peter, N.; Belon, C.; Dietlin, C.; Croutxé-Barghorn, C.; Vidal, L.; Rigolet, S. Macroporous organosilica films via a template-free photoinduced sol-gel process. *J. Mater. Chem.* **2010**, *20*, 9104–9112. [CrossRef]
29. Liu, B.; Wang, K.; Lu, H.; Huang, M.; Shen, Z.; Yang, J. Thermally responsive AIE-active polyurethanes based on a tetraaniline derivative. *RSC Adv.* **2020**, *10*, 41424–41429. [CrossRef]
30. Ma, C.; Han, T.; Niu, N.; Al-Shok, L.; Efstathiou, S.; Lester, D.; Huband, S.; Haddleton, D. Well-defined polyacrylamides with AIE properties via rapid Cu-mediated living radical polymerization in aqueous solution: Thermoresponsive nanoparticles for bioimaging. *Polym. Chem.* **2022**, *13*, 58–68. [CrossRef]

Disclaimer/Publisher's Note: The statements, opinions and data contained in all publications are solely those of the individual author(s) and contributor(s) and not of MDPI and/or the editor(s). MDPI and/or the editor(s) disclaim responsibility for any injury to people or property resulting from any ideas, methods, instructions or products referred to in the content.

Review

Research Progress on Quantum Dot-Embedded Polymer Films and Plates for LCD Backlight Display

Bin Xu [1,*], Jiankang Zhou [2], Chengran Zhang [2], Yunfu Chang [1] and Zhengtao Deng [2,*]

[1] Department of Electronic Information Engineering, School of Computer and Information Engineering, Nanjing Tech University, Nanjing 211816, China; changyunfu@njtech.edu.cn

[2] College of Engineering and Applied Sciences, Nanjing University, Nanjing 210023, China; 522023340080@smail.nju.edu.cn (J.Z.); zhangchengran@ntchanged.com (C.Z.)

* Correspondence: xubinnjtu@njtech.edu.cn (B.X.); dengz@nju.edu.cn (Z.D.)

Abstract: Quantum dot–polymer composites have the advantages of high luminescent quantum yield (PLQY), narrow emission half-peak full width (FWHM), and tunable emission spectra, and have broad application prospects in display and lighting fields. Research on quantum dots embedded in polymer films and plates has made great progress in both synthesis technology and optical properties. However, due to the shortcomings of quantum dots, such as cadmium selenide (CdSe), indium phosphide (InP), lead halide perovskite (LHP), poor water, oxygen, and light stability, and incapacity for large-scale synthesis, their practical application is still restricted. Various polymers, such as methyl methacrylate (PMMA), polyethylene terephthalate (PET), polystyrene (PS), polyvinylidene fluoride (PVDF), polypropylene (PP), etc., are widely used in packaging quantum dot materials because of their high plasticity, simple curing, high chemical stability, and good compatibility with quantum dot materials. This paper focuses on the application and development of quantum dot–polymer materials in the field of backlight displays, summarizes and expounds the synthesis strategies, advantages, and disadvantages of different quantum dot–polymer materials, provides inspiration for the optimization of quantum dot–polymer materials, and promotes their application in the field of wide-color-gamut backlight display.

Keywords: quantum dot; polymers; backlight display; display technology

Academic Editors: Tengling Ye and Bixin Li

Received: 30 December 2024
Revised: 13 January 2025
Accepted: 15 January 2025
Published: 17 January 2025

Citation: Xu, B.; Zhou, J.; Zhang, C.; Chang, Y.; Deng, Z. Research Progress on Quantum Dot-Embedded Polymer Films and Plates for LCD Backlight Display. *Polymers* **2025**, *17*, 233. https://doi.org/10.3390/polym17020233

Copyright: © 2025 by the authors. Licensee MDPI, Basel, Switzerland. This article is an open access article distributed under the terms and conditions of the Creative Commons Attribution (CC BY) license (https://creativecommons.org/licenses/by/4.0/).

1. Introduction

With the continuous progress of semiconductor technology, display technology has witnessed remarkable enhancements over the last several decades, particularly in aspects such as color gamut width, luminance, and lifespan. The backlight display technology currently employed in the market is mainly the photoluminescence (PL) type (Figure 1a). Photoluminescence technology is commonly seen in liquid crystal displays (LCD), where blue light LEDs and down-converting luminescent materials (such as phosphor or quantum dots) are utilized as the backlight source [1]. Among the multiple components of the display device, the luminescent material is the core factor determining the display performance. To offer superior display quality, the luminescent material must possess several key characteristics: high luminous efficiency—the material is capable of effectively converting input energy into light, thereby enhancing display luminance; high color purity—the light emitted by the material needs to have a distinct color to ensure clear and accurate displayed colors; and good stability—the material should maintain stable performance during long-term usage or under high-temperature conditions, preventing color distortion or luminance attenuation. These characteristics jointly determine the visual performance and service life of the display device [2,3]. Hence, the selection and optimization of luminescent materials are crucial for improving the quality of display technology.

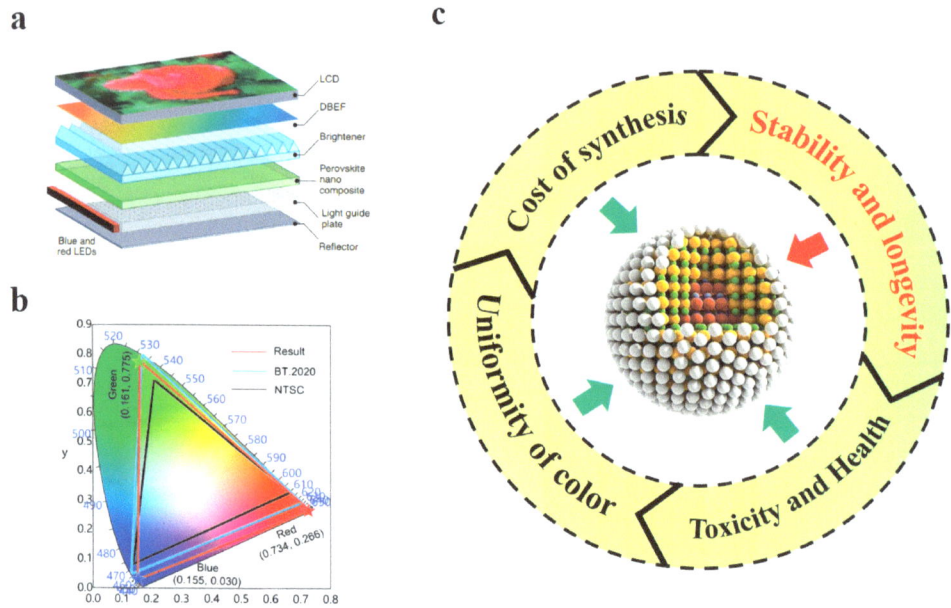

Figure 1. (**a**) Photoluminescent liquid crystal display structure schematic diagram. Reprinted with permission from ref. [1]. Copyright 2020 Springer Nature. (**b**) Gamut in CIE chromaticity diagram. Reprinted with permission from ref. [4]. Copyright 2023 Wiley. (**c**) Current problems and challenges faced by quantum dot materials.

Currently, most of the LCD backlight display materials available in the market employ blue light LED chips combined with a coating of yellow fluorescent materials (such as YAG:Ce^{3+}) to generate white light. Nevertheless, this approach confronts significant issues. Since the spectrum produced by blue light LED and yellow phosphor does not fully encompass the entire visible spectrum, particularly in the red and green regions, it leads to inaccurate color reproduction and low saturation [5]. The color gamut of this backlighting scheme is typically narrow, merely approximately ~72% of the National Television System Committee (NTSC) standard (Figure 1b), making it challenging to achieve higher standards of color performance. The light generated by the strategy of blue light LED and yellow fluorescent materials usually leans towards a cool hue (high color temperature), and thus the overall display effect might appear cold. To compensate for this, it is often necessary to adjust the color temperature of the backlight, but excessive adjustment might result in the distortion of other colors. The LCD panel itself has a certain degree of viewing angle dependence. When using a blue light LED backlight, the performance of color and luminance may vary at different viewing angles, especially when viewed at larger angles, and the problem of uneven color and luminance might be more pronounced. The spectrum of blue light LED contains a relatively larger proportion of short-wavelength blue light. Prolonged viewing might cause certain levels of eye fatigue, and some studies have even raised concerns about potential adverse effects on sleep quality, especially following excessive exposure to a strong blue light environment [6–11].

Quantum dots, as luminescent materials, can significantly enhance the color gamut and luminance of LED-backlit LCD displays with their outstanding color performance and high efficiency, thereby delivering more vivid and precise visual experiences. The application of this technology enables liquid crystal display televisions to present richer and more brilliant colors, particularly in the display of high dynamic range (HDR) content [12–16].

In 2013, Sony pioneered the launch of the world's first liquid crystal television (QD-LCD) equipped with quantum dot backlight technology, signifying the official advent of the quantum dot television era. In 2014, TCL showcased quantum dot LCD televisions at the International Consumer Electronics Show (CES) and the IFA exhibition in Berlin. This innovative technology promptly aroused extensive attention within the industry. Subsequently, renowned domestic and international television brands such as Samsung followed suit, and quantum dot televisions rapidly emerged as a significant new category in the television industry, highly favored by consumers [17].

However, the application of quantum dot materials in high-performance display devices still presents numerous issues that urgently require resolution (Figure 1c). Firstly, the problem of water and oxygen stability of quantum dots must be addressed [18–21]. Therefore, this article focuses on the research progress on quantum dot polymers in the field of backlight display, analyzes the important indicators of backlight display, contrasts the pros and cons of different types of quantum dots, conducts a detailed discussion on various encapsulation methods of quantum dots and polymers, and finally puts forward some challenges and opportunities in this domain, facilitating the development of the quantum dot backlight display field.

2. Important Indicators for Backlight Display and Related Parameters of Quantum Dots

2.1. Color Gamut of Backlight Display and FWHM

Color gamut refers to the range of colors that display devices (such as televisions, computer monitors, printers, etc.) are capable of displaying or reproducing. Different devices have varying color gamut sizes, and thus the types and saturation of colors they can display also differ. The CIE color space is a set of standardized color representation methods developed by the International Commission on Illumination for accurately describing and comparing colors [22,23]. The CIE color space is a model based on human visual perception, providing a unified and device-independent way to describe colors. The color gamut is a specific manifestation of the color range that can be reproduced by a device or the human eye within the color space [24]. On a color gamut chart, the triangular area enclosed by the three primary colors of red, green, and blue represents the color gamut of the display. The higher the color purity of the three primary colors, the closer their coordinates in the color gamut chart are to the spectral color curve (the boundary of the visible spectrum). This implies that the color range that the display can present is closer to the pure colors in the natural spectrum. When the coordinates of the three primary colors of the display are closer to the spectral color curve, the colors it can display will be more saturated and vivid. Furthermore, the larger the area of the color gamut triangle, the wider the color gamut covered by the display, indicating that the display is capable of presenting more colors. Therefore, the larger the color gamut, the richer the color performance of the display, and the more details and color gradations it can reproduce. The size of the color gamut of a display directly affects the accuracy and richness of its colors. Wide-color-gamut displays are typically more significant in applications such as image processing, professional design, and high-definition television because they can restore various colors more accurately and richly [23,25–28].

FWHM represents the width of the emission spectral line. Specifically, it is the distance between the two points on the emission intensity curve at half the peak intensity (half-peak). In the spectrum, the difference between the wavelength values corresponding to these two points is the FWHM, which can be used to quantify the spectral width or color purity of the luminescent material [29].

In display technology, multiple light sources with narrow FWHM (such as red light, green light, and blue light) are typically used to cover the entire color gamut. The FWHM of each light source is narrow, so the representation of each color is more precise and saturated. When combining these narrow FWHM light sources, their color gamuts will cover different areas in the color gamut chart, and these areas are often close and non-overlapping, thereby covering a wider color range. On the color gamut chart, the boundaries of the color gamut are usually defined by the most extreme colors that the display can accurately reproduce. Light sources with narrow FWHM can position the boundaries of the color gamut more precisely. For example, for red, green, and blue, the narrower the FWHM of each light source, the more accurately the boundaries of the color gamut can be restored, thereby expanding the color range that the display can display. A narrow FWHM enables the display to better restore the details of the spectrum and highly saturated colors without causing color blurring due to a broad color spectrum [30].

The FWHM of quantum dot materials is influenced by several key factors, including material composition and type, size distribution, synthesis methods, surface defects, and ligand modification. Specifically, different materials, such as CdSe, CdTe, InP, PbS, etc., exhibit distinct electronic structures and bandgap characteristics, leading to variations in emission spectral widths among various types of quantum dots. For example, perovskite quantum dots inherently possess a relatively narrow FWHM (<30 nm) [31–45], whereas $CuInS_2$ exhibits a broader FWHM (>100 nm) [46]. Moreover, a wide size distribution of quantum dots results in significant variations in emission wavelengths, thereby broadening the spectrum and increasing the FWHM. Reducing the size distribution, particularly minimizing its standard deviation, is critical for enhancing spectral quality and narrowing the FWHM. Surface defects and vacancies can also alter the electronic energy levels of quantum dots, causing spectral shifts or broadening. Surface modifications using organic ligands or inorganic shells can improve surface quality, reduce defects, and consequently narrow the emission spectrum [47,48]. Precise control over synthesis conditions, surface modification, and optimization of the reaction environment can effectively minimize the size distribution of quantum dots, resulting in narrower spectra and smaller FWHM, thereby enhancing the performance of quantum dot materials in display, optoelectronic, and other applications [41,45].

2.2. Backlight Display Brightness and Quantum Dot Photoluminescence Quantum Yield

The brightness of a backlight display is of paramount significance for display performance, particularly in aspects such as presentation of high dynamic range (HDR) content, contrast, and color manifestation in strong light environments. High brightness not only conspicuously enhances visibility, making the screen content more legible in bright settings, but also boosts the layering and detail presentation of images. It assists the display in presenting more elaborate details, especially in the dark and highlight sections, guaranteeing a fine and realistic picture [49]. Simultaneously, a higher brightness facilitates more precise color reproduction, elevating color saturation and vividness, thereby offering a more vivid and natural visual experience [50].

Photoluminescence quantum yield (PLQY) is a crucial parameter that describes the photoluminescence efficiency of materials. It is defined as the ratio of the number of photons re-emitted through radiative transitions (luminescence) to the number of photons absorbed by the material.

In backlight displays, fluorescent materials are typically employed to convert the light generated by blue or ultraviolet light sources into other colors (such as red and green), thereby achieving a complete color gamut and rich color performance. The higher the PLQY of the fluorescent material, the greater the portion of absorbed light energy that is converted into visible light, consequently enhancing the brightness of the display. Additionally, fluorescent materials with high PLQY typically imply fewer non-radiative losses (such as heat losses) [51]. High losses associated with low PLQY materials might lead to an increase in the temperature of the backlight system, thereby influencing the thermal management and efficiency of the system. With the enhancement of PLQY, not only is the brightness elevated, but the thermal effect is also mitigated, thereby improving the overall energy efficiency and stability of the display system [52–54].

2.3. Backlight Display Life and Quantum Dot Stability

The lifetime of a backlight display refers to the period during which the backlight source and fluorescent materials can operate normally while maintaining a specific brightness and performance. The lifetime of a backlight display is a critical factor influencing the overall service life, image quality, and user experience of the display. The lifetime of a backlight display is affected by multiple factors, including the quality of the light source and fluorescent materials, the driving current, the operating temperature, the heat dissipation design, the environmental conditions, the brightness setting, the power supply stability, and material aging. Reasonable design and utilization can prolong the lifetime of backlight displays, reduce the phenomenon of light attenuation, and maintain the display effect [55,56].

The stability of quantum dots pertains to the capacity of quantum dots to sustain their physical and chemical properties (such as optical properties, luminescence efficiency, color stability, etc.) over an extended period or under specific conditions. Quantum dots have significant applications in fields such as display technology, optoelectronics, and biological imaging, particularly being widely utilized in quantum dot displays (QLED) and quantum dot backlight displays. Nevertheless, the stability of quantum dots still constitutes a challenge in their commercial applications. High-stability quantum dots are of paramount importance for the persistence of display quality. They are required to maintain long-term brightness and color stability to avoid any impact on display quality due to light attenuation or color variations [57,58].

3. Development of Quantum Dot Backlight Display Materials

3.1. Group II–VI Quantum Dot Materials

Group II–VI quantum dot materials are semiconductor quantum dots formed by elements of Group II and Group VI in the periodic table. Group II–VI elements encompass elements such as zinc (Zn), cadmium (Cd), mercury (Hg), etc., while Group VI includes elements such as sulfur (S), selenium (Se), and tellurium (Te), etc. Table 1 presents the key parameters of II–VI quantum dot materials reported in recent years. The development of II–VI quantum dot materials in the field of backlight display has undergone a transformation from traditional cadmium-based quantum dots to cadmium-free alternative materials.

Table 1. Group II–VI quantum dot materials.

Materials	Emission Peak (nm)	FWHM (nm)	PLQY (%)	Ref.
CdSe	550	150	90	[59]
CdSe/CdS	470–570	30	60–80	[60]
CdSe/CdS/ZnS	620	/	75	[61]
ZnSe:Mn	580	50	90	[62]
ZnSe:Cu	508	18	/	[63]
ZnSe:Mn-Cu	490/585	/	13–17	[64]
(Zn,Se)Te/ZnSe/ZnS	463	63	95	[65]
CuInS$_2$/ZnS	920	217	65	[46]
(Cd,Zn)Se/ZnS	455–512	<20	81	[66]
CdTe/CdS/ZnS	561	47	73	[67]
CdTe:In	635	/	90	[68]
CdTe:Mn	496–542	<50	/	[69]

CdSe is the earliest II–VI quantum dot material that has been extensively studied and applied, primarily utilized in quantum dot backlight sources. It possesses a high quantum yield and a broad absorption spectral range (Figure 2a,b), and the emission wavelength can be precisely modulated [59,60]. CdS quantum dots are frequently employed as shell materials to form core–shell structures (such as CdSe/CdS) in conjunction with CdSe cores to enhance light stability and quantum efficiency [60,61]; As illustrated in Figure 2c, due to its relatively smaller bandgap, CdTe is more prevalently utilized in research on red light quantum dots [67–69]. These quantum dots are capable of providing high-color-gamut and high-brightness display effects. Nevertheless, the toxicity of cadmium and environmental pollution concerns have elicited apprehensions regarding CdTe usage [70–73]. The ROHS environmental protection standard of the European Union stipulates that the cadmium content (mass fraction) of quantum dot components should be less than 1×10^{-4}. From the perspective of the development trend of future quantum dot material technology, low-cadmium and cadmium-free materials will become the options for all end products. To address the environmental and health issues of cadmium-based quantum dots, researchers have actively developed low-cadmium and cadmium-free quantum dot materials [74–77].

Zinc-based quantum dots are non-toxic and comply with environmental protection requirements, serving as significant candidates for substituting cadmium-based quantum dots. ZnSe and ZnS are the principal alternative materials for cadmium-based quantum dots, and they have emerged as research hotspots for eco-friendly quantum dots due to their non-toxicity and favorable chemical stability: by doping manganese (Mn) or copper (Cu), the luminescent properties of zinc-based quantum dots can be regulated (as shown in Figure 2d), further enhancing their color purity and brightness [62,65]. However, the relatively lower quantum efficiency and brightness of zinc-based quantum dot materials have restricted their further development in the field of backlight display.

Copper indium sulfide (CuInS$_2$) quantum dots are a non-toxic and environmentally friendly type of quantum dot that have garnered attention in backlight displays in recent years [46]. They feature a wide optical bandgap and a large absorption coefficient, and the emission color can be adjusted by varying the composition ratio or the size of the quantum dots. Nevertheless, compared with cadmium-based quantum dots, CuInS$_2$ quantum dots have a broader spectral bandwidth (FWHM > 60 nm), and can only be applied in display applications with lower requirements for color purity. Moreover, the facile degradation of CuInS$_2$ quantum dots under conditions of high humidity and high temperature also constrains their application in the field of backlight display [78].

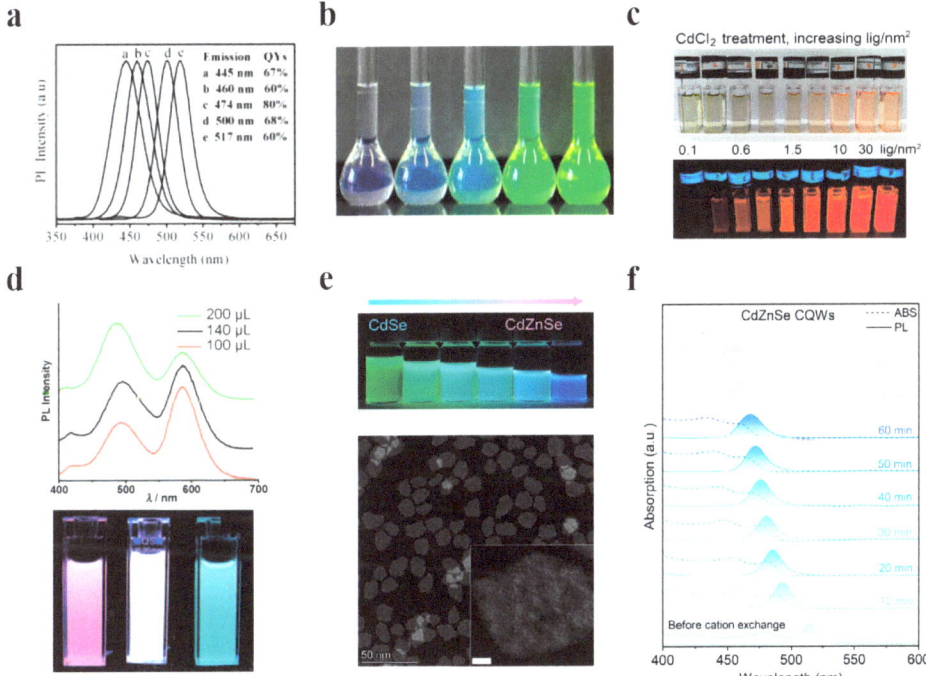

Figure 2. Group II–VI quantum dots. (**a**) PL spectra for CdSe/CdS core–shell nanocrystals with different core size and different shell thickness. (**b**) Photograph of solutions of CdSe/CdS core–shell nanocrystals with different core size and different shell thickness under normal indoor light without UV irradiation. Reprinted with permission from ref. [60]. Copyright 2005 Wiley. (**c**) Photographs under ambient room light (**top**) and UV light (**bottom**) showing effect of CdCl$_2$ treatment at increasing CdCl$_2$ concentrations on QD PL. Reprinted with permission from ref. [68]. Copyright 2018 American Chemical Society. (**d**) Cu:Mn-ZnSe-doped QD samples with different amounts of Cu precursors. Reprinted with permission from ref. [64]. Copyright 2011 Wiley. (**e**) Solutions of CQWs after different CE reaction times (10 to 60 min) under 365 nm UV light. (**f**) Normalized absorption and PL spectra of CdZnSe CQWs with respect to the CE reaction time. Reprinted with permission from ref. [66]. Copyright 2024 Wiley.

Cd$_{1-x}$Zn$_x$Se alloy quantum dots strike a balance between the high efficiency of cadmium-based quantum dots and the environmentally friendly nature of zinc-based quantum dots by adjusting the proportion of different elements (Figure 2e,f). However, the homogeneity and composition control of alloy quantum dots pose higher requirements for synthesis techniques, and the relatively high preparation cost of alloy quantum dots limits their application in large-scale production [66].

3.2. III–V Quantum Dot Materials

The development of III–V quantum dot materials in the field of backlight display has also made significant advancements, particularly as they typically exhibit non-toxicity and outstanding optical properties, serving as a crucial complement and alternative to II–VI quantum dots (refer to Table 2).

Table 2. Group III–V quantum dot materials.

Materials	Emission Peak (nm)	FWHM (nm)	PLQY (%)	Ref.
InP/Zn(Se,S)/ZnS	510	45	95	[79]
InP/Zn(Se,S)/ZnS	480–530	45	>90	[80]
InP/ZnSe/ZnS	535	35	90	[81]
InP/ZnS/ZnS	468	47	45	[82]
InP	650	/	24	[83]
InAs	700	/	11	[83]
GaP	400–520	75	35–40	[84]
$In_{1-x}Ga_xP$/ZnS	490–640	50	46	[85]
$In_{1-x}Ga_xAs$/ZnS	860	/	9.8	[85]
$In_{1-x}Ga_xP$/ZnS	486	46	65	[86]
$In_{1-x}Ga_xP$/ZnS	550–630	60	80	[87]

InP quantum dots are among the most focused upon types of III–V quantum dots. Due to their non-toxicity and high performance, they have gradually become an important candidate material for replacing cadmium-based quantum dots. InP quantum dots comply with the RoHS environmental protection requirements, making them highly suitable for consumer-grade backlight displays (Figure 3a,b). In recent years, the quantum efficiency and stability of InP quantum dots have been significantly enhanced through surface passivation and core–shell structure design. Nevertheless, as InP quantum dots tend to generate surface defects, which affect their optical properties and stability, more efficient surface passivation methods are required to suppress the occurrence of surface defects [79–82].

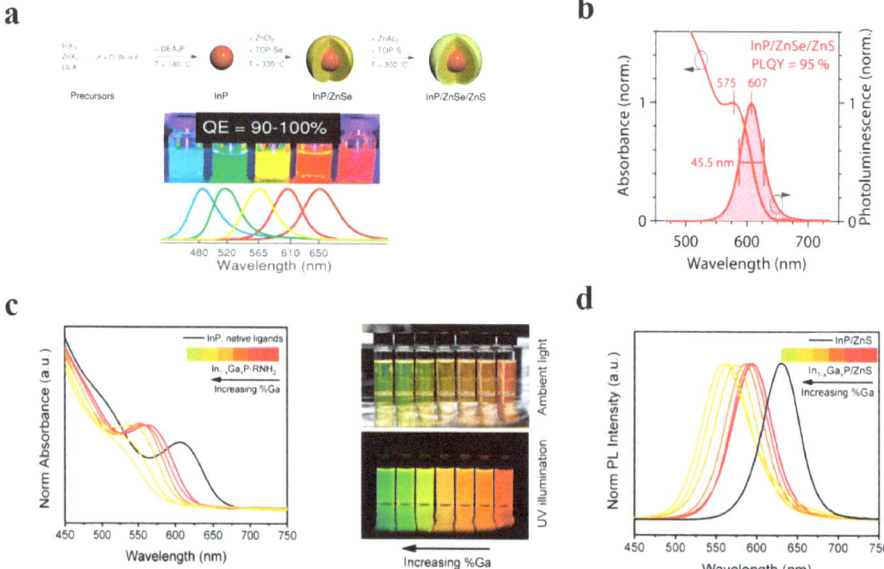

Figure 3. Group III–V quantum dots. (**a**) The synthesis strategy of InP/Zn(Se,S)/ZnS featuring a core–shell–shell structure and the attainment of multiple emission colors through the adjustment of the composition of the inner shell. (**b**) Absorbance and emission spectra of InP/ZnSe/ZnS QDs. Reprinted with permission from ref. [80]. Copyright 2022 American Chemical Society. (**c**) Absorption spectra of alloyed $In_{1-x}GaxP$ cores and the large range of emission colors produced by core–shell $In_{1-x}GaxP$/ZnS samples with varying gallium. (**d**) $In_{1-x}GaxP$/ZnS emission spectra. Reprinted with permission from ref. [87]. Copyright 2023 American Chemical Society.

InAs quantum dots are also a type of direct bandgap semiconductor material, and their emission wavelengths can be achieved by adjusting the size of the quantum dots, covering the full range from the visible-light to the infrared region. In backlight displays, InAs quantum dots are more frequently used in high-end display devices for extending the spectrum (particularly in the deep-red and near-infrared regions). Compared with other quantum dot materials, the preparation process of InAs quantum dots has strict requirements for growth conditions, presenting technical challenges in terms of uniformity and size control [83].

$In_{1-x}Ga_xP$ alloy quantum dots, as environmentally friendly red light materials, have received increasing attention in recent years. Alloyed III–V quantum dots can achieve flexible regulation of the emission wavelength by adjusting the material components (such as the ratios of In, Ga, and P) [85,86]. The emission wavelength range of $In_{1-x}Ga_xP$ quantum dots covers the red band of visible light (as shown in Figure 3c,d), making them extremely suitable for red light compensation in displays: the optimized $In_{1-x}Ga_xP$ quantum dots have high luminous efficiency and have been employed in high-end quantum dot backlight modules. However, compared to common III–V quantum dot materials, the preparation of alloy quantum dots is more complex, and the component control and uniformity of different elements pose higher demands on the synthesis process [87].

3.3. Perovskite Quantum Dot Materials

Perovskite quantum dots (PQDs) typically possess an ABX_3 crystal structure composed of corner-sharing $[BX_6]^{4-}$ octahedra. The A-site cation is confined within the cubic cage formed by the octahedra, and this distinctive crystal structure enables it to exhibit significant potential in the field of optoelectronics [1].

Common A-site cations include: inorganic cations, such as Cs^+ and Rb^+, and organic cations, such as methylammonium ($CH_3NH_3^+$, abbreviated as MA^+) and formamidinium cations ($CH(NH_2)_2)^+$, abbreviated as FA^+). The selection of the A-site cation has a direct influence on the size matching, tolerance factor, and overall stability of the crystal structure. The B-site ion is located at the center of the octahedron and is typically a divalent metal cation. Pb^{2+} is the most frequently utilized B-site ion, conferring high photoluminescence efficiency and outstanding optoelectronic properties to perovskite [88–93]. However, due to the high biotoxicity of Pb^{2+}, researchers have been seeking alternative metal ions, such as Sn^{2+}, Zn^{2+}, and Mn^{2+}, etc. These alternative metals can modify the lattice structure or regulate the optical properties, but many metal ions have difficulties in forming an ideal three-dimensional perovskite structure or may lead to bandgap characteristics that are not suitable for specific applications [94]; As depicted in Figure 4a,b, the X-site anion forms $[BX_6]^{4-}$ octahedra together with the B-site ion and constitutes the core component of the luminescence performance of perovskite materials. Regulation of the emission wavelength can be achieved by varying the halogen type, thereby covering the spectral range from ultraviolet to near-infrared [93].

Table 3 presents the key parameters of perovskite quantum dot materials reported in recent years. PQDs encompass almost all the advantages of luminescent materials, such as high brightness (their photoluminescence quantum yield (PLQY) approaches the theoretical limit), tunable emission (precise wavelength control (from ultraviolet to near-infrared) can be realized by altering the halogen type or doping), high color purity (the emission spectrum is narrow (typical FWHM is 20–40 nm) and the color saturation is extremely high), high light absorption coefficient (the light absorption capacity within the visible light range is extremely strong), high defect tolerance (even if there are defects in the crystal, its optical performance still exhibits excellence), and facile fabrication (mass production can be achieved through low-cost processes such as solution methods, spraying, spin coating, etc. [88,95]).

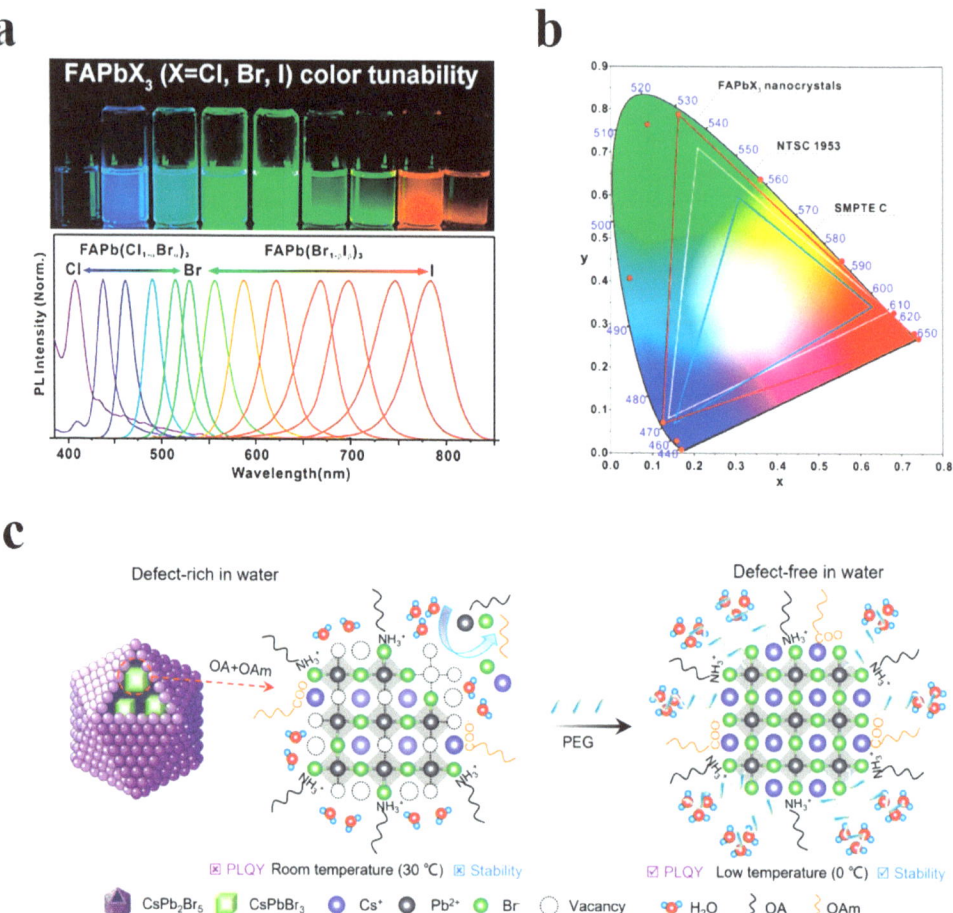

Figure 4. Perovskite quantum dots (**a**) FAPbX$_3$ nanocrystals dispersed in toluene under UV irradiation (λpeak = 365 nm) and PL emission spectra of FAPbX$_3$ nanocrystals. (**b**) Corresponding color gamut of FAPbX$_3$ nanocrystals displayed on the CIE diagram. Reprinted with permission from ref. [93]. Copyright 2017 American Chemical Society. (**c**) Scheme of synthesized aqueous-based CsPbBr$_3$/CsPb$_2$Br$_5$ PQDs using vacancy inhibitors of PEG, forming a defect-free surface in water (OA refers to oleic acid, and OAm represents oleylamine). Reprinted with permission from ref. [91]. Copyright 2023 Wiley.

Although perovskite (LHP) quantum dots have demonstrated tremendous potential and are widely regarded as a star material in the field of backlight applications, their stability issues, especially sensitivity to moisture, oxygen, light, and temperature, still constitute the main impediment restricting their wide application. As shown in Figure 4c, to solve this problem, researchers are enhancing the environmental stability of perovskite quantum dots through methods such as surface modification, optimization of the synthesis process, and encapsulation techniques [89–92]. The structural flexibility and outstanding optoelectronic properties of perovskite quantum dots have drawn significant attention in the fields of LED and backlight display technologies. With further improvements in stability and environmental performance, their potential in the future commercialization domain will be broader.

Table 3. Perovskite quantum dot materials.

Materials	Emission Peak (nm)	FWHM (nm)	PLQY (%)	Ref.
$CsPbCl_3$:Y	404	/	60	[96]
$CsPbCl_3$:Cd	381–410	/	60.5	[97]
$CsPbCl_3$	405	10.6	71	[98]
$CsPbCl_3$:Ni	408	/	96.5	[99]
$CsPbCl_3$:Mn	585	/	27	[100]
$CsPbCl_3$:Mn	579	80	54	[101]
$CsPbBr_3$	522	18	71.3	[102]
$CsPbBr_3$	517	18	94.6	[88]
$CsPbBr_3@PbSO_4$	522	16	99.8	[89]
$CsPbBr_3@Cs_4PbBr_6$	515	20	92	[90]
$CsPbBr_3@CsPb_2Br_5$	518	16	96	[91]
$MAPbBr_3@PbBr(OH)$	514	28	71.5	[92]
$FAPbBr_3$	530	22	85	[93]
$CsPbI_3$	640	/	88.2	[88]
$CsPbI_3$	660	/	88.1	[91]
$CsPb(Br_{0.4}, I_{0.6})_3$	641	32	32.4	[103]
$CsPbI_3$	690	31	100	[104]
$FA_{0.1}Cs_{0.9}PbI_3$	690	45	>70	[105]
$FAPbI_3$: Sr	735	/	100	[106]

4. Synthesis of Quantum Dot–Polymer Materials and Their Application in Backlight Display Field

The application of quantum dot–polymer materials in the field of backlight displays is rapidly emerging as a highly focused research hotspot. This type of material ingeniously combines the outstanding optical characteristics of QDs with the processing convenience of polymers, demonstrating immense potential, particularly in enhancing display performance, reducing production costs, and realizing flexible displays [107].

Table 4 presents advancements in research on quantum dot–polymer materials in the field of backlight displays in recent years. When quantum dots are combined with polymers, the advantages of both can be fully exploited. Quantum dots offer excellent optical properties, while polymers enhance the processing flexibility, mechanical strength, and environmental adaptability of the materials. As depicted in Figure 5a, the polymer substrate can effectively enwrap the quantum dots, protecting them from environmental factors such as moisture and oxygen, thereby significantly enhancing the stability and durability of the quantum dots [108–110]. Additionally, polymers can optimize the dispersion of quantum dots within the material, ensuring their homogeneous distribution, thereby avoiding aggregation and agglomeration phenomena and maintaining the efficient luminescence performance of the quantum dots [111].

As depicted in Figure 5a, through the combination of the properties of quantum dots and polymers, quantum dot–polymer materials can not only enhance the color gamut, brightness, and stability of backlight displays but also reduce production costs, facilitating large-scale manufacturing. With the in-depth progress of related research, quantum dot–polymer materials are anticipated to play a significant role in future display technologies and become an essential material foundation for next-generation displays, lighting equipment, and optoelectronic devices [112,113]. Currently, the common preparation methods of quantum dot polymers include pre-synthesis of quantum dots followed by encapsulation, in situ synthesis of quantum dots in polymers, and encapsulation of quantum dot composite materials in polymers. The pros and cons of the three methods are illustrated in Figure 5b. Hereinafter, we will conduct a detailed discussion on each synthesis method.

Table 4. Quantum dot–polymer materials for backlight displays.

QDs	Polymer	NTSC1953	PLQY	Light Stability	Ref.
$CsPbX_3$	PVDF	107%	/	>80 h	[31]
$MAPbBr_3$	PVDF	121%	94.6%	>400 h	[32]
$CsPbX_3$	PVDF	128%	70%	50 Day	[33]
$FAPbBr_3$	PVDF	118%	99%	>7 Day	[34]
$CsPbBrI_2$/Glass	PS	125%	36.9%	40 Day	[35]
(Zn,Cd)(Se,S)/ZnS	PP	120%	>75%	/	[36]
$CsPbX_3$/Glass	PP	110%	>90%	>1000 h	[37]
$Cs_{1-x}FA_xPbBr_3$	PMMA	/	99%	>90 Day	[38]
$CsPbBr_3$/SDDA	PMMA	102%	63%	/	[39]
$CsPb(Br/I)_3$/Glass	PMMA	120%	53%	>100 h	[40]
Cs_4PbBr_6/$CsPbBr_3$	PMMA	131%	85%	/	[41]
CdSe/CdS & (Cd,Zn)(Se,S)/ZnS	PDMS	133%	98%/97%	800 h	[42]
CdSe/ZnS/CdSZnS & CdSe/CdS/ZnS/(Cd,Zn)S	Silicone Encapsulant	100%	100%	>2200 h	[43]
CdSe/$Zn_xCd_{1-x}S$/ZnS & CdSe/CdS/ZnS	Silicone glue	118%	81%	>150 h	[44]
$CsPbBr_3$/Glass	Silicone glue	126%	86%	/	[45]

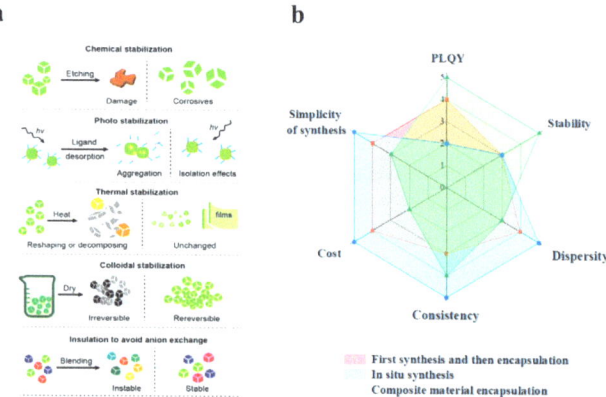

Figure 5. (**a**) Schematic diagram of functions by encapsulation illustrated by $CsPbX_3$ QDs. Reprinted with permission from ref. [20]. Copyright 2019 Wiley. (**b**) Radar map of the advantages and disadvantages of three methods of combining QDs with polymers.

4.1. Quantum Dots First Synthesized and Then Packaged

Synthesis followed by purification and then encapsulation of quantum dots is currently the most common synthesis approach for quantum dot–polymer materials. The advantage of this method lies in the relatively independent processes of quantum dot synthesis, purification, and polymer encapsulation, which can guarantee the high quality and stability of quantum dots, thereby enhancing the overall performance of the polymer composite materials. Nevertheless, although this process can achieve high purity and good luminescent performance of quantum dots, it still confronts several challenges. For instance, the synthesis, purification, and encapsulation processes are complex, requiring multiple steps. There exist interface issues between the polymer and quantum dots with poor compatibility. Quantum dots are prone to aggregation during the encapsulation process. Moreover, during the encapsulation, the polymer matrix material may not completely cover the quantum dots or the encapsulation may be incomplete, resulting in the exposure of quantum dots to the external environment and subsequently reducing their stability.

As depicted in Figure 6a–e, Lu et al. synthesized high-quality $Cs_{1-x}FA_xPbBr_3$ quantum dots using a dual-solvent-assisted reprecipitation method [38]. Subsequently, the purified quantum dots were mixed with MMA to fabricate quantum dot–polymer films. The $Cs_{0.2}FA_{0.8}PbBr_3$ quantum dot–polymer film maintained 98% of its initial strength under normal temperature and humidity conditions. After being exposed to 60 °C/90% relative humidity for 300 h, it retained 75% of its initial strength, demonstrating outstanding environmental stability. The white light-emitting device fabricated using the synthesized PQD, $K_2SiF_6:Mn^{4+}$ phosphor, and blue LED chip achieved a Rec. 2020 color gamut coverage rate of 96.7%. The quantum dot–polymer material synthesized by this method possesses high efficiency, environmental reliability, and a wide NTSC 2020 coverage range, and is regarded as a promising candidate in display devices.

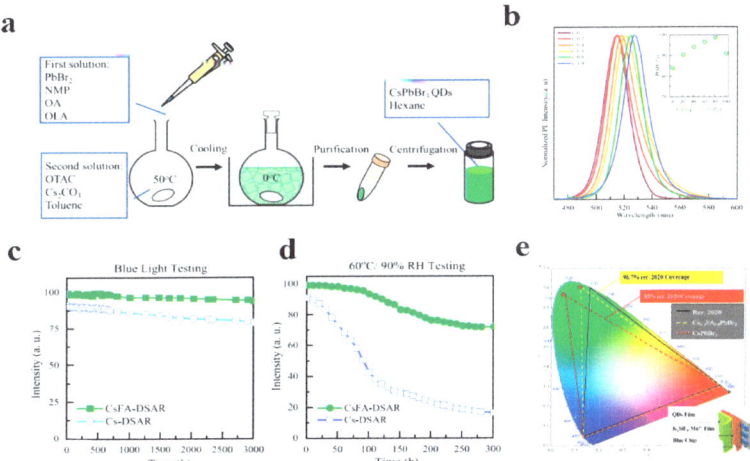

Figure 6. (a) Flowchart of the dual-solvent assisted reprecipitation (DSAR) technique. (b) Photoluminescence spectra (the inset shows photoluminescence quantum yield). (c) Blue light stability test. (d) The 60 °C/90%RH stability test. (e) Color gamut of the fabricated devices using the Cs-DSAR and CsFA-DSAR PQDs. Reprinted with permission from ref. [38]. Copyright 2023 Elsevier.

4.2. Quantum Dots Synthesized In Situ in Polymers

In situ synthesis enables the direct generation of quantum dots within the polymer matrix, ensuring their homogeneous distribution within the polymer material. This circumvents the issues of quantum dot agglomeration or uneven dispersion that might arise in conventional approaches, guaranteeing the consistency and stability of the material's optical and mechanical properties.

Unlike the pre-synthesis of quantum dots using solution or solid-state methods followed by polymer encapsulation, in situ synthesis permits precise control over the growth process of quantum dots during the polymer synthesis, achieving more accurate size control and a more uniform distribution, while maximizing the retention of their outstanding optical characteristics. Compared to physical mixing methods, in situ synthesis enhances the interaction between quantum dots and the polymer through a chemical synthesis process, reducing interfacial defects or stress concentration, thereby enhancing the stability of the composite material [114].

During in situ synthesis, the interface between the quantum dots and the polymer is formed naturally through chemical reactions and intermolecular interactions, rendering the compatibility between the quantum dots and the polymer matrix more intimate. This direct binding contributes to improving the interface contact between the two, thereby enhancing the photoelectric performance of the quantum dots and strengthening the mechanical properties of the polymer material. Moreover, in situ synthesis enables the direct embedding of quantum dots during the synthesis of the polymer matrix, eliminating the need for additional post-treatment steps, reducing the complexity and cost of the manufacturing process. This is particularly significant for large-scale production, enabling low-cost and straightforward production procedures. Additionally, in situ synthesis obviates the requirement for additional dispersants or surface modifiers, avoiding potential chemical contamination or instability and enhancing the simplicity and controllability of the process.

As illustrated in Figure 7a–d, Zhang et al. dissolved MAX, PbX_2 (x = Cl, Br, I), and PVDF simultaneously in DMF to prepare a precursor solution. Subsequently, they utilized a reduced-pressure environment to rapidly remove the solvent and in situ-synthesized $MAPbX_3$–PVDF polymer films [32]. The interaction between the -CF_2- groups in PVDF and the organic A-site MA^+ led to uniform size and spatial distribution of the fabricated quantum dots in the composite films. These quantum dot–polymer films exhibited superior PL performance, with a PLQY as high as 94.6 ± 1%. The combination of green-emitting quantum dot–polymer films with red-emitting $K_2SiF_6:Mn^{4+}$ resulted in a backlight display device with high luminous efficiency (up to 109 lm W^{-1} at 20 mA) and a wide color gamut (121% of the NTSC standard), providing possibilities for quantum dot–polymer LCD backlight displays.

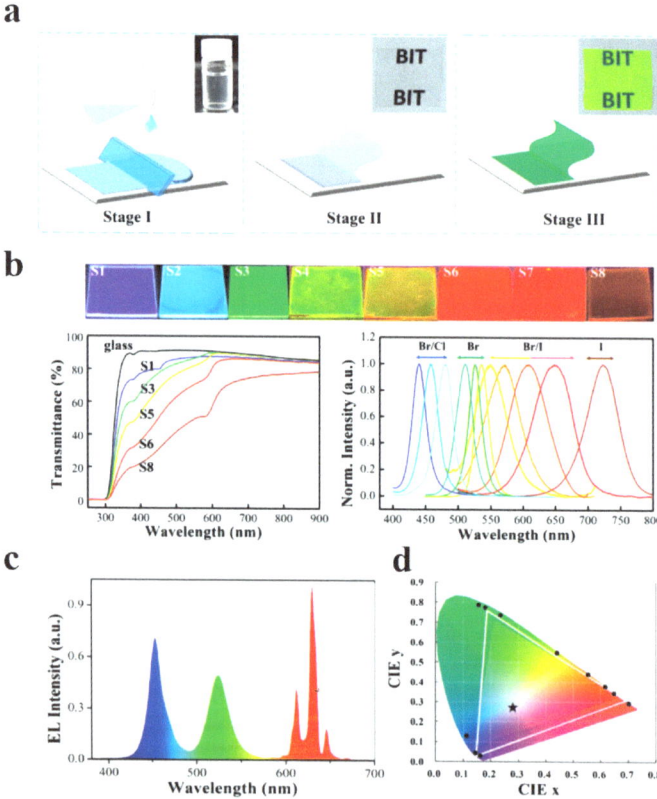

Figure 7. (**a**) Schematic illustration of the in situ fabrication of $MAPbBr_3$ NCs-embedded PVDF composite films. (**b**) Optical images under a UV lamp (365 nm) of color-tunable $MAPbX_3$–PVDF composite films with different halogen constitutions on glass substrates. (**c**) Emission spectrum of the white LED using green emissive $MAPbBr_3$–PVDF composite films and red emissive phosphor. (**d**) The color coordinate (star) and the white triangle (white line) of obtained white LED in CIE 1931 diagram. Reprinted with permission from ref. [32]. Copyright 2016 Wiley.

4.3. Quantum Dots Encapsulated In Polymers

Quantum dot composite materials refer to composites formed by combining quantum dots with other materials (such as glass, metals, inorganic materials, etc.). By integrating the excellent optical properties of quantum dots with the characteristics of the matrix materials, quantum dot composite materials can exhibit distinctive advantages in multiple fields. Currently, common quantum dot composite materials include quantum dot–glass composites and quantum dot core–shell composites.

Quantum dot–glass composites are formed by growing quantum dots in situ at high temperatures in glass, confining the quantum dots within the dense glass network structure to restrict their size. Additionally, the excellent waterproof performance of glass can significantly enhance the thermal and water stability of quantum dots. As depicted in Figure 8, Lin et al. first mixed quantum dot raw materials and borosilicate glass components, and synthesized quantum dot–glass composites using the melt-quenching method followed by annealing treatment. Subsequently, the quantum dot–glass composites were ground into powder and fabricated into a large-area yellow single-layer quantum dot–polymer film encapsulated with PP through industrial melting extrusion and rolling methods [37]. The photoluminescence quantum yield (PLQY) of the quantum dot–polymer film was as high as 92%, with a narrow FWHM of 19 nm (green) and 33 nm (red). Significantly, the quantum dot plate could pass the stringent aging test at 85 °C/85% RH and achieve a working T90 lifetime of over 1000 h. Finally, by coupling the yellow single-layer quantum dot–polymer film with a blue light guide plate, a white backlight unit was designed. The constructed prototype display possesses superior color rendering performance, with narrow green/red emissions and a color gamut reaching 110% of the National Television System Committee (NTSC).

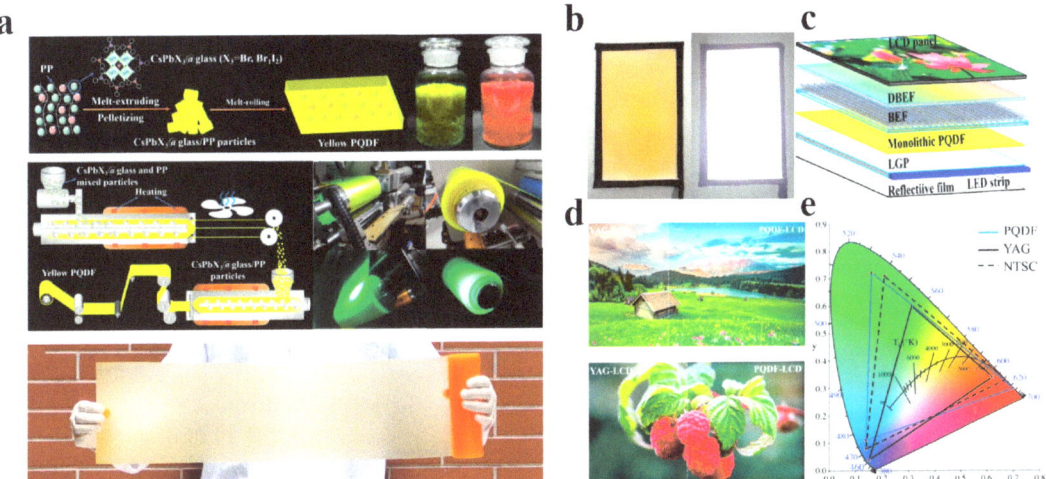

Figure 8. (**a**) Schematic diagrams of the preparation procedure for yellow PQDF via melt extruding-rolling method. Photographs of the as-prepared PP-encapsulated CsPbX$_3$@glass composite particles and the corresponding green and yellow PQDFs. (**b**) Photograph of yellow monolithic PQDF-based backlit unit and luminescent image of the backlit unit at an operating voltage of 12 V. (**c**) Schematic structure of an LCD prototype using yellow PQDF as a light converter. (**d**) Comparison of the display performance of a YAG-based LCD and PQDF-based LCD. (**e**) Color gamut of PQDF-based LCD (blue solid triangle), NSTC 1953 standard (black dashed triangle), and YAG-based commercial LCD (black solid triangle). Reprinted with permission from ref. [37]. Copyright 2024 Wiley.

Early researchers discovered that defects on the surface of quantum dots, particularly surface vacancies and unsaturated bonds, frequently led to the occurrence of non-radiative recombination processes, which significantly reduced the luminescence efficiency and optical stability of quantum dots [114–118]. These defects not only result in the energy loss of light but also may cause the photodegradation of quantum dots, severely influencing their luminescence performance during long-term usage, especially in environmental conditions such as variations in humidity and temperature. Hence, how to effectively suppress the surface defects of quantum dots and enhance their luminescence efficiency and stability has emerged as a significant challenge for researchers [47,48].

To address this issue, quantum dot core–shell composite materials have emerged. By coating the surface of quantum dots with a shell layer possessing excellent properties (typically semiconductor or metal materials), researchers can effectively seal or repair the surface defects and reduce the occurrence of non-radiative recombination, thereby significantly enhancing the fluorescence quantum efficiency, optical stability, and environmental adaptability of quantum dots. The core–shell structure not only protects the core of quantum dots from the external environment but also further improves their optical characteristics, such as tunable emission wavelengths, high brightness, and longer photodegradation time, through the rational design of the combination of core and shell materials [117,118]. Therefore, quantum dot core–shell composite materials exhibit extensive application prospects in fields such as display technology, optoelectronic devices, sensors, biomedicine, etc., and have become one of the effective approaches to address the issue of optical performance degradation of quantum dots.

As depicted in Figure 9, Jang et al. synthesized green-emitting CdSe/ZnS/CdSZnS and red-emitting CdSe/CdS/ZnS/CdSZnS multilayer core–shell quantum dot composite materials using the thermal injection method and subsequently mixed them with organosilicon to prepare quantum dot–polymer films [43]. Due to the passivation of the quantum dot surface by the multi-core–shell structure, the luminescence efficiency of these quantum dots reached 100%. The quantum dots were encapsulated in blue LEDs and used as green and red color converters to fabricate white LEDs for backlight display. The equivalent quantum effects of green and red QD-LEDs reached 72% and 34%, respectively, and the QD-LEDs maintained their initial efficiency for over 2200 h in a normal working environment. The white QD-LED, adjusted to (0.24, 0.21) in CIE 1931 for backlight applications, had a luminous efficiency of 41 lm/W, and the color reproducibility was 100% compared to the NTSC color space. Moreover, the white QD-LED backlight was successfully integrated onto a 46-inch liquid crystal television panel, demonstrating excellent color gamut.

As shown in Figure 9, Kang et al. synthesized rod-shaped CdSe/$Zn_xCd_{1-x}S$/ZnS and CdSe/CdS/ZnS multi-core–shell structure quantum dot materials with a 1D structure using the thermal injection method, addressing the PLQY quenching and stability issues of rod-shaped quantum dot materials [44]. The introduction of the outer gradient $Zn_xCd_{1-x}S$ shell layer and ZnS shell layer in the CdSe/$Zn_xCd_{1-x}S$/ZnS gradient alloy quantum rods provided a smoother confinement potential, reduced non-radiative energy transfer, and achieved a solid-state PLQY of 81%. The white light quantum rod polymer prepared by mixing with silica gel achieved an astonishing luminous efficiency of 149 lmW-1, with a color gamut of 118% NTSC and 90% BT2020, possessing high thermal stability and optical stability and being highly suitable for LCD backlight applications. Additionally, it showed higher energy efficiency compared to the most advanced LED backlighting devices, with a maximum efficiency of an astonishing 200 lm/W. Although the efficiency decreased to 120 lm/W at higher currents (50 mA), this is still much higher than that of phosphor-based LEDs (50 lm/W).

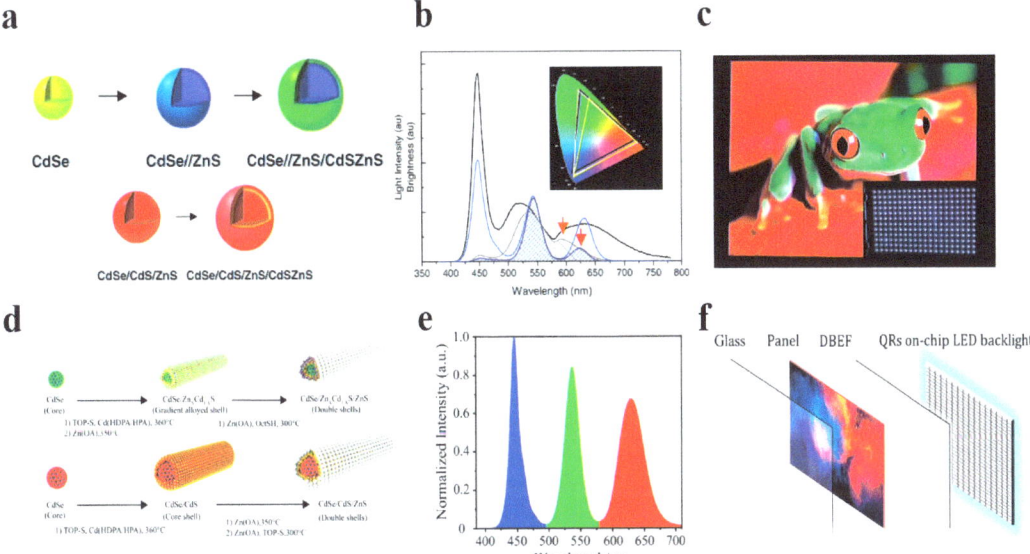

Figure 9. (**a**) Schematic structures of the growth of green and red QDs. (**b**) Light intensity spectra (solid line) and brightness (hatched area) of QD-LED (blue) and phosphor-LED (gray). Inset: color triangles of QD-LED (white) and phosphor-LED (yellow) compared to NTSC1931 (black). (**c**) Display image of a 46-inch LCD TV panel and a quarter of the white QD-LED backlights (inset). Reprinted with permission from ref. [43]. Copyright 2010 Wiley. (**d**) Schematic of syntheses of the ZnS modified green and red QRs. (**e**) The spectrum of the fabricated QRWLED consists of three emission band peaks at 450, 527, and 624 nm. (**f**) Schematic of a QRs on-chip backlight with dual-brightness-enhancement film (DBEF) design for displays. Reprinted with permission from ref. [44]. Copyright 2021 Wiley.

5. Conclusions and Prospects

Quantum dot–polymer materials (QDPs) have exhibited tremendous application potential in the field of backlight display, especially in enhancing color performance, brightness, efficiency of displays, and reducing costs. Quantum dot materials, with their outstanding optical properties such as high luminance, tunable emission wavelengths, and high color purity, have offered a novel impetus for the innovation of backlight display technology. Polymer materials in turn, due to their favorable processability, tunability, flexibility, and advantages in large-scale production, have addressed the stability and processing difficulties that quantum dot materials might encounter in practical applications. By combining quantum dots and polymers, it is possible to maintain high optical performance while endowing the materials with better processability and durability.

At present, significant advancements have been achieved in the application of quantum dot–polymer materials in backlight display, particularly in quantum dot-enhanced liquid crystal displays (QLED) and quantum dot backlight units (QBLUs), which can offer a broader color gamut, higher luminance, and longer service life [119–122]. Through optimizing the size, composition, and surface modification of quantum dots, as well as the structure and performance of polymer substrates, researchers have continuously enhanced the performance of these materials, making their application in high-end display devices more mature and widespread [123,124].

In the future, research on quantum dot–polymer materials in the field of backlight display will evolve towards more efficient, stable, and economical directions. Specifically, the following areas will be the key points of future research.

(1) Material stability and environmental adaptability: Despite the fact that quantum dot–polymer composite materials have achieved favorable results in laboratories, in practical applications, environmental factors such as temperature, humidity, and ultraviolet radiation remain critical in influencing their long-term stability. Hence, in the future, more materials with high environmental stability need to be developed, and their durability can be further enhanced through innovative surface modification or encapsulation techniques.

(2) High-efficiency light conversion technology: Although quantum dots themselves possess excellent optical properties, how to enhance the photoelectric conversion efficiency of quantum dots and reduce energy loss remains a crucial aspect for future development. Exploring new quantum dot materials, such as perovskite quantum dots and two-dimensional quantum dots, might provide new breakthroughs for high-efficiency display materials.

(3) Flexible display technology: With the rise of flexible display technology, the application of quantum dot–polymer materials in bendable and stretchable display devices holds broad prospects. The combination of the flexibility of polymers and the optical characteristics of quantum dots can bring about a brand-new display experience. Therefore, future research needs to explore more quantum dot–polymer composite materials suitable for flexible substrates while ensuring that their performance is not compromised during the flexibilization process.

(4) Low-cost, large-scale production: Currently, the production cost of quantum dot–polymer materials is relatively high, restricting their large-scale application in the consumer electronics field. Future research will focus on developing low-cost and straightforward preparation methods, such as solution methods and large-area coating techniques, to meet the requirements of large-scale production and commercialization.

In conclusion, quantum dot–polymer materials have broad application prospects and significant market potential in the field of backlight display. With the continuous optimization of material performance and the advancement of production technologies, they will play a vital role in a wider range of display technologies in the future, contributing to the realization of more efficient, higher-quality, and more innovative display devices.

Author Contributions: Conceptualization, Z.D. and B.X.; writing—original draft preparation, B.X. and J.Z.; writing—review and editing, C.Z. and Y.C.; supervision, Z.D.; funding acquisition, B.X. All authors have read and agreed to the published version of the manuscript.

Funding: This research was funded by the National Natural Science Foundation of China (22075129), Nantong Major Scientific and Technological Achievements Transformation Plan Project (XA2022017), and Nanjing University Technology Development Project (2022200996).

Acknowledgments: The authors are grateful to Tao Zhang from Nanjing University for his valuable discussions.

Conflicts of Interest: The authors declare no conflicts of interest.

References

1. Park, A.; Goudarzi, A.; Yaghmaie, P.; Thomas, V.J.; Maine, E. Rapid response through the entrepreneurial capabilities of academic scientists. *Nat. Nanotechnol.* **2022**, *17*, 802–807. [CrossRef]
2. Srivastava, A.K.; Zhang, W.; Schneider, J.; Halpert, J.E.; Rogach, A.L. Luminescent Down-Conversion Semiconductor Quantum Dots and Aligned Quantum Rods for Liquid Crystal Displays. *Adv. Sci.* **2019**, *6*, 1901345. [CrossRef] [PubMed]
3. Fang, M.-H.; Leaño, J.L.; Liu, R.-S. Control of Narrow-Band Emission in Phosphor Materials for Application in Light-Emitting Diodes. *ACS Energy Lett.* **2018**, *3*, 2573–2586. [CrossRef]

4. Fan, X.; Wang, S.; Yang, X.; Zhong, C.; Chen, G.; Yu, C.; Chen, Y.; Wu, T.; Kuo, H.C.; Lin, Y.; et al. Brightened Bicomponent Perovskite Nanocomposite Based on Forster Resonance Energy Transfer for Micro-LED Displays. *Adv. Mater.* **2023**, *35*, e2300834. [CrossRef] [PubMed]
5. Peng, M.; Sun, S.; Xu, B.; Deng, Z. Polymer-Encapsulated Halide Perovskite Color Converters to Overcome Blue Overshoot and Cyan Gap of White Light-Emitting Diodes. *Adv. Funct. Mater.* **2023**, *33*, 2300583. [CrossRef]
6. Gianluca, T.; Ferguson, I.; Tsubota, K. Effects of blue light on the circadian system and eye physiology. *Molecular* **2016**, *22*, 61. [PubMed] [PubMed Central]
7. Ouyang, X.; Yang, J.; Hong, Z.; Wu, Y.; Xie, Y.; Wang, G. Mechanisms of blue light-induced eye hazard and protective measures: A review. *Biomed. Pharmacother.* **2020**, *130*, 110577. [CrossRef]
8. Zhong, J.; Zhuo, Y.; Hariyani, S.; Zhao, W.; Wen, J.; Brgoch, J. Closing the Cyan Gap Toward Full-Spectrum LED Lighting with NaMgBO$_3$:Ce^{3+}. *Chem. Mater.* **2019**, *32*, 882–888. [CrossRef]
9. Shibuya, T.; Akiba, T.; Iwanaga, T. Assessment of the Blue Light Hazard for Light Sources with Non-Uniform Luminance. *Leukos* **2020**, *17*, 205–209. [CrossRef]
10. Francon, A.; Behar-Cohen, F.; Torriglia, A. The blue light hazard and its use on the evaluation of photochemical risk for domestic lighting. An in vivo study. *Environ. Int.* **2024**, *184*, 108471. [CrossRef]
11. Zielinska-Dabkowska, K.M. Make lighting healthier. *Nature* **2018**, *553*, 274–276. [CrossRef]
12. Sarma, D.D.; Kamat, P.V. 2023 Nobel Prize in Chemistry: A Mega Recognition for Nanosized Quantum Dots. *ACS Energy Lett.* **2023**, *8*, 5149–5151. [CrossRef]
13. Moscatelli, A. Physically unclonable functions fight forgery. *Nat. Nanotechnol.* **2022**, *17*, 818. [CrossRef] [PubMed]
14. Huang, C.Y.; Li, H.; Wu, Y.; Lin, C.H.; Guan, X.; Hu, L.; Kim, J.; Zhu, X.; Zeng, H.; Wu, T. Inorganic Halide Perovskite Quantum Dots: A Versatile Nanomaterial Platform for Electronic Applications. *Nanomicro Lett.* **2022**, *15*, 16. [CrossRef]
15. Zhou, W.; Coleman, J.J. Semiconductor quantum dots. *Curr. Opin. Solid. State Mater. Sci.* **2016**, *20*, 352–360. [CrossRef]
16. Moon, H.; Lee, C.; Lee, W.; Kim, J.; Chae, H. Stability of Quantum Dots, Quantum Dot Films, and Quantum Dot Light-Emitting Diodes for Display Applications. *Adv. Mater.* **2019**, *31*, e1804294. [CrossRef] [PubMed]
17. Zhu, R.; Luo, Z.; Chen, H.; Dong, Y.; Wu, S.T. Realizing Rec. 2020 color gamut with quantum dot displays. *Opt Express* **2015**, *23*, 23680–23693. [CrossRef] [PubMed]
18. Yang, W.; Fei, L.; Gao, F.; Liu, W.; Xu, H.; Yang, L.; Liu, Y. Thermal polymerization synthesis of CsPbBr$_3$ perovskite-quantum-dots@copolymer composite: Towards long-term stability and optical phosphor application. *Chem. Eng. J.* **2020**, *387*, 124180. [CrossRef]
19. Kim, D.; Yun, T.; An, S.; Lee, C.L. How to improve the structural stabilities of halide perovskite quantum dots: Review of various strategies to enhance the structural stabilities of halide perovskite quantum dots. *Nano Converg.* **2024**, *11*, 4. [CrossRef] [PubMed]
20. Lv, W.; Li, L.; Xu, M.; Hong, J.; Tang, X.; Xu, L.; Wu, Y.; Zhu, R.; Chen, R.; Huang, W. Improving the Stability of Metal Halide Perovskite Quantum Dots by Encapsulation. *Adv. Mater.* **2019**, *31*, e1900682. [CrossRef]
21. Wei, Y.; Cheng, Z.; Lin, J. An overview on enhancing the stability of lead halide perovskite quantum dots and their applications in phosphor-converted LEDs. *Chem. Soc. Rev.* **2019**, *48*, 310–350. [CrossRef]
22. Chen, H.W.; Zhu, R.D.; He, J.; Duan, W.; Hu, W.; Lu, Y.Q.; Li, M.C.; Lee, S.L.; Dong, Y.J.; Wu, S.T. Going beyond the limit of an LCD's color gamut. *Light. Sci. Appl.* **2017**, *6*, e17043. [CrossRef]
23. Wang, X.; Bao, Z.; Chang, Y.-C.; Liu, R.-S. Perovskite Quantum Dots for Application in High Color Gamut Backlighting Display of Light-Emitting Diodes. *ACS Energy Lett.* **2020**, *5*, 3374–3396. [CrossRef]
24. Fryc, I.; Listowski, M.; Supronowicz, R. Going beyond the 20th century color space to evaluate LED color consistency. *Opt. Express* **2023**, *31*, 38666–38687. [CrossRef] [PubMed]
25. Zhao, X.; Wang, N.; Liu, K.; Yao, R.; Guo, Z.; Zhao, J.; Liu, Q. Enhancing Optical Properties of Zn-Mn Solid Solution Hybrid Halides for Wide Color Gamut Backlight Displays. *Small* **2024**, *20*, e2405137. [CrossRef]
26. Li, X.; Lou, B.; Chen, X.; Wang, M.; Jiang, H.; Lin, S.; Ma, Z.; Jia, M.; Han, Y.; Tian, Y.; et al. Deep-blue narrow-band emissive cesium europium bromide perovskite nanocrystals with record high emission efficiency for wide-color-gamut backlight displays. *Mater. Horiz.* **2024**, *11*, 1294–1304. [CrossRef] [PubMed]
27. Liao, H.; Zhao, M.; Zhou, Y.; Molokeev, M.S.; Liu, Q.; Zhang, Q.; Xia, Z. Polyhedron Transformation toward Stable Narrow-Band Green Phosphors for Wide-Color-Gamut Liquid Crystal Display. *Adv. Funct. Mater.* **2019**, *29*, 1901988. [CrossRef]
28. Zhao, M.; Liao, H.; Ning, L.; Zhang, Q.; Liu, Q.; Xia, Z. Next-Generation Narrow-Band Green-Emitting RbLi(Li$_3$SiO$_4$)$_2$:Eu^{2+} Phosphor for Backlight Display Application. *Adv. Mater.* **2018**, *30*, e1802489. [CrossRef]
29. Liu, F.; Cheng, Z.; Wan, L.; Gao, L.; Yan, Z.; Hu, D.; Ying, L.; Lu, P.; Ma, Y. Anthracene-based emitters for highly efficient deep blue organic light-emitting diodes with narrow emission spectrum. *Chem. Eng. J.* **2021**, *426*, 131351. [CrossRef]
30. Zhou, Y.; Ming, H.; Zhang, S.; Deng, T.; Song, E.; Zhang, Q. Unveiling Mn^{4+} substitution in oxyfluoride phosphor Rb2MoO2F4:Mn^{4+} applied to wide-gamut fast-response backlight displays. *Chem. Eng. J.* **2021**, *415*, 128974. [CrossRef]

31. Wu, Z.; Zhang, Y.; Du, D.; Yang, K.; Wu, J.; Dai, T.; Dong, C.; Xia, J.; Wu, A.; Zhao, Z. Disordered metasurface-enhanced perovskite composite films with ultra-stable and wide color gamut used for backlit displays. *Nano Energy* **2022**, *100*, 107436. [CrossRef]
32. Zhou, Q.; Bai, Z.; Lu, W.G.; Wang, Y.; Zou, B.; Zhong, H. In Situ Fabrication of Halide Perovskite Nanocrystal-Embedded Polymer Composite Films with Enhanced Photoluminescence for Display Backlights. *Adv. Mater.* **2016**, *28*, 9163–9168. [CrossRef] [PubMed]
33. He, K.; Chen, D.; Yuan, L.; Xu, J.; Xu, K.; Hu, J.; Liang, S.; Zhu, H. Crystallization engineering in PVDF enables ultrastable and highly efficient CsPbBr$_3$ quantum dots film for wide color gamut Mini-LED backlight. *Chem. Eng. J.* **2024**, *480*, 148066. [CrossRef]
34. Yang, C.; Niu, W.; Chen, R.; Pang, T.; Lin, J.; Zheng, Y.; Zhang, R.; Wang, Z.; Huang, P.; Huang, F.; et al. In Situ Growth of Ultrapure Green-Emitting FAPbBr$_3$-PVDF Films via a Synergetic Dual-Additive Strategy for Wide Color Gamut Backlit Display. *Adv. Mater. Technol.* **2022**, *7*. [CrossRef]
35. Huang, Q.; Chen, W.; Liang, X.; Lv, C.; Xiang, W. Ag Nanoparticles Optimized the Optical Properties and Stability of CsPbBrI$_2$ Glass for High Quality Backlight Display. *ACS Sustain. Chem. Eng.* **2023**, *11*, 9773–9781. [CrossRef]
36. Fang, F.; Wen, Z.; Chen, W.; Wang, Z.; Sun, J.; Liu, H.; Tang, H.; Hao, J.; Liu, P.; Xu, B.; et al. Thermally Processed Quantum-Dot Polypropylene Composite Color Converter Film for Displays. *ACS Appl. Mater. Interfaces* **2022**, *14*, 31160–31169. [CrossRef] [PubMed]
37. Lin, J.; Chen, S.; Ye, W.; Zeng, Y.; Xiao, H.; Pang, T.; Zheng, Y.; Zhuang, B.; Huang, F.; Chen, D. Ultra-Stable Yellow Monolithic Perovskite Quantum Dots Film for Backlit Display. *Adv. Funct. Mater.* **2024**, *34*, 2314795. [CrossRef]
38. Lu, S.-A.; Meena, M.L.; Gupta, K.K.; Lu, C.-H. Reprecipitation synthesis and spectroscopic characterization of Cs$_{1-x}$FA$_x$PbBr$_3$ nanocrystals for backlight display devices. *Appl. Surf. Sci.* **2024**, *643*, 158576. [CrossRef]
39. Zhang, X.; Wang, H.-C.; Tang, A.-C.; Lin, S.-Y.; Tong, H.-C.; Chen, C.-Y.; Lee, Y.-C.; Tsai, T.-L.; Liu, R.-S. Robust and Stable Narrow-Band Green Emitter: An Option for Advanced Wide-Color-Gamut Backlight Display. *Chem. Mater.* **2016**, *28*, 8493–8497. [CrossRef]
40. Liu, X.; Tong, Y.; Wang, Q.; Liang, X.; Zhang, Z.; Fan, H.; Xiang, W. The direct water quenching process in the preparation of broad wavelength tunable CsPb(Br/I)$_3$ NCs@glass for backlight display. *Mater. Today Nano* **2023**, *21*, 100288. [CrossRef]
41. Li, X.; Wen, Z.; Ding, S.; Fang, F.; Xu, B.; Sun, J.; Liu, C.; Wang, K.; Sun, X.W. Facile In Situ Fabrication of Cs$_4$PbBr$_6$/CsPbBr$_3$ Nanocomposite Containing Polymer Films for Ultrawide Color Gamut Displays. *Adv. Opt. Mater.* **2020**, *8*, 2000232. [CrossRef]
42. Onal, A.; Eren, G.O.; Melikov, R.; Kaya, L.; Nizamoglu, S. Quantum Dot Enabled Efficient White LEDs for Wide Color Gamut Displays. *Adv. Mater. Technol.* **2023**, *8*, 2201799. [CrossRef]
43. Jang, E.; Jun, S.; Jang, H.; Lim, J.; Kim, B.; Kim, Y. White-light-emitting diodes with quantum dot color converters for display backlights. *Adv. Mater.* **2010**, *22*, 3076–3080. [CrossRef]
44. Kang, C.; Prodanov, M.F.; Gao, Y.; Mallem, K.; Yuan, Z.; Vashchenko, V.V.; Srivastava, A.K. Quantum-Rod On-Chip LEDs for Display Backlights with Efficacy of 149 lm W^{-1}: A Step toward 200 lm W^{-1}. *Adv. Mater.* **2021**, *33*, e2104685. [CrossRef] [PubMed]
45. Li, J.; Fan, Y.; Xuan, T.; Zhang, H.; Li, W.; Hu, C.; Zhuang, J.; Liu, R.S.; Lei, B.; Liu, Y.; et al. In Situ Growth of High-Quality CsPbBr$_3$ Quantum Dots with Unusual Morphology inside a Transparent Glass with a Heterogeneous Crystallization Environment for Wide Gamut Displays. *ACS Appl. Mater. Interfaces* **2022**, *14*, 30029–30038. [CrossRef] [PubMed]
46. He, L.; Cao, S.; Li, Q.; Bi, Y.; Song, Y.; Ji, W.; Zou, B.; Zhao, J. Achieving near-unity quantum yield in blue ZnSeTe quantum dots through NH$_4$F molecular-assisted synthesis for highly efficient light-emitting diodes. *Chem. Eng. J.* **2024**, *489*, 151347. [CrossRef]
47. Kim, H.; Park, J.H.; Kim, K.; Lee, D.; Song, M.H.; Park, J. Highly Emissive Blue Quantum Dots with Superior Thermal Stability via In Situ Surface Reconstruction of Mixed CsPbBr$_3$–Cs$_4$PbBr$_6$ Nanocrystals. *Adv. Sci.* **2022**, *9*, e2104660. [CrossRef] [PubMed]
48. Lin, L.; Liu, A.A.; Zhao, W.; Yang, Y.; Zhu, D.L.; Dong, B.R.; Ding, F.; Ning, D.; Zhu, X.; Liu, D.; et al. Multihierarchical Regulation To Achieve Quantum Dot Nanospheres with a Photoluminescence Quantum Yield Close to 100. *J. Am. Chem. Soc.* **2024**, *146*, 21348–21356. [CrossRef]
49. Li, S.; Tian, R.; Yan, T.; Guo, Y.; Liu, Y.; Zhou, T.-L.; Wang, L.; Xie, R.-J. Small-sized nitride phosphors achieving mini-LED backlights with superhigh brightness and ultralong durability. *Mater. Today* **2023**, *70*, 82–92. [CrossRef]
50. Mantel, C.; Sogaard, J.; Bech, S.; Korhonen, J.; Pedersen, J.M.; Forchhammer, S. Modeling the Quality of Videos Displayed with Local Dimming Backlight at Different Peak White and Ambient Light Levels. *IEEE Trans. Image Process* **2016**, *25*, 3751–3761. [CrossRef] [PubMed]
51. Zhang, S.; Li, Z.; Fang, Z.; Qiu, B.; Pathak, J.L.; Sharafudeen, K.; Saravanakumar, S.; Li, Z.; Han, G.; Li, Y. A high-performance metal halide perovskite-based laser-driven display. *Mater. Horiz.* **2023**, *10*, 3499–3506. [CrossRef] [PubMed]
52. Zhang, H.; Li, H.; Liu, C.; Jiang, H.; Li, J.; Liu, Y.; He, J.; Wang, R.; Hu, W.; Zhu, J. Manipulating cationic ordering toward highly efficient and zero-thermal-quenching cyan photoluminescence. *Chem. Eng. J.* **2024**, *490*, 151727. [CrossRef]
53. Tang, J.; Zhang, X.; Liao, S.; Zhu, Y.; Han, Y.; Su, H.; Qiu, Z.; Lian, S.; Zhang, J. Killing Three Birds with One Stone: Energy Transfer Inducing Efficient, Zero Thermal Quenching, and Emission-Color Tunable Phosphors. *Adv. Opt. Mater.* **2024**, *12*, 2401811. [CrossRef]

54. Zhao, Y.; Riemersma, C.; Pietra, F.; Koole, R.; Donegá, C.d.M.; Meijerink, A. High-Temperature Luminescence Quenching of Colloidal Quantum Dots. *ACS Nano* **2012**, *6*, 9058–9067. [CrossRef] [PubMed]
55. Yang, F.; Li, B.; Li, Y.; Duan, Y.; Ding, Y.; Xiong, Y.; Guo, S. One-step fast fabrication of multi-layer quantum dot diffusion plate for stable display and ultra-long life, a novel quantum dot packaging strategy. *Chem. Eng. J.* **2024**, *481*, 148386. [CrossRef]
56. Liu, Y.; Wang, C.; Chen, G.; Wang, S.; Yu, Z.; Wang, T.; Ke, W.; Fang, G. A generic lanthanum doping strategy enabling efficient lead halide perovskite luminescence for backlights. *Sci. Bull.* **2023**, *68*, 1017–1026. [CrossRef]
57. Wang, Q.; Tong, Y.; Yang, M.; Ye, H.; Liang, X.; Wang, X.; Xiang, W. ZnO induced self-crystallization of $CsPb(Br/I)_3$ nanocrystal glasses with improved stability for backlight display application. *J. Mater. Sci. Technol.* **2022**, *121*, 140–147. [CrossRef]
58. Wang, P.; Wang, B.; Liu, Y.; Li, L.; Zhao, H.; Chen, Y.; Li, J.; Liu, S.F.; Zhao, K. Ultrastable Perovskite-Zeolite Composite Enabled by Encapsulation and In Situ Passivation. *Angew. Chem. Int. Ed. Engl.* **2020**, *59*, 23100–23106. [CrossRef] [PubMed]
59. Delikanli, S.; Yu, G.; Yeltik, A.; Bose, S.; Erdem, T.; Yu, J.; Erdem, O.; Sharma, M.; Sharma, V.K.; Quliyeva, U.; et al. Ultrathin Highly Luminescent Two-Monolayer Colloidal CdSe Nanoplatelets. *Adv. Funct. Mater.* **2019**, *29*, 1901028. [CrossRef]
60. Pan, D.; Wang, Q.; Jiang, S.; Ji, X.; An, L. Synthesis of Extremely Small CdSe and Highly Luminescent CdSe/CdS Core–Shell Nanocrystals via a Novel Two-Phase Thermal Approach. *Adv. Mater.* **2005**, *17*, 176–179. [CrossRef]
61. Deka, S.; Quarta, A.; Lupo, M.G.; Falqui, A.; Boninelli, S.; Giannini, C.; Morello, G.; De Giorgi, M.; Lanzani, G.; Spinella, C.; et al. CdSe/CdS/ZnS double shell nanorods with high photoluminescence efficiency and their exploitation as biolabeling probes. *J. Am. Chem. Soc.* **2009**, *131*, 2948–2958. [CrossRef] [PubMed]
62. Irvine, S.E.; Staudt, T.; Rittweger, E.; Engelhardt, J.; Hell, S.W. Direct light-driven modulation of luminescence from Mn-doped ZnSe quantum dots. *Angew. Chem. Int. Ed. Engl.* **2008**, *47*, 2685–2688. [CrossRef] [PubMed]
63. Kim, J.S.; Kim, S.H.; Lee, H.S. Energy spacing and sub-band modulation of Cu doped ZnSe quantum dots. *J. Alloys Compd.* **2022**, *914*, 165372. [CrossRef]
64. Panda, S.K.; Hickey, S.G.; Demir, H.V.; Eychmuller, A. Bright white-light emitting manganese and copper co-doped ZnSe quantum dots. *Angew. Chem. Int. Ed. Engl.* **2011**, *50*, 4432–4436. [CrossRef]
65. Lim, L.J.; Zhao, X.; Tan, Z.K. Non-Toxic $CuInS_2$/ZnS Colloidal Quantum Dots for Near-Infrared Light-Emitting Diodes. *Adv. Mater.* **2023**, *35*, e2301887. [CrossRef] [PubMed]
66. Zhu, Y.; Lu, X.; Qiu, J.; Bai, P.; Hu, A.; Yao, Y.; Liu, Q.; Li, Y.; Yu, W.; Li, Y.; et al. High-Performance Green and Blue Light-Emitting Diodes Enabled by CdZnSe/ZnS Core/Shell Colloidal Quantum Wells. *Adv. Mater.* **2024**, e2414631. [CrossRef] [PubMed]
67. Wei, F.; Lu, X.; Wu, Y.; Cai, Z.; Liu, L.; Zhou, P.; Hu, Q. Synthesis of highly luminescent CdTe/CdS/ZnS quantum dots by a one-pot capping method. *Chem. Eng. J.* **2013**, *226*, 416–422. [CrossRef]
68. Kirkwood, N.; Monchen, J.O.V.; Crisp, R.W.; Grimaldi, G.; Bergstein, H.A.C.; du Fossé, I.; van der Stam, W.; Infante, I.; Houtepen, A.J. Finding and Fixing Traps in II–VI and III–V Colloidal Quantum Dots: The Importance of Z-Type Ligand Passivation. *J. Am. Chem. Soc.* **2018**, *140*, 15712–15723. [CrossRef]
69. Tynkevych, O.; Karavan, V.; Vorona, I.; Filonenko, S.; Khalavka, Y. Synthesis and Properties of Water-Soluble Blue-Emitting Mn-Alloyed CdTe Quantum Dots. *Nanoscale Res. Lett.* **2018**, *13*, 132. [CrossRef] [PubMed]
70. Gomes, S.I.L.; Costa, J.M.S.; Amorim, M.J.B. Aging in animals–Individuals decline and the impacts on toxicity–Hazard of Cd in Enchytraeus crypticus. *Ecol. Indic.* **2024**, *165*, 112231. [CrossRef]
71. Godt, J.; Scheidig, F.; Grosse-Siestrup, C.; Esche, V.; Brandenburg, P.; Reich, A.; Groneberg, D.A. The toxicity of cadmium and resulting hazards for human health. *J. Occup. Med. Toxicol.* **2006**, *1*, 22. [CrossRef]
72. Clemens, S.; Aarts, M.G.; Thomine, S.; Verbruggen, N. Plant science: The key to preventing slow cadmium poisoning. *Trends Plant Sci.* **2013**, *18*, 92–99. [CrossRef] [PubMed]
73. Baba, H.; Tsuneyama, K.; Yazaki, M.; Nagata, K.; Minamisaka, T.; Tsuda, T.; Nomoto, K.; Hayashi, S.; Miwa, S.; Nakajima, T.; et al. The liver in itai-itai disease (chronic cadmium poisoning): Pathological features and metallothionein expression. *Mod. Pathol.* **2013**, *26*, 1228–1234. [CrossRef]
74. Xu, G.; Zeng, S.; Zhang, B.; Swihart, M.T.; Yong, K.T.; Prasad, P.N. New Generation Cadmium-Free Quantum Dots for Biophotonics and Nanomedicine. *Chem. Rev.* **2016**, *116*, 12234–12327. [CrossRef]
75. Ranjbar-Navazi, Z.; Omidi, Y.; Eskandani, M.; Davaran, S. Cadmium-free quantum dot-based theranostics. *TrAC Trends Anal. Chem.* **2019**, *118*, 386–400. [CrossRef]
76. Babkin, I.A.; Udepurkar, A.P.; Van Avermaet, H.; de Oliveira-Silva, R.; Sakellariou, D.; Hens, Z.; Van den Mooter, G.; Kuhn, S.; Clasen, C. Encapsulation of Cadmium-Free InP/ZnSe/ZnS Quantum Dots in Poly(LMA-co-EGDMA) Microparticles via Co-Flow Droplet Microfluidics. *Small Methods* **2023**, *7*, e2201454. [CrossRef] [PubMed]
77. Jin, L.; Selopal, G.S.; Tong, X.; Perepichka, D.F.; Wang, Z.M.; Rosei, F. Heavy-Metal-Free Colloidal Quantum Dots: Progress and Opportunities in Solar Technologies. *Adv. Mater.* **2024**, *36*, e2402912. [CrossRef] [PubMed]
78. Zaiats, G.; Kinge, S.; Kamat, P.V. Origin of Dual Photoluminescence States in $ZnS–CuInS_2$ Alloy Nanostructures. *J. Phys. Chem. C* **2016**, *120*, 10641–10646. [CrossRef]

79. Liu, P.; Lou, Y.; Ding, S.; Zhang, W.; Wu, Z.; Yang, H.; Xu, B.; Wang, K.; Sun, X.W. Green InP/ZnSeS/ZnS Core Multi-Shelled Quantum Dots Synthesized with Aminophosphine for Effective Display Applications. *Adv. Funct. Mater.* **2021**, *31*, 2008453. [CrossRef]

80. Van Avermaet, H.; Schiettecatte, P.; Hinz, S.; Giordano, L.; Ferrari, F.; Nayral, C.; Delpech, F.; Maultzsch, J.; Lange, H.; Hens, Z. Full-Spectrum InP-Based Quantum Dots with Near-Unity Photoluminescence Quantum Efficiency. *ACS Nano* **2022**, *16*, 9701–9712. [CrossRef] [PubMed]

81. Li, Y.; Hou, X.; Dai, X.; Yao, Z.; Lv, L.; Jin, Y.; Peng, X. Stoichiometry-Controlled InP-Based Quantum Dots: Synthesis, Photoluminescence, and Electroluminescence. *J. Am. Chem. Soc.* **2019**, *141*, 6448–6452. [CrossRef]

82. Zhang, W.; Ding, S.; Zhuang, W.; Wu, D.; Liu, P.; Qu, X.; Liu, H.; Yang, H.; Wu, Z.; Wang, K.; et al. InP/ZnS/ZnS Core/Shell Blue Quantum Dots for Efficient Light-Emitting Diodes. *Adv. Funct. Mater.* **2020**, *30*, 2005303. [CrossRef]

83. Kim, T.G.; Zherebetskyy, D.; Bekenstein, Y.; Oh, M.H.; Wang, L.W.; Jang, E.; Alivisatos, A.P. Trap Passivation in Indium-Based Quantum Dots through Surface Fluorination: Mechanism and Applications. *ACS Nano* **2018**, *12*, 11529–11540. [CrossRef] [PubMed]

84. Choi, Y.; Choi, C.; Bae, J.; Park, J.; Shin, K. Synthesis of gallium phosphide quantum dots with high photoluminescence quantum yield and their application as color converters for LEDs. *J. Ind. Eng. Chem.* **2023**, *123*, 509–516. [CrossRef]

85. Srivastava, V.; Kamysbayev, V.; Hong, L.; Dunietz, E.; Klie, R.F.; Talapin, D.V. Colloidal Chemistry in Molten Salts: Synthesis of Luminescent $In_{1-x}Ga_xP$ and $In_{1-x}Ga_xAs$ Quantum Dots. *J. Am. Chem. Soc.* **2018**, *140*, 12144–12151. [CrossRef] [PubMed]

86. Kim, Y.; Yang, K.; Lee, S. Highly luminescent blue-emitting $In_{1-x}Ga_xP$@ZnS quantum dots and their applications in QLEDs with inverted structure. *J. Mater. Chem. C* **2020**, *8*, 7679–7687. [CrossRef]

87. Gupta, A.; Ondry, J.C.; Lin, K.; Chen, Y.; Hudson, M.H.; Chen, M.; Schaller, R.D.; Rossini, A.J.; Rabani, E.; Talapin, D.V. Composition-Defined Optical Properties and the Direct-to-Indirect Transition in Core-Shell $In_{1-x}Ga_xP$/ZnS Colloidal Quantum Dots. *J. Am. Chem. Soc.* **2023**, *145*, 16429–16448. [CrossRef]

88. Yin, J.; Zhang, J.; Wu, Z.; Wu, F.; Li, X.; Dai, J.; Chen, C. Origin of Water-Stable $CsPbX_3$ Quantum Dots Assisted by Zwitterionic Ligands and Sequential Strategies for Enhanced Luminescence Based on Crystal Evolution. *Small* **2024**, *20*, e2307042. [CrossRef] [PubMed]

89. Yin, J.; Wu, F.; Dai, J.; Chen, C. $CsPbBr_3$@$PbSO_4$ nanocomposites with near-unity photoluminescence and ultrastability via in-water in situ embedding synthesis strategy. *Chem. Eng. J.* **2024**, *499*, 156066. [CrossRef]

90. Quan, L.N.; Quintero-Bermudez, R.; Voznyy, O.; Walters, G.; Jain, A.; Fan, J.Z.; Zheng, X.; Yang, Z.; Sargent, E.H. Highly Emissive Green Perovskite Nanocrystals in a Solid State Crystalline Matrix. *Adv. Mater.* **2017**, *29*, 1605945. [CrossRef] [PubMed]

91. Lian, H.; Zhang, W.; Zou, R.; Gu, S.; Kuang, R.; Zhu, Y.; Zhang, X.; Ma, C.G.; Wang, J.; Li, Y. Aqueous-Based Inorganic Colloidal Halide Perovskites Customizing Liquid Scintillators. *Adv. Mater.* **2023**, *35*, e2304743. [CrossRef]

92. Liu, K.-K.; Liu, Q.; Yang, D.-W.; Liang, Y.-C.; Sui, L.-Z.; Wei, J.-Y.; Xue, G.-W.; Zhao, W.-B.; Wu, X.-Y.; Dong, L.; et al. Water-induced MAPbBr3@PbBr(OH) with enhanced luminescence and stability. *Light Sci. Appl.* **2020**, *9*, 44. [CrossRef] [PubMed]

93. Minh, D.N.; Kim, J.; Hyon, J.; Sim, J.H.; Sowlih, H.H.; Seo, C.; Nam, J.; Eom, S.; Suk, S.; Lee, S.; et al. Room-Temperature Synthesis of Widely Tunable Formamidinium Lead Halide Perovskite Nanocrystals. *Chem. Mater.* **2017**, *29*, 5713–5719. [CrossRef]

94. López-Fernández, I.; Valli, D.; Wang, C.Y.; Samanta, S.; Okamoto, T.; Huang, Y.T.; Sun, K.; Liu, Y.; Chirvony, V.S.; Patra, A.; et al. Lead-Free Halide Perovskite Materials and Optoelectronic Devices: Progress and Prospective. *Adv. Funct. Mater.* **2023**, *34*, 2307896. [CrossRef]

95. Chen, H.; Pina, J.M.; Hou, Y.; Sargent, E.H. Synthesis, Applications, and Prospects of Quantum-Dot-in-Perovskite Solids. *Adv. Energy Mater.* **2021**, *12*, 2100774. [CrossRef]

96. Ahmed, G.H.; El-Demellawi, J.K.; Yin, J.; Pan, J.; Velusamy, D.B.; Hedhili, M.N.; Alarousu, E.; Bakr, O.M.; Alshareef, H.N.; Mohammed, O.F. Giant Photoluminescence Enhancement in $CsPbCl_3$ Perovskite Nanocrystals by Simultaneous Dual-Surface Passivation. *ACS Energy Lett.* **2018**, *3*, 2301–2307. [CrossRef]

97. Zhang, Y.; Cheng, X.; Tu, D.; Gong, Z.; Li, R.; Yang, Y.; Zheng, W.; Xu, J.; Deng, S.; Chen, X. Engineering the Bandgap and Surface Structure of $CsPbCl_3$ Nanocrystals to Achieve Efficient Ultraviolet Luminescence. *Angew. Chem. Int. Ed. Engl.* **2021**, *60*, 9693–9698. [CrossRef] [PubMed]

98. Zhang, C.; Wan, Q.; Ono, L.K.; Liu, Y.; Zheng, W.; Zhang, Q.; Liu, M.; Kong, L.; Li, L.; Qi, Y. Narrow-Band Violet-Light-Emitting Diodes Based on Stable Cesium Lead Chloride Perovskite Nanocrystals. *ACS Energy Lett.* **2021**, *6*, 3545–3554. [CrossRef]

99. Yong, Z.J.; Guo, S.Q.; Ma, J.P.; Zhang, J.Y.; Li, Z.Y.; Chen, Y.M.; Zhang, B.B.; Zhou, Y.; Shu, J.; Gu, J.L.; et al. Doping-Enhanced Short-Range Order of Perovskite Nanocrystals for Near-Unity Violet Luminescence Quantum Yield. *J. Am. Chem. Soc.* **2018**, *140*, 9942–9951. [CrossRef]

100. Das Adhikari, S.; Dutta, S.K.; Dutta, A.; Guria, A.K.; Pradhan, N. Chemically Tailoring the Dopant Emission in Manganese-Doped $CsPbCl_3$ Perovskite Nanocrystals. *Angew. Chem. Int. Ed. Engl.* **2017**, *56*, 8746–8750. [CrossRef] [PubMed]

101. Liu, H.; Wu, Z.; Shao, J.; Yao, D.; Gao, H.; Liu, Y.; Yu, W.; Zhang, H.; Yang, B. $CsPb_xMn_{1-x}Cl_3$ Perovskite Quantum Dots with High Mn Substitution Ratio. *ACS Nano* **2017**, *11*, 2239–2247. [CrossRef]

102. Dong, Y.; Tang, X.; Zhang, Z.; Song, J.; Niu, T.; Shan, D.; Zeng, H. Perovskite Nanocrystal Fluorescence-Linked Immunosorbent Assay Methodology for Sensitive Point-of-Care Biological Test. *Matter* **2020**, *3*, 273–286. [CrossRef]
103. Song, Y.H.; Choi, S.H.; Yoo, J.S.; Kang, B.K.; Ji, E.K.; Jung, H.S.; Yoon, D.H. Design of long-term stable red-emitting CsPb(Br$_{0.4}$, I$_{0.6}$)$_3$ perovskite quantum dot film for generation of warm white light. *Chem. Eng. J.* **2017**, *313*, 461–465. [CrossRef]
104. Liu, F.; Zhang, Y.; Ding, C.; Kobayashi, S.; Izuishi, T.; Nakazawa, N.; Toyoda, T.; Ohta, T.; Hayase, S.; Minemoto, T.; et al. Highly Luminescent Phase-Stable CsPbI$_3$ Perovskite Quantum Dots Achieving Near 100% Absolute Photoluminescence Quantum Yield. *ACS Nano* **2017**, *11*, 10373–10383. [CrossRef] [PubMed]
105. Protesescu, L.; Yakunin, S.; Kumar, S.; Bar, J.; Bertolotti, F.; Masciocchi, N.; Guagliardi, A.; Grotevent, M.; Shorubalko, I.; Bodnarchuk, M.I.; et al. Dismantling the "Red Wall" of Colloidal Perovskites: Highly Luminescent Formamidinium and Formamidinium-Cesium Lead Iodide Nanocrystals. *ACS Nano* **2017**, *11*, 3119–3134. [CrossRef]
106. Gualdrón-Reyes, A.F.; Macias-Pinilla, D.F.; Masi, S.; Echeverría-Arrondo, C.; Agouram, S.; Muñoz-Sanjosé, V.; Rodríguez-Pereira, J.; Macak, J.M.; Mora-Seró, I. Engineering Sr-doping for enabling long-term stable FAPb$_{1-x}$Sr$_x$I$_3$ quantum dots with 100% photoluminescence quantum yield. *J. Mater. Chem. C* **2021**, *9*, 1555–1566. [CrossRef]
107. Tang, Y.; Zhang, X.; Liao, K.; Qiu, L.; Du, N.; Xiao, L.; Ma, J.; Wu, B.; Wu, Z.; Wang, G. Ultra-Stable Perovskite Quantum Dot Polymer Films. *Adv. Opt. Mater.* **2024**. Early View. [CrossRef]
108. Jeon, H.; Wajahat, M.; Park, S.; Pyo, J.; Seol, S.K.; Kim, N.; Jeon, I.; Jung, I.D. 3D Printing of Luminescent Perovskite Quantum Dot–Polymer Architectures. *Adv. Funct. Mater.* **2024**, *34*, 2400594. [CrossRef]
109. Yoon, C.; Yang, K.P.; Kim, J.; Shin, K.; Lee, K. Fabrication of highly transparent and luminescent quantum dot/polymer nanocomposite for light emitting diode using amphiphilic polymer-modified quantum dots. *Chem. Eng. J.* **2020**, *382*, 122792. [CrossRef]
110. Gong, Y.; Shen, J.; Zhu, Y.; Yang, X.; Zhang, L.; Li, C. Stretch induced photoluminescence enhanced perovskite quantum dot polymer composites. *J. Mater. Chem. C* **2020**, *8*, 1413–1420. [CrossRef]
111. Sung, C.-H.; Huang, S.-D.; Kumar, G.; Lin, W.-C.; Lin, C.-C.; Kuo, H.-C.; Chen, F.-C. Highly luminescent perovskite quantum dots for light-emitting devices: Photopatternable perovskite quantum dot–polymer nanocomposites. *J. Mater. Chem. C* **2022**, *10*, 15941–15947. [CrossRef]
112. Yoon, H.C.; Lee, S.; Song, J.K.; Yang, H.; Do, Y.R. Efficient and Stable CsPbBr$_3$ Quantum-Dot Powders Passivated and Encapsulated with a Mixed Silicon Nitride and Silicon Oxide Inorganic Polymer Matrix. *ACS Appl. Mater. Interfaces* **2018**, *10*, 11756–11767. [CrossRef]
113. Peng, X.; Hu, L.; Sun, X.; Lu, Y.; Chu, D.; Xiao, P. Fabrication of High-Performance CsPbBr$_3$ Perovskite Quantum Dots/Polymer Composites via Photopolymerization: Implications for Luminescent Displays and Lighting. *ACS Appl. Nano Mater.* **2022**, *6*, 646–655. [CrossRef]
114. He, S.; Kumar, N.; Beng Lee, H.; Ko, K.-J.; Jung, Y.-J.; Il Kim, J.; Bae, S.; Lee, J.-H.; Kang, J.-W. Tailoring the refractive index and surface defects of CsPbBr$_3$ quantum dots via alkyl cation-engineering for efficient perovskite light-emitting diodes. *Chem. Eng. J.* **2021**, *425*, 130678. [CrossRef]
115. Duncan, T.V.; Bajaj, A.; Gray, P.J. Surface defects and particle size determine transport of CdSe quantum dots out of plastics and into the environment. *J. Hazard. Mater.* **2022**, *439*, 129687. [CrossRef] [PubMed]
116. Kilina, S.V.; Tamukong, P.K.; Kilin, D.S. Surface Chemistry of Semiconducting Quantum Dots: Theoretical Perspectives. *Acc. Chem. Res.* **2016**, *49*, 2127–2135. [CrossRef]
117. Gwak, N.; Shin, S.; Yoo, H.; Seo, G.W.; Kim, S.; Jang, H.; Lee, M.; Park, T.H.; Kim, B.J.; Lim, J.; et al. Highly Luminescent Shell-Less Indium Phosphide Quantum Dots Enabled by Atomistically Tailored Surface States. *Adv. Mater.* **2024**, *36*, e2404480. [CrossRef] [PubMed]
118. Hu, L.; Duan, L.; Yao, Y.; Chen, W.; Zhou, Z.; Cazorla, C.; Lin, C.H.; Guan, X.; Geng, X.; Wang, F.; et al. Quantum Dot Passivation of Halide Perovskite Films with Reduced Defects, Suppressed Phase Segregation, and Enhanced Stability. *Adv. Sci.* **2022**, *9*, e2102258. [CrossRef]
119. Zeng, H. QLED goes to be both bright and efficient. *Sci. Bull.* **2019**, *64*, 464–465. [CrossRef] [PubMed]
120. Li, X.; Hu, B.; Zhang, M.; Wang, X.; Chen, L.; Wang, A.; Wang, Y.; Du, Z.; Jiang, L.; Liu, H. Continuous and Controllable Liquid Transfer Guided by a Fibrous Liquid Bridge: Toward High-Performance QLEDs. *Adv. Mater.* **2019**, *31*, e1904610. [CrossRef] [PubMed]
121. Kim, D.C.; Seung, H.; Yoo, J.; Kim, J.; Song, H.H.; Kim, J.S.; Kim, Y.; Lee, K.; Choi, C.; Jung, D.; et al. Intrinsically stretchable quantum dot light-emitting diodes. *Nat. Electron.* **2024**, *7*, 365–374. [CrossRef]
122. Yoo, J.; Lee, K.; Yang, U.J.; Song, H.H.; Jang, J.H.; Lee, G.H.; Bootharaju, M.S.; Kim, J.H.; Kim, K.; Park, S.I.; et al. Highly efficient printed quantum dot light-emitting diodes through ultrahigh-definition double-layer transfer printing. *Nat. Photonics* **2024**, *18*, 1105–1112. [CrossRef]

123. Kim, T.; Kim, K.H.; Kim, S.; Choi, S.M.; Jang, H.; Seo, H.K.; Lee, H.; Chung, D.Y.; Jang, E. Efficient and stable blue quantum dot light-emitting diode. *Nature* **2020**, *586*, 385–389. [CrossRef] [PubMed]
124. Nomura, M.; Arakawa, Y. Shaking quantum dots. *Nat. Photonics* **2011**, *6*, 9–10. [CrossRef]

Disclaimer/Publisher's Note: The statements, opinions and data contained in all publications are solely those of the individual author(s) and contributor(s) and not of MDPI and/or the editor(s). MDPI and/or the editor(s) disclaim responsibility for any injury to people or property resulting from any ideas, methods, instructions or products referred to in the content.

Review

Application of Perovskite Nanocrystals as Fluorescent Probes in the Detection of Agriculture- and Food-Related Hazardous Substances

Wei Zhao [1,*], Jianguo Zhang [1], Fanjun Kong [3] and Tengling Ye [2,4,*]

[1] Maize Research Institute, Heilongjiang Academy of Agricultural Sciences, Harbin 150086, China; zhangjianguo72@163.com
[2] State Key Laboratory of Luminescent Materials and Devices, South China University of Technology, Guangzhou 510640, China
[3] Harbin Technician College, Harbin 150500, China
[4] School of Chemistry and Chemical Engineering, Harbin Institute of Technology, Harbin 150001, China
* Correspondence: wei825@126.com (W.Z.); ytl@hit.edu.cn (T.Y.)

Abstract: Halide perovskite nanocrystals (PNCs) are a new kind of luminescent material for fluorescent probes. Compared with traditional nanosized luminescent materials, PNCs have better optical properties, such as high fluorescence quantum yield, tunable band gap, low size dependence, narrow emission bandwidth, and so on. Therefore, they have broad application prospects as fluorescent probes in the detection of agriculture- and food-related hazardous substances. In this paper, the structure and basic properties of PNCs are briefly described. The water stabilization methods, such as polymer surface coating, ion doping, surface passivation, etc.; are summarized. The recent advances of PNCs such as fluorescent probes for detecting hazardous substances in the field of agricultural and food are reviewed, and the detection effect and mechanism are discussed and analyzed. Finally, the problems and solutions faced by PNCs as fluorescent probes in agriculture and food were summarized and prospected. It is expected to provide a reference for further application of PNCs as fluorescent probes in agriculture and food.

Keywords: perovskite; nanocrystal; fluorescent probe; agriculture; food

1. Introduction

With the rapid development of Chinese agriculture and food in recent decades, the excessive and indiscriminate use of some hazardous substances (such as pesticides, antibiotics, industrial pollutants, harmful ion pollutants, etc.) is also increasing. As a result, these hazardous substances may cause harm to human beings through agricultural products and food. In order to ensure that the residues of these related hazardous substances do not exceed the standard, fluorescent probes are used as indicators under the excitation of a certain wavelength of light. Through the detection of the generated fluorescence, qualitative or quantitative analysis of the detected substances can be realized. Fluorescent probes play an important role in agriculture and food because of their advantages of simple operation, high sensitivity, and non-destructive visualization [1–3].

In the past few decades, fluorescent probes based on semiconductor quantum dots (QDs) have gained wide attention due to their convenience, speed, low cost, and low limit of detection (LOD). To date, a large number of QD-based fluorescent probes have been developed to detect a variety of agricultural- and food-related contaminants, such as pesticides, harmful ion contaminants, chemical dyes, etc. Currently, QDs are generally composed of IV, II-VI, IV-VI, or II-V elements, such as CdTe, ZnO, and ZnS QDs [4]. However, due to the low quantum yield of these QDs, the sensitivity of fluorescent probes prepared by surface modification is low. Therefore, it is necessary to develop new semiconductor fluorescent probes with excellent photoluminescence (PL) performance and high fluorescence quantum yield. Halide PNCs have become a new promising fluorescent probe material in recent years due to their excellent properties, such as high fluorescence quantum yield, tunable band gap, low size dependence, narrow emission bandwidth, and so on [5,6].

In this paper, as shown in Figure 1, the structure and basic properties of halide PNC luminescent materials and the methods to improve the water stability of PNCs are briefly introduced. On this basis, the research progress of PNCs as fluorescent probes for detecting hazardous substances in the field of agricultural and food is reviewed, and the detection effect and mechanism are discussed and analyzed. Finally, the problems and solutions faced by PNCs as fluorescent probes in agriculture and food were summarized and prospected. It is expected to provide a reference for further application of PNCs as fluorescent probes in agricultural and food.

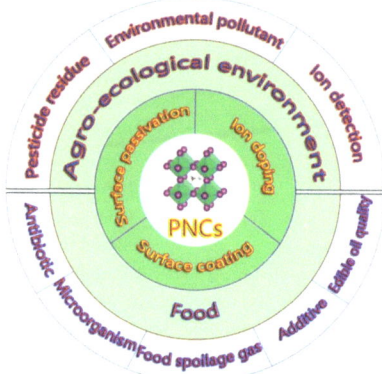

Figure 1. Schematic illustration of the outline.

1.1. Structure and Basic Properties of PNCs

Halide perovskite is a new type of semiconductor material with the molecular formula ABX_3, where A mainly represents alkali metal cation cesium (Cs^+), methylamine ion (MA, $CH_3NH_3^+$), or formamidine ion (FA, $CH(NH_2)_2^+$); B represents metal cations such as Pb^{2+}, Sn^{2+}, etc.; and X stands for the halide anion (Cl, Br, I, or their combination). As shown in Figure 2, the lattice structure of perovskite, such as $CaTiO_3$, has a three-dimensional (3D) angular shared $[BX_6]$ octahedron, which is formed by the coordination of the B site in the center and six X site anions. Each octahedron is connected through the shared angle to form an extended 3D network structure, while A is embedded in the gap of the network structure [7]. Halide perovskite can have three different crystalline structures: rhombic, tetragonal, and cubic phases. The cubic phase can exist stably at high temperatures [8]. When A is an organic group such as MA or FA, it is called an organic–inorganic hybrid perovskite, whereas when A is Cs^+, it is called an all-inorganic perovskite.

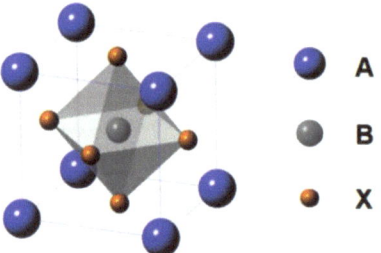

Figure 2. Typical ABX_3 perovskite crystal structure. Reproduced from ref. [7] Copyright (2019), with permission from Royal Society of Chemistry.

However, the low exciton binding energy of bulk phase perovskite materials makes carriers easy to be trapped by defects in the lattice, leading to slow radiation recombination of carriers in bulk phase perovskite. To address this problem, efforts have been made to reduce the size of perovskite crystals or reduce its structural dimension to enhance carrier restriction within the perovskite lattice [9,10]. Nano-structured perovskites, such as PNCs, perovskite nanowires, and perovskite nanosheets, can improve the photoluminescent quantum yield (PLQY) of materials due to the quantum domain effect. Among them, PNCs have three dimensions below 100 nm, and perovskite quantum dots (PQDs) have three dimensions below 20 nm, exhibiting a clear quantum limiting effect and high PLQY, which is the most widely studied. The band gap and color of halide perovskite can be continuously adjusted from near-infrared to the violet region by modifying the halide elements and their respective composition, covering the entire visible spectral region. Moreover, the narrow spectral distribution of PNCs is crucial for easy detection in fluorescent probes. Numerous simple and effective synthetic methods are available for obtaining PNCs, such as hot injection [11], room-temperature reprecipitation [12], solvothermal synthesis [13], microwave method [14], and ultrasonication [15], among others.

1.2. Methods to Improve the Stability of PNCs

Although PNCs possess excellent properties, as ionic crystals they have a large surface energy and are susceptible to external influences, leading to instability and degradation of their structures. For instance, water, light radiation, and temperature changes can all cause fluorescence quenching and structural degradation [16]. In applications such as agriculture and food, where hazardous substance detection requires a water environment, stability in water is crucial for PNCs to function effectively as fluorescent probes. Fortunately, recent studies have developed a range of strategies to address the limitations of PNCs, including surface coating, metal ion doping, and surface passivation.

1.2.1. Surface Coating by Polymer, Inorganic Oxide, and Porous Materials

The surface of the PNCs is coated and isolated from the environment, which greatly improves the stability of the PNCs, especially the water resistance. At present, PNCs are coated with polymer, inorganic oxide, and porous materials. The polymer forms a dense network on the NC surface, which can effectively prevent environmental erosion. Researchers usually dissolve or expand the polymer in toluene, N, N-dimethylformamide, and other solvents, realize the encapsulation protection of PNCs through in situ growth or adsorption, and then remove the solvent under vacuum and high temperature to obtain polymer–PNCs composite materials [17,18]. As shown in Figure 3, Xuan et al. used divinylbenzene, ethyl acetate, and azodiisobutyronitrile (AIBN) as reaction raw materials to prepare superhydrophobic organic polymers through solvothermal methods. High-quality PQDs were prepared by hot injection, and the cyclohexane solution of PQDs was mixed with the organic polymer to fully absorb the QDs into the pores on the surface of the polymer, forming a complex (CPB@SHFW). Fluorescence quantum efficiency remained at 91% after immersion in water for one month, and the contact angle of the complex was maintained at 150°. The pores of the polymer provided a superhydrophobic environment to improve the water stability of the QDs, while also acting as a framework to maintain the excellent luminous performance of PQDs. This study successfully achieved the preparation of "water-resistant" perovskite QDs [19].

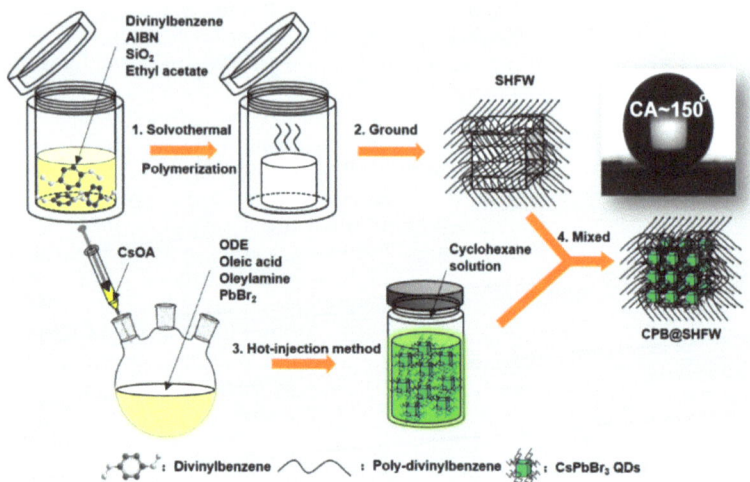

Figure 3. Schematic illustration for preparation of CPB@SHFW composites. Reproduced from ref. [19] Copyright (2019), with permission from American Chemical Society.

Although the hydrophobic strategy used to coat PNCs has improved their stability, the application of fluorescent probes often requires PNCs to be dispersed in a water environment. To address this challenge, the introduction of amphiphilic polymers that are both hydrophilic and hydrophobic has been shown to be an effective solution. Avugadda et al. reported the creation of $CsPbBr_3$ NCs in capsules that maintain a PLQY of approximately 60% for over two years in water, using a few scalable fabrication steps that involve an automated routine. For this purpose, an encapsulating amphiphilic polymer, Polystyrene-block-poly(acrylic acid) (PS-b-PAA), with low molecular weight (34 kDa) was selected. As illustrated in Figure 4, this polymer, in the presence of NCs in toluene and with the addition of methanol (MeOH) in a single-phase system, facilitates the formation of polymer capsules containing the NCs. This method does not require prior surface modification of the as-synthesized NCs and is insensitive to their surface coating. $CsPbBr_3$@PS-b-PAA is soluble in water, and its emission remains stable in a saline solution [20].

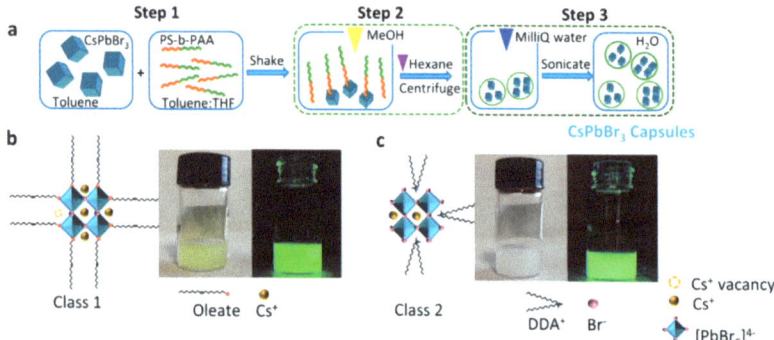

Figure 4. (**a**) Schematic illustration of the capsules embedding the $CsPbBr_3$ NCs. (**b**,**c**) Photographs of aged samples of (**b**) Cs-oleate and (**c**) DDAB-coated $CsPbBr_3$ NCs dispersed in water under normal and UV light. Reproduced from ref. [20] Copyright (2022), with permission from American Chemical Society.

Compared to polymers, inorganic nano oxide coatings can also play a role in stabilizing PNCs, with the advantage of better water dispersion. Nano SiO_2 has become a typical representative of inorganic materials used for coating PNCs due to its high heat resistance, stability, and transparency. Li et al. have developed a simple method for producing water-resistant PQDs@SiO_2 nanodots (wr-PNDs) that can be easily dispersed in water and maintain their emission for up to six weeks, demonstrating an unprecedented water-resistance capability. The synthetic process for creating wr-PNDs is illustrated in Figure 5. Initially, pristine PQDs are dispersed in a mixed solvent of toluene and (3-mercaptopropyl)trimethoxysilane ($C_6H_{15}O_3Si$-SH, MPTMS), followed by the addition of de-ionized water to initiate the hydrolysis of MPTMS. MPTMS molecules can be tightly adsorbed onto the PQDs due to the formation of Pb-S bonding. These bound molecules reduce the interfacial energy between the PQDs and silica by providing additional steric stabilization, thereby preventing phase separation between the PQDs and silica. Silica-encapsulated PQDs-Pb-S-SiO_2-SH nanodots are produced as hydrolysis and condensation of MPTMS are triggered by the added water [21].

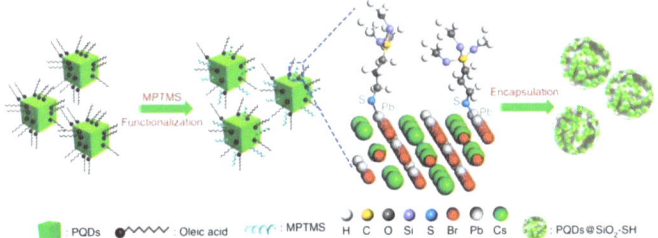

Figure 5. Synthesis of wr-PNDs by encapsulating into SiO_2. Reproduced from ref. [21] Copyright (2020), with permission from Springer Nature.

Porous materials provide a high degree of porosity that can be used to prevent self-aggregation of PNCs and reduce environmental erosion, making it another viable method for enhancing the stability of PNCs coated with inorganic materials. A universal synthesis method of PNCs with pore domain limiting-shell isolation dual protection was developed by You et al.; as depicted in Figure 6. In this approach, perovskite precursor was injected into mesoporous silicon, and PNCs were grown in situ in the mesoporous silicon pores through heating and annealing. Next, dense Al_2O_3 shells were deposited on the mesoporous silicon surface through atomic deposition technology, which effectively isolates the impact of the environment on PNCs, owing to the abundant -OH on the mesoporous silicon surface. The resulting $mSiO_2$-$CsPbBr_3$@AlO_xNCs can remain stably dispersed in water for more than 90 days and can effectively shield the ion exchange effect, thereby significantly enhancing the stability of PNCs and expanding their practical application range [22]. Likewise, Zhang et al. synthesized ceramic-like stable and highly luminescent $CsPbBr_3$ NCs by embedding $CsPbBr_3$ in silica derived from molecular sieve templates at high temperature (600–900 °C). The resulting $CsPbBr_3$-SiO_2 powder exhibits high PLQY (~71%) and excellent stability, which can be compared with the ceramic Sr_2SiO_4:Eu^{2+} green phosphor [23].

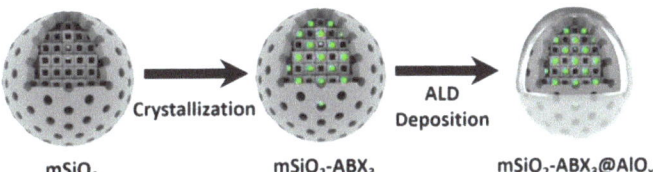

Figure 6. Synthetic procedure of the $mSiO_2$-ABX_3@AlO_x (A = MA, FA, Cs, B = Pb, Mn, X = Cl, Br, I) NCs. Reproduced from ref. [22] Copyright (2020), with permission from Elsevier.

Metal–organic framework compounds (MOFs) are a novel class of porous materials that are assembled by inorganic metals and organic ligands. Due to their high porosity and large specific surface area, MOFs are often utilized as carriers for nanoparticles, drugs, and other materials. Incorporating PNCs into the pores of MOFs can significantly enhance their stability. However, in order to function as a carrier for PNCs, the MOF design must address several challenges. Firstly, the MOF matrix must remain stable during PNCs incorporation and fluorescence detection. As many bivalent metal-based MOFs are sensitive to water or ambient moisture, high-valent metal-based MOFs are ideal hosts that provide stable pore structures for encapsulation. Secondly, the aperture of MOFs must be sufficiently large to accommodate PNCs. While micropore MOFs with apertures smaller than 2 nm are frequently used as encapsulation materials for perovskite nanoparticles, the limited space can hinder QD growth and damage the micropore structure. Therefore, mesoporous MOFs with pore sizes greater than 4 nm are more suitable for PNC fixation. In the case of $CsPbBr_3$ nanocrystals, QDs with complete crystal structures are more suitable for fluorescence detection applications to ensure efficient light absorption and prevent dense surface recombination. Simultaneously, the mesoporous cage can effectively separate the QDs and realize the quantum limited domain of perovskite. As illustrated in Figure 7, Qiao et al. encapsulated $CsPbBr_3$ nanocrystals into a stable ferric organic skeleton (MOF) with mesoporous cages (5.5 and 4.2 nm) via a sequential deposition pathway to obtain a perovskite–MOF composite CSPbBr$_3$@PN-333 (Fe). The $CsPbBr_3$ nanocrystals are stabilized by the restriction of the MOF cage without aggregation or leaching [24]. Wu et al. employed a sequential deposition method to encapsulate $MAPbI_3$ in cheap iron-based MOFs (PSN-221) to obtain the composite $MAPbI_3$@PCN-221(Fe$_x$) (x = 0–1), which they used for photocatalytic CO_2 reduction. Due to the protection offered by PSN-221 (Fe$_x$), this composite catalytic system exhibits ultra-high stability and can operate continuously in water for more than 80 h. Additionally, the maximum efficiency of photocatalytic CO_2 is 38 times higher than that of pure PCN-221(Fe$_x$) [25].

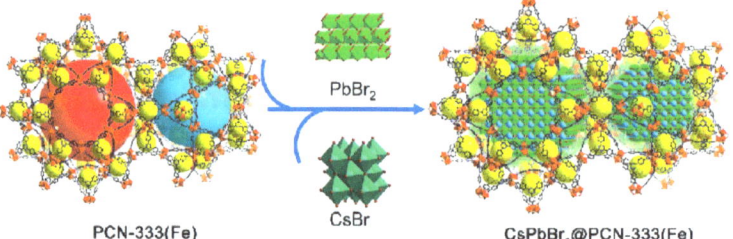

Figure 7. Schematic illustration of the preparation of CsPbBr$_3$@PCN-333(Fe). Reproduced from ref. [24] Copyright (2021), with permission from American Chemical Society.

1.2.2. Ion Doping

Ion doping is a widely used and effective method for introducing heteroatoms into the lattice, making it one of the most important approaches for modifying semiconductors. This method is valuable because it can maintain the structure and properties of the original main crystal to a large extent. Perovskite structures have a high tolerance for impurity doping, enabling various impurity atoms to be doped into the lattice of PQDs. This doping strategy can significantly improve the stability of perovskite and regulate its photoelectric characteristics [26]. The typical perovskite structure (ABX_3) consists of corner-sharing $[BX_6]^{4-}$ octahedra, with A-site cations confined within the cuboctahedra cages. In the ABX_3 structure, the ionic radii of A-site, B-site, and X-site ions must correspond to the Goldschmidt tolerance factor t [27]:

$$t = \frac{r_A + r_X}{\sqrt{2}(r_B + r_X)}$$

The Goldschmidt tolerance factor is defined by the formula $t = (r_A + r_X)/\sqrt{2}(r_B + r_X)$, where r_A, r_B, and r_X are the ionic radii of the A-site, B-site, and X-site elements, respectively [26]. To maintain a stable perovskite structure, t should fall between 0.813 and 1.107, with a range of $0.9 \leq t \leq 1$ generally considered optimal for the cubic phase [7]. The $[BX_6]^{4-}$ framework confines the A-site cations, and those that are either too large or too small can cause the cubic-like crystal structure to distort, warp, and eventually break down. MA (2.70 Å) is generally the most suitable size for the cubic perovskite structure, whereas FA (2.79 Å) and Cs (1.88 Å) are slightly too large or small to serve as ideal A-sites [28], leading to deviations from the equilibrium interplanar distances. A mixed A-site cation system that combines larger and smaller ions can help to stabilize the 3D perovskite framework and overcome these issues. For instance, Protesescu et al. replaced some Cs^+ ions with FA^+ ions to improve the structural stability of $CsPbI_3$ PQDs, resulting in a higher t value relative to pure $CsPbI_3$ PQDs. The resulting $FA_{0.1}Cs_{0.9}PbI_3$ PQDs display a 3D orthogonal structure, stable red-light emission (685 nm), a PLQY over 70%, and can maintain their PLQY in solution (with less than 5% relative decrease) after several months of storage at ambient conditions, owing to the restricted transformation from 3D cubic perovskite structure to 1D polycrystals [29]. Triple cations such as (Cs, FA)PbI_3 or (Cs, MA, FA)PbI_3 have also demonstrated improved stability by promoting phase structure stability, and multi-cationic doping of A sites has emerged as a key method for obtaining high-quality perovskite films [30]. However, it is rarely reported in the field of PNCs [31]. In addition, Wang et al. synthesized ultra-stable quasi-2D PQDs by introducing butadiamine cation (BA^+) into methylamine lead bromide perovskite ($MAPbBr_3$), as shown in Figure 8. The quasi-2D perovskite $(BA)_2(MA)_{x-1}Pb_xBr_{3x+1}$, which has reduced dimensionality, exhibits higher luminous efficiency and better environmental stability than traditional 3D perovskite due to the higher exciton binding energy and formation energy of the reduced dimension perovskite [32].

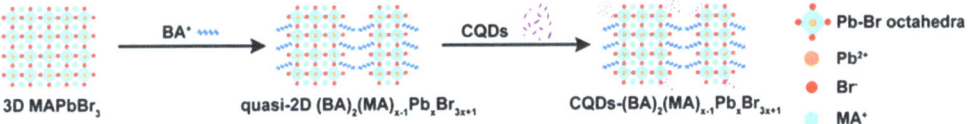

Figure 8. Schematic representation of metal halide perovskites with 3D and quasi-2D structures. Reproduced from ref. [32] Copyright (2021), with permission from Royal Society of Chemistry.

The incorporation of dopants to replace B-site Pb^{2+} in perovskite structures has been shown to significantly enhance their stability or modify their optical properties, examples of which include Zn^{2+}, Ni^{2+}, and RE ions [33,34]. The smaller ionic radius of Zn^{2+} (0.74 Å) compared to Pb^{2+} (1.19 Å) causes the perovskite lattice to contract, leading to an increase in exciton binding energy and promoting radiative recombination within $CsPbI_3$ PQDs [35]. In addition, Mn^{2+} doping also increases the Goldschmidt tolerance factor (t), enhancing structural stability [36]. Similarly, the introduction of Ni^{2+} ions into $CsPbI_3$ PQDs results in lattice contraction due to the smaller ionic radius of Ni^{2+} (≈ 0.69 Å) relative to Pb^{2+} (≈ 1.19 Å), resulting in a blue shift in the PL and a higher quantum efficiency, as shown in Figure 9a,b. Moreover, Ni^{2+} doping improves the stability of PNCs by increasing t relative to $CsPbI_3$ PQDs, leading to a more stable perovskite structure. The stability of Ni:$CsPbBr_3$ PNCs with distinct Ni/Pb ratios against moisture and UV light was tested and shown in Figure 9c,d, providing compelling evidence for the beneficial effects of Ni^{2+} dopants on emission performance [37].

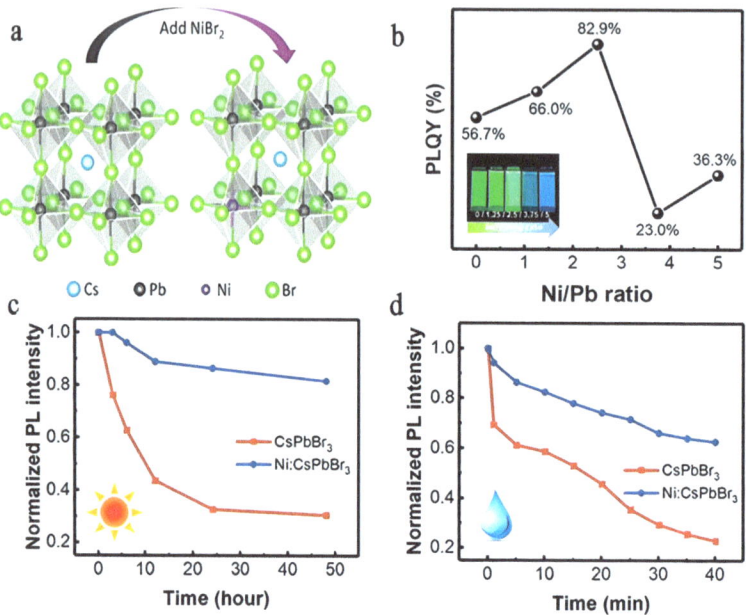

Figure 9. (**a**) Schematic of CsPbBr$_3$ PNCs with and without Ni^{2+} substitution. (**b**) PLQY with Ni/Pb feed molar ratios of 0, 1.25, 2.5, 3.75, and 5; insets show photographs of Ni:CsPbBr$_3$ PNCs in hexane under 365 nm UV irradiation. Normalized PL intensity of CsPbBr$_3$ and Ni:CsPbBr$_3$ PNCs (Ni/Pb = 2.5), (**c**) under constant UV radiation (365 nm, 4 W), and (**d**) in deionized water. Reproduced from ref. [37] Copyright (2021), with permission from Wiley-VCH.

1.2.3. Surface Passivation

The large specific surface area of PNCs results in a high concentration of surface atoms, which in turn leads to an abundance of dangling bonds and defects. These defects can trap electrons in the conduction band and cause non-radiative energy loss, thereby reducing the stability of nanocrystals. Surface passivation using appropriate ligands is an effective strategy to improve the stability of PNCs. However, commonly used ligands like oleic acid and oleylamine are dynamic and can easily desorb in the purification process, leading to PNC aggregation and loss of optical properties. Thus, ligands that bind more firmly to the surface of PNCs are required for better passivation, reducing the impact of water and oxygen and improving stability. Polydentate ligands, containing multiple functional groups, are shown to bind more strongly to nanocrystals and provide improved stability. For instance, Figure 10 illustrates a strategy for introducing polydentate ligand of AHDA to synthesis stable PNCs. The polyamine chelating ligand N′-(2-aminoethyl)-N′-hexadecylethane-1, 2-diamine (AHDA) can anchor the PNC surface lattice with a higher binding energy (2.36 eV) than that of the commonly used oleylammonium ligands (1.47 eV). The strong chelation effect achieves ultra-stable CsPbI$_3$ PNCs with a high binding energy, exhibiting excellent stability even under harsh conditions such as repeated purification, polar solvents, heat, and light [38]. The use of mercaptan materials such as 2-aminoethylmercaptan (AET) can also lead to tightly bound ligand layers on the surface of PNCs, resulting in high stability, even in water. The AET-CsPbI$_3$ QDs (in both solutions and films) retained more than 95% of the initial PL intensity in water after 1 h [39]. These strong chelating ligand strategies offer promising avenues for achieving stable and high-performance PNCs for a range of applications.

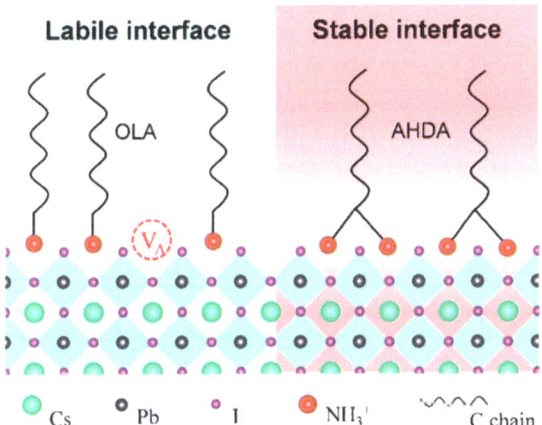

Figure 10. Multi-amine AHDA ligands stabilize the CsPbI$_3$ surface. Reproduced from ref. [38] Copyright (2022), with permission from American Chemical Society.

In recent years, zwitterionic organic ligands have emerged as a promising approach for surface engineering of PNCs and defect elimination, owing to their multifunctional zwitterionic structure. As shown in Figure 11, Krieg et al. proposed a surface passivation strategy for PQDs using long-chain zwitterionic ligands such as sulfobetaine and phosphocholine, which comprise both cationic and anionic groups. These zwitterionic ligands can coordinate with both surface cations and anions and are firmly attached to the surface of PQDs through the chelation effect, thereby obtaining PQDs with high stability. For example, 3-(N,Ndimethyloctadecylammonio) propanesulfonate can result in much improved chemical durability. The sulfobetaine-capped CsPbBr$_3$ NCs retained PLQYs in the range of 70–90% for 28–50 days under ambient conditions [40]. Moreover, they demonstrated that the strong binding between the amphoteric lecithin and PQDs provides CsPbBr$_3$ PQDs solutions with superior colloidal stability over a wide range of concentrations [41]. In another study, Wang et al. used polymers containing multiple sulfobetaine amphoteric groups to achieve desirable surface coatings and CsPbBr$_3$ PQDs that maintain bright PL emission even after 1.5 years of storage in polar solvents such as acetone and ethanol [42].

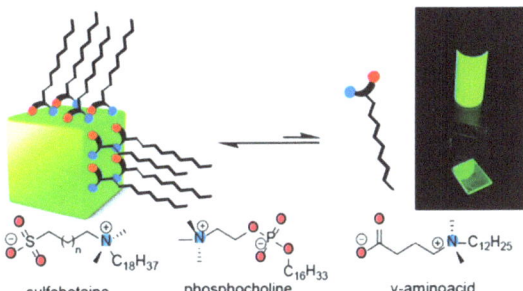

Figure 11. Surface engineering of PNCs with zwitterionic organic ligands, including 3-(N,N-dimethyloctadecylammonio) propanesulfonate (n = 1), N-hexadecylphosphocholine, and N,N-dimethyldodecylammoniumbutyrate. Reproduced from ref. [39] Copyright (2019), with permission from Wiley-VCH.

In addition, the presence of long alkyl chain organic ligands in PQD films severely impedes charge transport and interlayer coupling, calling for the development of more conductive organic semiconductor ligands to prepare PNCs with favorable optoelectrical properties. Pan et al. demonstrated that the incorporation of organic semiconductor molecules (rhodamine B derivative, COM) into $CsPbBr_3$ NCs resulted in COM-$CsPbBr_3$ NCs with a high PLQY of 82% and exceptional stability under harsh commercial accelerated operating stability tests such as high temperature (85 °C) and high humidity (85%). The product exhibits remarkable endurance against high temperature and high humidity, retaining 84% of the initial PL intensity value for 300 h. The stability of $CsPbBr_3$ NCs can be significantly improved by the strong interaction between organic semiconductor molecules and the quasi-type II heterostructure formed by $CsPbBr_3$ and moderate photocarrier transfer. Furthermore, the diversity and versatility of organic semiconductors present a broad prospect for their combined application in the field of luminescence and display [43].

Table 1 summarizes methods to improve the stability of PNCs and their stability performance. Surface coating is a way from outside to inside, playing the main role of encapsulation. Metal ion doping and surface passivation are ways to improve the intrinsic structural stability from inside to outside. They are not absolute by classification and may work together to improve the stability of PNCs sometimes.

Table 1. Summary of methods to improve the stability of PNCs and their stability performance.

Methods	Modification Material		Water-Stabilized PNCs	Stability of PNCs	Ref.
Surface coating	Polymer	Divinylbenzene, ethyl acetate, and AIBN	CPB@SHFW	PLQY remained at 91%, 1 month in water	[19]
		PS-b-PAA	$CsPbBr_3$@PS-b-PAA	PLQY remained at 60%, 2 years in water	[20]
	Inorganic oxide	$C_6H_{15}O_3Si$-SH	PQDs-Pb-S-SiO_2-SH	PLQY remained at 80%, 13 h in water	[21]
	Porous materials	Mesoporous silicon	$mSiO_2$-$CsPbBr_3$@AlO_x	Over 20% PL emission intensity, 90 days in water	[22]
		Ferric organic skeleton MOFs	$CsPbBr_3$@PN-333 (Fe)	The Li–O_2 battery can be cycled stably for more than 200 h at 0.01 mA cm^{-2} under illumination	[24]
			$MAPbI_3$@PCN-221(Fe_x)	The catalytic system can operate continuously in water for more than 80 h	[25]
Metal ion doping	A site	Cs^+ ions with FA^+	$FA_{0.1}Cs_{0.9}PbI_3$ PQDs	PLQY remained at 95%, several months in solution	[29]
	B site	Ni^{2+}	Ni:$CsPbBr_3$ PNCs	The stability of Ni:$CsPbBr_3$ PNCs are higher than $CsPbBr_3$ PNCs	[37]
Surface passivation	Strong chelating ligands	The polyamine chelating ligand: AHDA	AHDA-$CsPbI_3$ PNCs	PLQY remained at 60%, after 15 purification cycles	[38]
		SH ligands: 2-aminoethylmercaptan (AET)	AET-$CsPbI_3$-QDs	PLQY remained at 95%, 1 h in water	[39]
	Zwitterionic organic ligands	Sulfobetaine or phosphocholine	Sulfobetaine-capped $CsPbBr_3$ NCs	PLQY remained at 70–90% for 28–50 days under ambient conditions	[40]
	Organic semiconductor ligands	Rhodamine B derivative (COM)	COM-$CsPbBr_3$ NCs	84% of the initial PL intensity, 300 h under HT 85 °C and HH 85%	[43]

2. Application of PNCs as Fluorescent Probes in the Detection of Agriculture- and Food-Related Hazardous Substances

2.1. The Detection of Agriculture-Related Hazardous Substances

2.1.1. Pesticide Residue Detection

Pesticides are crucial in minimizing crop losses, ensuring food security, preventing the emergence and spread of plant diseases and insect pests, and improving health conditions. However, their use also leads to environmental pollution, pesticide residues, and other issues. Pesticide residues refer to the presence of trace amounts of pesticide precursor, toxic metabolites, degradants, and impurities that remain in grains, vegetables, fruits, livestock products, aquatic products, soil, and water bodies even after the application of pesticides in agricultural production. Currently, the annual production of chemical pesticides worldwide is nearly 4.1 million tons, and these synthetic compounds were utilized as pesticides, fungicides, algaecides, deworming agents, defoliants, and other pesticides [44]. The substantial use of pesticides, particularly organic pesticides, has resulted in severe pesticide pollution. Ingesting food containing high levels of highly toxic pesticide residues can lead to acute poisoning in both humans and animals. While long-term consumption of agricultural products with excessive pesticide residues may not cause acute poisoning, it can result in chronic poisoning in humans, leading to the development of diseases, inducing cancer, and even affecting future generations [45]. Therefore, the real-time and rapid detection of pesticide residues in agricultural products is of great significance for global food safety and environmental safety. Researchers have conducted significant research on rapid detection technologies for pesticide residues, especially for organophosphorus, organochlorine, triazine herbicides, and other pesticides, yielding some remarkable achievements.

Organophosphorus Pesticide

Organophosphorus pesticides are a type of pesticide that contains organic compounds with phosphorus. They are widely used in agricultural production to control plant diseases and insect pests but can leave residues in crops. Ensuring food safety requires rapid and accurate detection of organophosphorus pesticides.

Dimethoate oxide (OMT) is an organophosphorus insecticide with a strong imbibition insecticidal effect. It can be absorbed into the plant stem and leaves and transmitted to other parts of the plant to kill harmful pests. Huang et al. successfully synthesized a new MIPs@CsPbBr$_3$ QDs composite using imprinting technology with a sol–gel reaction, as shown in Figure 12. APTES-capped CsPbBr$_3$ QDs were used as a fluorescent carrier, and TMOS was used as the cross-linker to gradually form a silica matrix, improving the stability of perovskites, and synthesized water-soluble PQDs. OMT was selected as the template molecule due to its hydrogen bonding interaction with APTES. Loading the MIPs with OMT significantly quenched the fluorescence of MIPs@CsPbBr$_3$ QDs, with a linear range of OMT from 50 to 400 ng/mL and a LOD of 18.8 ng/mL. After removing the OMT template via solvent extraction, the resulting MIP@CsPbBr$_3$ QD composites showed selective recognition ability towards specific template-shaped molecules such as dimethoate, with an imprinting factor of 3.2, indicating excellent specificity of the MIPs for inorganic metal halide (IMH) perovskites [4].

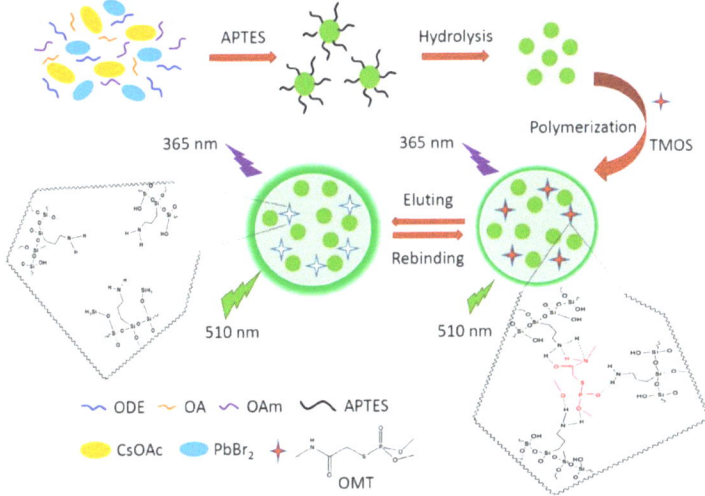

Figure 12. Schematic illustration of the preparation of the MIPs@CsPbBr$_3$ QD sensor. Reproduced from ref. [4] Copyright (2018), with permission from American Chemical Society.

2,2-dichlorovinyl dimethyl phosphate (DDVP) is a highly effective and wide-spectrum organophosphorus insecticide commonly used to control various pests on crops such as cotton, fruit trees, vegetables, sugarcane, tobacco, tea, and mulberry. However, on 27 October 2017, the World Health Organization's International Agency for Research on Cancer classified DDVP as a group 2B carcinogen. In a study by Huang et al.; molecularly imprinted mesoporous silica (MIMS) with in situ perovskite $CsPbBr_3$ quantum dots (QDs) was prepared and applied for the fluorescence detection of DDVP, as shown in Figure 13. The QDs stability was greatly improved by encapsulating them in mesoporous silica, resulting in higher analytical performance. The design of PQDs-encapsulated MIMS is based on PQDs being encapsulated into mesoporous silica and acting as a transducer for recognition signal fluorescent readout, while molecular imprinting on the surface of mesoporous silica provides selectivity. $CsPbBr_3$ QDs with uniform size and high fluorescence intensity were first grown directly in the channels of mesoporous silica SBA-15. Then, molecular imprinting was applied to the surface of QDs-encapsulated mesoporous silica using a sol–gel method. The mesoporous silica provided a solid support for molecular imprinting and acted as a medium for the in situ synthesis of QDs. The QDs' fluorescent properties were expected to be affected once the template molecules were captured in the binding sites, reducing the distance between the template and the QDs. Under optimized conditions, the QDs-encapsulated MIMS showed a linear relationship between DDVP concentration of 5–25 μg/L, and the LOD was 1.27 μg/L. The recovery of DDVP increased from 87.4% to 101% [46].

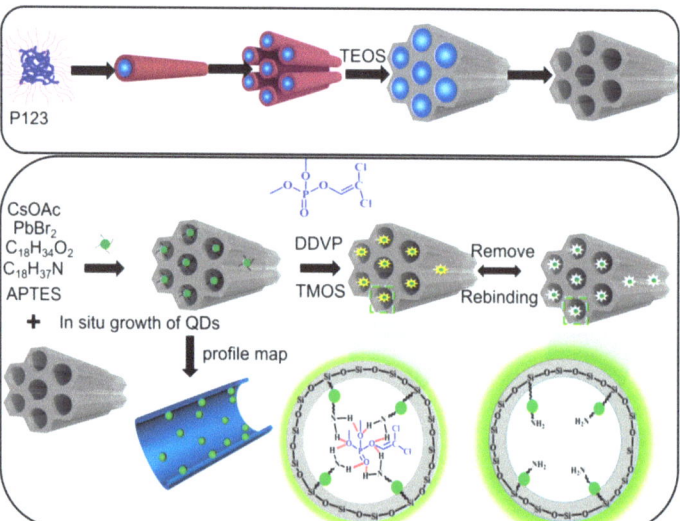

Figure 13. Schematic illustration of the preparation of QDs-encapsulated MIMS. Reproduced from ref. [46] Copyright (2020), with permission from Elsevier.

Phoxim is a highly effective broad-spectrum organophosphorus insecticide, exhibiting properties of stomach toxicity, contact pesticide, and long-lasting effect under dark conditions. To detect phoxim specifically and sensitively in samples, Tan et al. synthesized perovskite $CsPbBr_3$ QDs embedded in molecularly imprinted polymers (MIP) by thermal injection. As depicted in Figure 14, $CsPbBr_3$ QDs with high-brightness cores were encapsulated by MIPs using a sol–gel approach initiated by trace moisture in the air. To overcome the disadvantage of using siloxane monomers with simple organic groups, a multifunctional monomer, BUPTEOS, was synthesized with phenyl and hydrogen bond donors. During pre-polymerization, BUPTEOS and phoxim formed a complex through hydrogen bond interactions and π-π accumulation, leading to the formation of specific imprinted cavities in the MIP/QDs composites upon slow sol-gel molecular imprinting process and removal of templates. The resulting MIP/QDs complex demonstrated excellent selectivity to phoxim, with a blotting factor of 3.27. Compared to previous studies on detection of organophosphorus pesticides, the MIP/QDs fluorescent probe exhibited high sensitivity and specificity. Under optimized conditions, the fluorescence quenching of MIP/QDs showed a good linear correlation with the concentration range of 5–100 ng/mL, and the LOD was 1.45 ng/mL. By encapsulating the QDs of the novel MIP/perovskite CsPbBr$_3$ QDs fluorescent probe in the imprinted silica matrix, the stability of the QDs in phoxim detection was significantly improved. In this study, siloxane functional monomers with two functional groups were synthesized by in situ slow hydrolysis of organosilicon monomers [47].

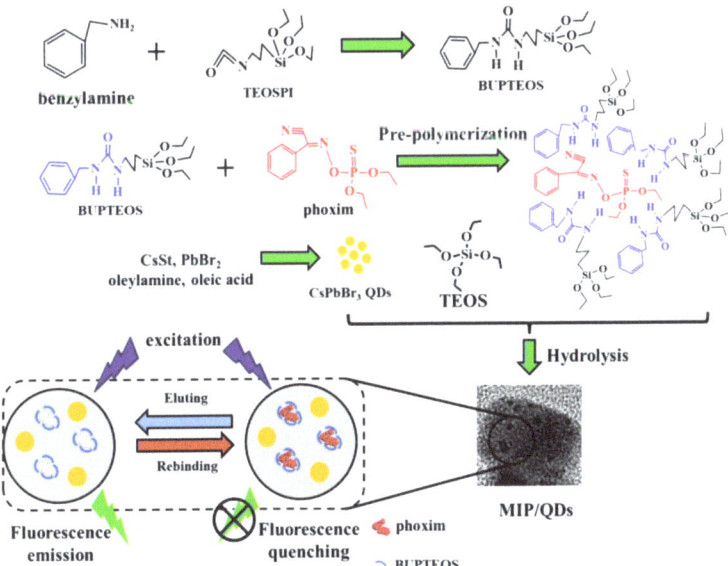

Figure 14. Schematic illustration of the preparation of MIP/CsPbBr$_3$ QDs composites. Reproduced from ref. [47] Copyright (2019), with permission from Elsevier.

Triazine Herbicides

Triazine herbicides can be classified into symmetrical triazines (1,3,5-triazine) and asymmetric triazines (1,2,4-triazine). The main commercial herbicide is symmetrical triazine, which can be further divided into thiomethyl-s-triazine (e.g., prometryn, ametryn, and terbutryn), simazine, terbuthylazine, fluoroalkyl-s-triazine (e.g., indaziflam and triaziflam), and methoxytriazine (e.g., prometon and terbumeton) [48].

Prometryn is commonly used for pre- and post-bud weeding in cotton and bean fields as well as for algae removal in aquatic products. Zhang et al. developed a novel molecularly imprinted electrochemical luminescence (MIECL) sensor by using perovskite quantum dots (PQDs) coated with a molecularly imprinted silica layer (MIP/CsPbBr$_3$-QDs) as recognition and response elements. As shown in Figure 15, the MIP/CsPbBr$_3$-QDs layer was immobilized onto the surface of a glassy carbon electrode (GCE) via one-pot electropolymerization to enhance water resistance, improve the stability of CsPbBr$_3$-QDs, and reduce film shedding. The highly selective and ultrasensitive MIECL sensor was successfully constructed for prometryn analysis in fish and seawater samples. The LODs for fish and seawater samples were 0.010 µg/kg and 0.050 µg/L, respectively. The recoveries of fish and seawater samples were 88.0% to 106.0%, with a relative standard deviation lower than 4.2%. The MIECL sensor based on MIP/CsPbBr$_3$-QDs shows good stability, accuracy, and precision and can sensitively detect prometryn in aquatic products and environmental samples [49].

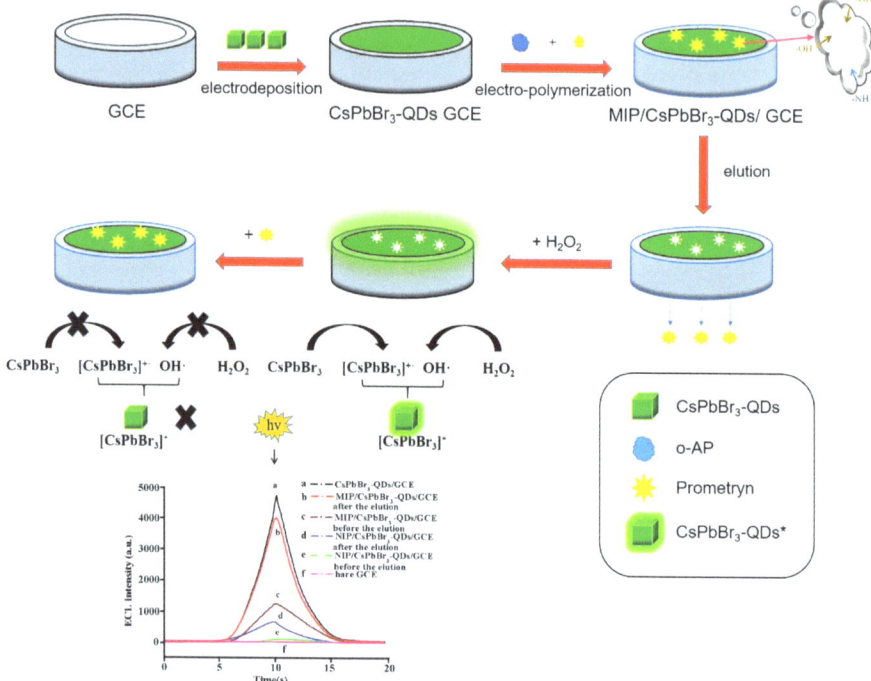

Figure 15. Schematic illustration for the fabrication process and prometryn determination principle of ultrasensitive MIECL sensor. Reproduced from ref. [49] Copyright (2022), with permission from Elsevier.

Simazine is a highly effective selective herbicide that has been listed as a category 3 carcinogen by the International Agency for Research on Cancer (IARC) on 27 October 2017. To detect simazine in aquatic products, Pan et al. developed a novel MIECL sensor based on the molecular-imprinted polymer perovskite (MIP-CsPbBr$_3$) and Ru(bpy)$_3^{2+}$ luminescent molecules. Under optimal experimental conditions, MIP-CsPbBr$_3$-GCEs were exposed to different concentrations of simazine solution (0.1–500 μg/L) and analyzed using the MIECL sensor. The ECL strength showed a linear relationship with the logarithm of simazine concentration within the range of 0.1–500.0 μg/L, and the limit of detection (LOD) was 0.06 μg/L. The MIECL sensor method was validated by analyzing fish and shrimp samples with recovery rates of 86.5–103.9% and relative standard deviations lower than 1.6%. The developed MIECL sensor demonstrated excellent selectivity, sensitivity, reproducibility, accuracy, and precision for the determination of simazine in actual aquatic samples [50].

Organochlorine Pesticides

Organochlorine pesticides are a group of organic compounds containing organochlorine elements, commonly used to prevent and control plant diseases and insect pests. They can be categorized into two main types: benzene-based and cyclopentadiene-based. Yang et al. demonstrated that $CH_3NH_3PbBr_3$ quantum dots (MAPB-QDs) undergo a blue shift in fluorescence spectra upon exposure to polar organochlorine pesticides (OCPs), providing a method for the detection of these compounds. The presence of polar OCPs alters the structure and emission properties of MAPB-QDs by desorbing the capping ligands OA and OAm from their surface, resulting in defect sites that can be occupied by polar OCPs. This allows the chlorine element in OCPs to fully dope into the QDs, increasing the band gap of the MAPB-QDs and causing a blue shift in their wavelength. To address the insufficient stability of MAPB-QDs in the presence of moisture, the researchers mixed MAPB-QDs with PDMS to create a colorimetric card that can be used for the rapid detection of OCPs in actual samples. This study represents the first use of MAPB-QDs for the detection of polar OCPs (Figure 16) [51].

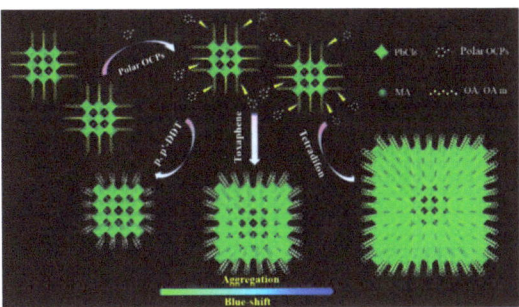

Figure 16. Schematic illustration of the reaction mechanism for MAPB-QDs and polar OCPs. Reproduced from ref. [51] Copyright (2020), with permission from Royal Society of Chemistry.

Other Pesticides

Clodinafop is a herbicide belonging to the aryloxy phenoxy propionate class, soluble in various organic solvents but decomposes under strong acidic or basic conditions. Vajubhai et al. synthesized $CsPbI_3$ PQDs using a microwave radiation method and utilized them for detecting clodinafop in samples via fluorescence spectrometry. Figure 17 illustrates the simple synthesis method of $CsPbI_3$ PQDs utilizing PbI_2, Cs_2CO_3, OAm, OA, and ODE as reagents. The resulting PQDs showed a strong red emission at 686 nm and were well-dispersed in hexane. They were utilized as a probe for fluorescence detection of clodinafop in an organic phase (hexane), where the strong red emission at 686 nm was quenched by clodinafop. The study found a good linear relationship between the quenching intensity and the concentration of clodinafop in the range of 0.1–100 μM, with a LOD of 34.70 nM. Furthermore, the fluorescent probes were integrated into liquid–liquid microextraction for the fluorescence analysis of clodinafop in fruits, vegetables, and grains. The developed probes showed a good linearity to the concentration of clodinafop in the range of 0.1–5.0 μM, with a LOD of 34.70 nM, and exhibited a good recovery rate (97–100%) and low relative standard deviation. The results indicate that $CsPbI_3$ PQDs can be used as a sensor for the quantitative detection of clodinafop in samples [1].

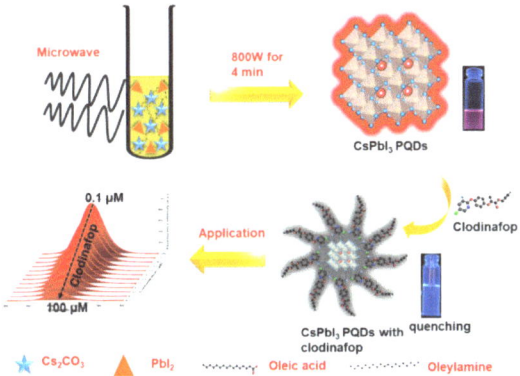

Figure 17. Schematic illustration of microwave synthesis of $CsPbI_3$ PQDs for sensing of clodinafop. Reproduced from ref. [1] Copyright (2022), with permission from American Chemical Society.

Propanil is a highly selective contact herbicide that contains amides, primarily used in rice fields or seedling fields to control barnyard grass and other gramineous and dicotyledonous weeds. The MIP-QDs, under optimized parameters, showed good linearity in the concentration range of paspalum from 1.0 µg/L to 2.0×10^4 µg/L. The developed MIP-QD-based fluorescent probe exhibited a good recovery rate ranging from 87.2% to 112.2%, with relative standard deviations of less than 6.0% for fish and seawater samples. The LOD of paspalum in fish and seawater was 0.42 µg/kg and 0.38 µg/L, respectively. Fluorescence test strips based on MIP-QD also showed satisfactory recovery rate ranging from 90.1% to 111.1%, and the LOD of paspalum in seawater samples was 0.6 µg/L. The developed fluorescent probe and test strip were successfully used for the detection of paspalum in environmental and aquatic products [52].

As shown in Table 2, PNCs demonstrate good stability, high sensitivity, low LOD, and high recovery rate for conventional pesticides. Furthermore, they can be developed into test strips for rapid detection with broad and convenient application prospects.

Table 2. Summary of PNC-based fluorescent probes for pesticide residue detection.

Pesticide Type	Target Analyte	Molecular Structure	Fluorescent Probe	Linearity Range	LOD	Recovery Rate	Ref.
Organophosphorus	OMT		MIPs@CsPbBr$_3$	50–400 ng/mL	18.8 ng/mL		[4]
	DDVP		CsPbBr$_3$ QDs	5–25 µg/L	1.27 µg/L	87.4–101%	[46]
	Phoxim		CsPbBr$_3$ QDs	5–100 ng/mL	1.45 ng/mL		[47]
Triazine herbicides	Prometryn		MIP/CsPbBr$_3$-QDs		0.010 µg/kg, 0.050 µg/L	88.0–106.0%	[49]
	Simazine		MIP-CsPbBr$_3$	0.1–500.0 µg/L	0.06 µg/L	86.5–103.9%	[50]
Organochlorine	OCPs		MAPB-QDs				[51]
Other pesticides	Clodinafop		CsPbI$_3$ PQDs	0.1–5.0 µM	34.70 nM	97–100%	[1]
	Propanil		MIP-QDs	1.0 µg/L– 2×10^4 µg/L	0.42 µg/kg, 0.38 µg/L	87.2–112.2%	[52]

2.1.2. Environmental Pollutant Detection

Uranium is a radioactive element found in various chemical forms in the environment, with uranyl ion (UO_2^{2+}) being the most common valence state under normal conditions. Inhalation or ingestion of U can lead to irreversible kidney damage, destruction of biomolecules, and DNA damage. According to World Health Organization guidelines, the concentration of UO_2^{2+} in drinking water should be less than 30 ppb. As most uranium complexes are highly soluble in water and mobile in aquatic environments, detecting its presence is essential for monitoring agroecological systems.

Halali et al. synthesized all-inorganic CsPbBr$_3$ PQDs with bright PL through thermal injection and used them for UO_2^{2+} sensing. As shown in Figure 18, the first step involved the adsorption of UO_2^{2+} on $CsPbBr_3$ PQDs, and zeta potential measurements confirmed the process of electrostatic attraction and adsorption, facilitating quenching. Dynamic light scattering (DLS) revealed that $CsPbBr_3$ PQDs treated with uranyl ions had significantly larger hydrodynamic radii (1244 nm) than the untreated probe (47.9 nm), indicating adsorption of UO_2^{2+} on $CsPbBr_3$ PQDs. Fluorescence quenching resulted from the adsorbed UO_2^{2+} ions on the surface of $CsPbBr_3$ PQDs. The ultra-low LOD of the probe was 83.33 nM (19.83 ppb), and it demonstrated rapid detection of UO_2^{2+} even in non-polar solvents such as toluene. This study suggests new avenues for designing probes (such as PQDs) and analytes (such as metal ions) for trace level sensing of metal ions [53].

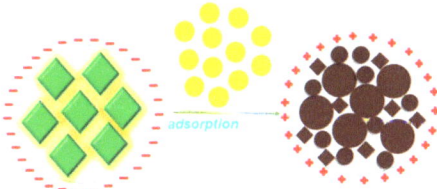

Figure 18. Adsorption mechanism UO_2^{2+} ions on $CsPbBr_3$ PQDs due to the zeta potential process. Reproduced from ref. [53] Copyright (2020), with permission from Elsevier.

O-nitrophenol (ONP) is a member of the phenolic compound family and is widely used as an intermediate in organic synthesis, dyes, medicines, and other products. ONP is also employed as a photographic developer, pH indicator, and photosensitive material, and as a rubber antioxidant. However, ONP is highly toxic and is considered a priority pollutant for environmental analysis and monitoring. Inhaling ONP can cause adverse effects such as headache, dizziness, and nausea, and serious exposure can lead to damage to the central nervous system or the liver/kidney function. Deng et al. developed a molecularly imprinted fluorescent probe using a combination of highly luminescent PQDs and molecular imprinting technology to rapidly detect ONP (see Figure 19). To enhance the hydrophobic properties of the PQDs, a superhydrophobic porous organic polymer framework (SHFW) was created by combining a superhydrophobic material with a pore-forming agent. This SHFW protected the PQDs from contact with water vapor in the air and increased their stability. The $CsPbBr_3$@SHFW fluorescent probe was designed using SHFW-modified $CsPbBr_3$ QDs to detect ONP in river silt. The probe exhibited high selectivity for ONP, and under optimal detection conditions, the linear range of $CsPbBr_3$@SHFW for ONP detection was 0–280 µM, with a limit of detection of 7.69×10^{-3} µM [54].

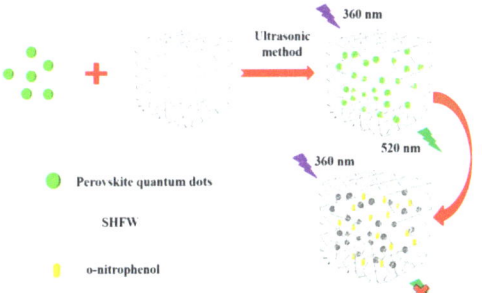

Figure 19. The schematic diagram of the synthetic $CsPbBr_3$@SHFW for fluorescence detection of ONP. Reproduced from ref. [54] Copyright (2021), with permission from Royal Society of Chemistry.

Table 3 summarizes the PNC-based fluorescent probes for environmental pollutant detection. High fluorescence performance of $CsPbBr_3$ PQDs proved them to be efficient fluorescent probes for the detection of pollutants U and ONP.

Table 3. Summary of PNC-based fluorescent probes for environmental pollutant detection.

Target Analyte	Molecular Structure	Fluorescent Probe	Linearity Range	LOD	Ref.
U	U	CsPbBr$_3$ PQD	0–3300 nM (3.3 μM)	0–3300 nM (3.3 μM)	[53]
O-nitrophenol (ONP)		CsPbBr$_3$@SHFW	0–280 μM	7.69 × 10^{-3} μM	[54]

2.1.3. Ion Detection

Cation

As industrialization and urbanization continue to progress, harmful ion pollutants have become a major source of land and water pollution, with highly toxic ions posing a global concern. Conventional detection methods are beset with issues such as high cost, long detection cycle, low accuracy, and low efficiency. Thus, the development of simple and selective sensors is crucial for detecting harmful ion pollutants. Compared to other methods, the fluorescence-based method has advantages such as low cost, high sensitivity, fast detection, and ease of use. Using PNCs to develop trace detection methods represents a breakthrough in new methods.

Liu et al. synthesized all-inorganic CsPbBr$_3$ PQDs via the thermal injection method as a fluorescent probe for the detection of Cu^{2+} in organic phase. The cubic-phase CsPbBr$_3$-NC exhibits a quantum yield as high as 90%. The PL intensity of CsPbBr$_3$ PQD was quenched significantly within seconds of adding Cu^{2+} owing to effective electron transfer from PQD to Cu^{2+}, as revealed by absorption spectra and transient PL lifetime experiments. The sensor detected Cu^{2+} in the range of 0 to 100 nM, with a low LOD of 0.1 nM, demonstrating excellent sensitivity and selectivity for Cu^{2+} in hexane [55].

Li et al. established a method for detecting Cu^{2+} in aqueous solution based on CsPbBr$_3$ (CPB) QDs. A strong organic ligand (oleylamine, OAm) was added, and due to the strong interaction between the amine ligand and metal ions, OAm in cyclohexane captured Cu^{2+} from water, forming an OAm-Cu^{2+} complex at the cyclohexane/water interface (Figure 20). These Cu^{2+} complexes rapidly diffused into cyclohexane, ultimately quenching the PL of CPB QDs. The fluorescence probe based on CPB QDs exhibited a wide linear range (10^6 M–10^2 M), a short response time (1 min), and good selectivity for Cu^{2+}. The metal ion phase transfer strategy avoided direct application of unstable PQDs in a water environment [56].

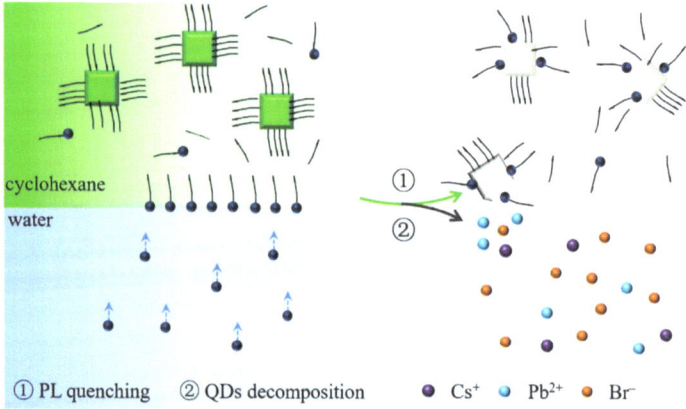

Figure 20. Schematic diagram for phase transfer of Cu^{2+} from water to cyclohexane induced PL quenching (①) and decomposition (②) of CPB QDs. Reproduced from ref. [56] Copyright (2021), with permission from Elsevier.

Mercury ion pollution is a significant issue that impacts human health and environmental safety. To address this problem, Lu et al. developed a novel fluorescence nanosensor based on the surface ion exchange mechanism for rapid, highly sensitive, and selective visual detection of mercury ions (Hg^{2+}), using $CH_3NH_3PbBr_3$ PQDs. $CH_3NH_3PbBr_3$ PQDs were synthesized using ligand-assisted reprecipitation (LARP) technology, without adding any modifier. These PQDs exhibit strong green fluorescence with a high quantum yield of 50.28%. The main principle of the sensor is based on the fact that the strong green fluorescence of $CH_3NH_3PbBr_3$ QDs at 520 nm is significantly quenched by Hg^{2+}, and the blue shift occurs with the increase of Hg^{2+} fluorescence due to surface ion exchange, which differs from most other detection mechanisms shown in Figure 21. The LOD of the sensor is 0.124 nM (24.87 ppt) within the range of 0 nM to 100 nM. The fluorescence intensity of PQDs remains unaffected by interfering ions such as Cd^{2+}, Pb^{2+}, Cu^{2+}, etc.; indicating high selectivity of the fluorescent probe towards Hg^{2+}. The detection of Hg^{2+} using $CH_3NH_3PbBr_3$ PQDs has several advantages, including simple operation, small sample size, high efficiency, and visual detection [57].

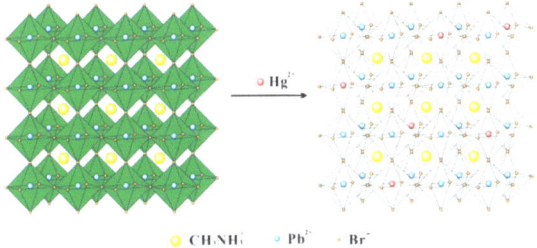

Figure 21. An illustration of the ion exchange process on the surface of $CH_3NH_3PbBr_3$ QDs. Reproduced from ref. [57] Copyright (2017), with permission from Elsevier.

Chen et al. developed a novel method for detecting Fe(III) by encapsulating $CsPbBr_3$ PQDs into poly(styrene/acrylamide) nanospheres using an improved expansion and contraction strategy, as illustrated in Figure 22. The CPB PQDs were dissolved in toluene, which also served as a swelling agent for PSAA polymers, while isopropanol was used as a dispersant for PSAA spheres. After ultrasonic swelling treatment, the composites were rapidly extracted from n-hexane. The resulting CPB@PSAA composites were well-dispersed in an aqueous solution and maintained their bright fluorescence for over 12 months. The fluorescence of the solution could be selectively quenched by the presence of iron ions, enabling the selective sensing of Fe(III). The results demonstrated a good linear relationship between 5 and 150 μM under optimal conditions, with an LOD of approximately 2.2 μM. Moreover, Fe(III) sensing was successfully applied to Yangtze River water, human serum, and tea with satisfactory results [58].

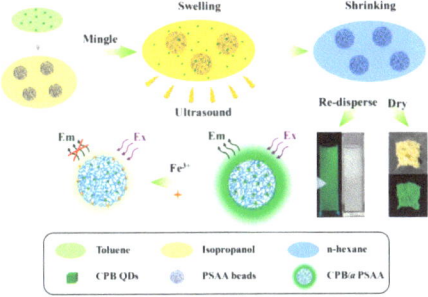

Figure 22. Synthetic diagram of the preparation of CPB@PSAA composites for the detection of iron ion. Reproduced from ref. [58] Copyright (2020), with permission from Elsevier.

In 2019, Wang et al. developed a novel composite material, carbon quantum dots (CQDs) doped MAPbBr$_3$ PQDs encapsulated with ultra-stable SiO$_2$, for potential applications in fluorescence analysis. The composite material was prepared by using carboxy-rich CQDs and MAPbBr$_3$ to form hydrogen-bond interactions, as depicted in Figure 23, which significantly improved the thermal stability of CQDs-MAPbBr$_3$. Furthermore, the composite material was encapsulated with highly dense SiO$_2$ via in situ growth, which greatly enhanced its water resistance and stability. The composite material exhibited excellent water and thermal stability for over 9 months in aqueous solution and maintained strong fluorescence emission even after annealing at 150 °C. The composite material was then employed for fluorescence analysis to detect Zn^{2+} and Ag$^+$ [59].

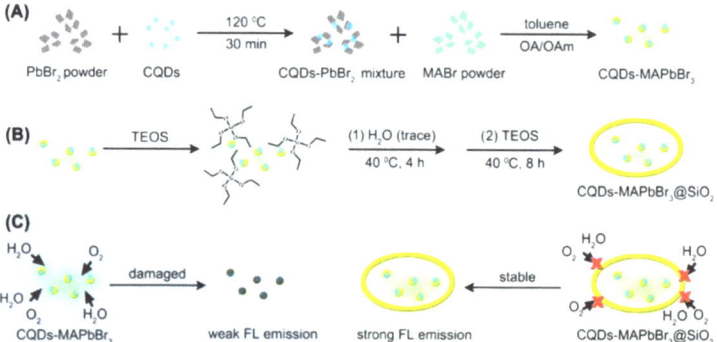

Figure 23. Schematic illustration of the synthesis process of (**A**) CQDs-MAPbBr$_3$, (**B**) CQDs-MAPbBr$_3$@SiO$_2$, and (**C**) Comparison of CQDs-MAPbBr$_3$ and CQDs-MAPbBr$_3$@SiO$_2$ under air and moisture conditions. Reproduced from ref. [59] Copyright (2019), with permission from American Chemical Society.

Yan et al. developed a novel fluorescence detection platform for selective detection of Pb^{2+} ions based on the excellent luminescence properties of lead halide perovskite CH$_3$NH$_3$PbBr$_3$ (MAPbBr$_3$) and CsPbBr$_3$. As shown in Figure 24, a high concentration of CH$_3$NH$_3$Br (MABr) solution was employed as a fluorescent probe. Upon adding PbBr$_2$, a rapid chemical reaction occurred to form MAPbBr$_3$, which exhibited a significant luminous response under ultraviolet light. The fluorescence probe mechanism for Pb^{2+} is attributed to the excellent PL properties of MAPbBr$_3$ in MAPbBr$_3$@MABr solution. Hence, the non-fluorescent MABr demonstrated a sensitive and selective luminescence response to Pb^{2+} under UV irradiation. Moreover, the reaction of MABr and PbBr$_2$ facilitated the conversion of Pb^{2+} to MAPbBr$_3$, enabling the extraction of Pb^{2+} from waste products. Figure 24 shows the detailed experimental process and results of this study [60].

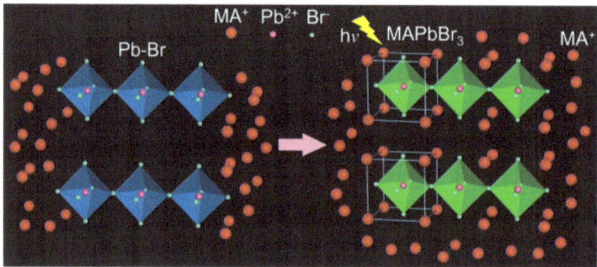

Figure 24. Schematic illustration for the luminescent response of the MABr solution to Pb^{2+}. Reproduced from ref. [60] Copyright (2019), with permission from Springer Nature.

When MABr encounters Pb (II) bound to sulfhydryl groups, it leads to the growth of high-fluorescence (MAPbBr$_3$) perovskite in situ in sulfhydryl-functionalized mesoporous alumina films. This sulfhydryl modification significantly enhances the extraction ability of mesoporous alumina membranes for Pb (II), as reported by Wang et al. Figure 25 illustrates the selection of a commercial meso-Al$_2$O$_3$ film with a pore size of 20 nm as a prototype for 3-mercaptopropyltriethoxysilane (MPTS) modification, due to the strong Pb-S bonding. The hydroxyl groups on the surface of the meso-Al$_2$O$_3$ film facilitate the sulfydryl modification through silane hydrolysis, as shown in Figure 25c. The meso-Al$_2$O$_3$-SH film efficiently enriches Pb(II) in aqueous solutions, which is subsequently dried in an oven. Excess MABr solution is added to the film, and the solvent is evaporated in an oven at 60 °C. Upon drying, MAPbBr$_3$ nanocrystals form on the skeleton or in the holes of the meso-Al$_2$O$_3$-SH film, exhibiting green fluorescence emission under the excitation of 365 nm ultraviolet light (Figure 25a). The PL intensity of the MAPbBr$_3$ nanocrystals increases linearly with the increase of Pb(II) concentration in the sample solution, enabling the visual determination of Pb(II) (Figure 25b). Under optimal conditions, the adsorption capacity of Pb (II) by sulfhydryl-functionalized membranes reaches 94.9 µg/g. Due to the strong extraction ability of Pb (II), the fluorescence opening determination of Pb (II) is realized by forming MAPbBr$_3$ perovskite at the site of sulfhydryl-functional alumina film. The LOD is as low as 5×10^{-3} µg/mL. This method integrates the extraction and determination of Pb (II) without background, effectively addressing the problem of low concentration detection and promoting the entire analysis process [61].

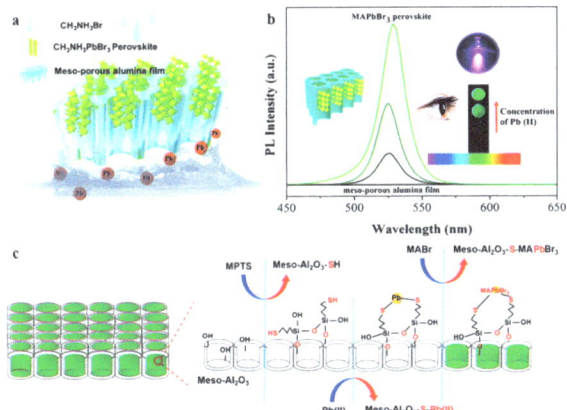

Figure 25. (**a**) Schematic of Pb(II) extraction and MAPbBr$_3$ perovskite formation. (**b**) Schematic of fluorescence turn-on and Pb(II) determination. (**c**) Schematic of sulfydryl functionalization, Pb(II) enrichment, and on-site conversion to MAPbBr$_3$ perovskite. Reproduced from ref. [61] Copyright (2021), with permission from Elsevier.

Anion

Chlorine and iodine residual trace elements during tap water disinfection can pose significant potential harm to human health. Park et al. synthesized CsPbBr$_3$ perovskite quantum dots (PQDs) cellulose composites using a thermal injection method, as depicted in Figure 26. The composites showed excellent stability, durability, and photoluminescence (PL) properties under various humidity conditions. Initially, cellulose fibers were stabilized with PbBr$_2$ precursors to adsorb ions between fibers (Figure 26a). Subsequently, Cs-oleate was added to the PbBr$_2$/cellulose composite at a higher temperature of 200 °C, leading to the nucleation of CsPbBr$_3$ PQDs and their immediate physical aggregation with cellulose fibers. The successful synthesis of CsPbBr$_3$ PQDs and their integration with cellulose nanofibers were confirmed through green PL under UV field (Figure 26c). A portable sensor for early diagnosis was developed based on this composite, which used the rapid anion exchange strategy between halide anions in cellulose fiber PQDs with a high porous structure to detect iodide and chloride ions in real water samples, with detection values of 2.56 mM and 4.11 mM, respectively. CsPbBr$_3$ PQDs/cellulose composites have the advantages of using low-cost materials for one-pot continuous synthesis, high sensitivity, reasonable detection time, simple and portable visual detection, flexibility of practical application, and a hydrophilic sensing platform. The response detection time can be shortened to change the color, indicating the possibility of real-time and on-site detection [62].

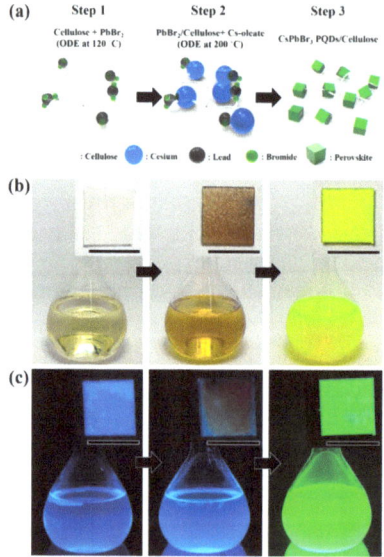

Figure 26. (**a**) Scheme showing the stepwise process for generating CsPbBr$_3$ PQDs/cellulose composites. (**b**) Color change under visible light and (**c**) photoluminescence under UV light. Reproduced from ref. [62] Copyright (2020), with permission from American Chemical Society.

Table 4 summarizes the PNC-based fluorescent probes developed for ion detection. The probes have demonstrated high sensitivity, rapid response times, and ease of use in detecting Cu^{2+}, Hg^{2+}, Fe^{3+}, Zn^{2+}, Ag$^+$, Pb^{2+}, I$^-$, and Cl$^-$ anions in environmental samples. These features make PNC-based fluorescent probes promising candidates for visual detection of harmful ions with great potential for practical applications.

Table 4. Summary of PNC-based fluorescent probes for ion detection.

Ion Type	Target Analyte	Fluorescent Probe	Linearity Range	LOD	Ref.
Cation	Cu^{2+}	CsPbBr$_3$ (PQD)	0–100 nM	0.1 nM	[55]
	Cu^{2+}	CsPbBr$_3$ (CPB)	10^6 M–10^2 M		[56]
	Hg^{2+}	CH$_3$NH$_3$PbBr$_3$ (QDs)	0–100 nM	0.124 nM (24.87 ppt)	[57]
	Fe(III)	CPB@PSAA	5–150 μM	2.2 μM	[58]
	Zn^{2+}, Ag$^+$	CQD-MAPbBr$_3$@SiO$_2$			[59]
	Pb^{2+}	CH$_3$NH$_3$PbBr$_3$ (MAPbBr$_3$)			[60]
	Pb^{2+}	MAPbBr$_3$		5×10^{-3} μg/mL	[61]
Anion	I$^-$, Cl$^-$	CsPbBr$_3$ PQDs		2.56 mM, 4.11 mM	[62]

2.2. The Detection of Food-Related Hazardous Substances

2.2.1. Antibiotic Detection

The overuse of antibiotics can lead to the proliferation of resistant bacteria, which can have harmful effects on human health and cause allergic reactions. Furthermore, the presence of antibiotic residues in food, due to their usage in livestock, is a significant concern. Hence, a simple and highly sensitive technique is required for detecting antibiotics in food samples. Roxithromycin (ROX) is a new generation of macrolide antibiotics that mainly targets Gram-positive bacteria, anaerobic bacteria, chlamydia, and mycoplasma. Han et al. have developed a novel $CsPbBr_3$-loaded PHEMA MIP nanogel with high stability against water and oxidation, using multifunctional MIP nanogel synthesized from four 2-(hydroxyethyl)methacrylate (HEMA) derivatives with different functions. First, MIP antioxidant-nanogels were prepared with ROX as a template, and then perovskite nanoparticles were loaded into nanogels via in situ synthesis through the hot-injection method. The developed $CsPbBr_3$-loaded MIP antioxidant-nanogels exhibited excellent stability to air/water and enhanced stability to water-based solvents. Finally, these nanogels were successfully applied for the selective and sensitive detection of ROX antibiotics in animal-derived food products, as shown in Figure 27. The limit of selectivity and sensitivity to ROX was 1.7×10^{-5} µg/mL (20.6 pM), and the detection results demonstrated good recovery, indicating the excellent performance of the developed MIP antioxidant-nanogel loaded with $CsPbBr_3$ [63].

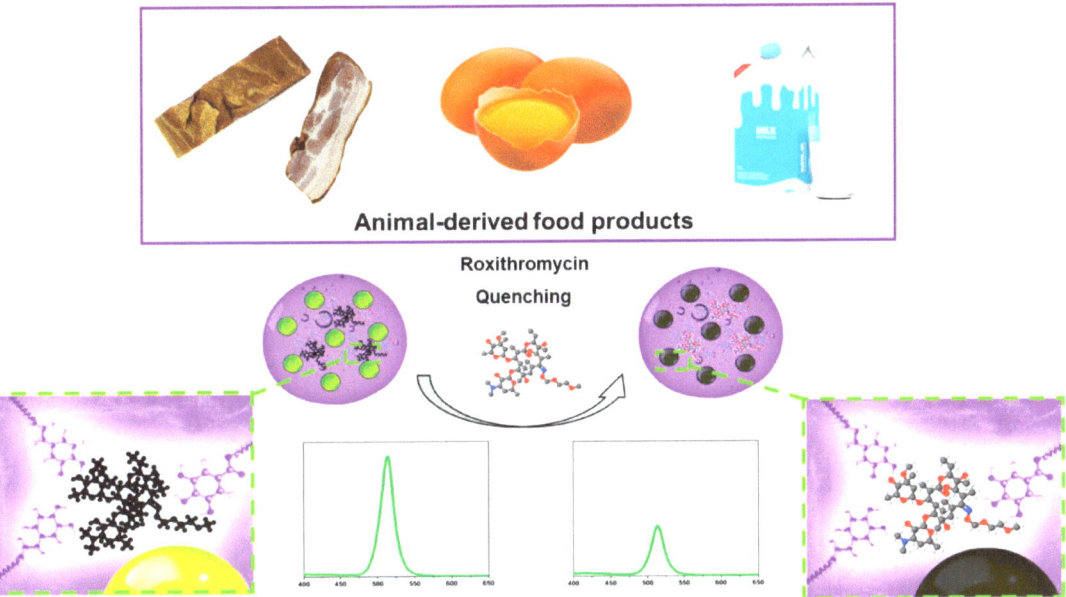

Figure 27. Schematic illustration of ROX sensing in animal-derived food using $CsPbBr_3$-loaded MIP antioxidant-nanogels. Reproduced from ref. [63] Copyright (2022), with permission from Springer Nature.

Tetracycline (TC) is a class of broad-spectrum antibiotics with phenanthrene nuclei discovered in the 1940s. This class of antibiotics is widely used to treat infections caused by Gram-positive and negative bacteria, intracellular mycoplasma, chlamydia, and Rickettsia. In addition, tetracycline is often used as a growth promoter for animals in some countries, including the United States.

In Figure 28, Jia et al. developed a perovskite-composited ratiometric fluorescent nanosensor with good water stability for detecting tetracycline. The sensor was inspired by the respective advantages of perovskite and rare earth elements. The team prepared rod-like MAPbBr$_3$@PbBr(OH) perovskite using the hydration method, with the PbBr(OH) surface layer providing dual advantages of preventing water from entering PNCs and dispersing PNCs in the matrix, resulting in high stability of MAPbBr$_3$@PbBr(OH) in water. Under UV irradiation, the perovskite nanorods emitted stable green fluorescence, which served as the fluorescence internal standard for ratiometric fluorescent nanosensors. The silica shell was used to cover the surface of PQDs, which facilitated the subsequent surface modification of europium compounds. Upon the addition of TC, the green fluorescence intensity of MAPbBr$_3$@PbBr(OH)@SiO$_2$-Cit-Eu decreased slowly, while the energy transfer from TC to Eu^{3+} led to the gradual enhancement of red fluorescence. The detection system exhibited a multi-color visual transformation from green to red fluorescence, which could be applied to the visual rapid detection of TC in actual liquid foods and have great potential application value. The sensor has a good linear response in the range of 0–25 μM, and the detection limit is 11.15 nM. Qualitative and semi-quantitative visual sensing of tetracycline on the surface of a variety of foods (e.g., eggs, apples, bananas, pears, and oranges) can be achieved with the help of color analysis software on a smartphone [64].

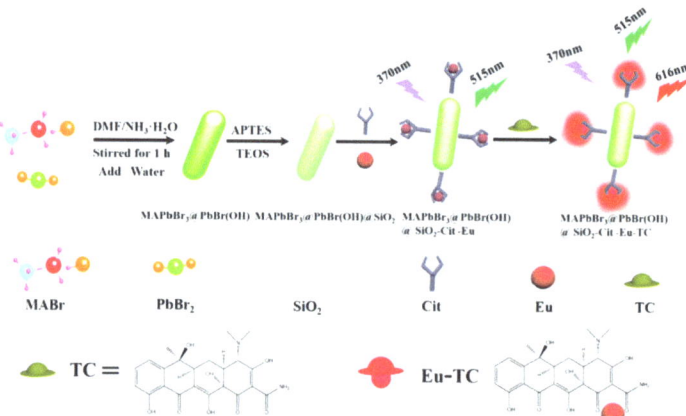

Figure 28. Preparation strategy of MAPbBr$_3$@PbBr(OH)@SiO$_2$-Cit-Eu ratio fluorescent probe and visual sensing schematic of TC. Reproduced from ref. [64] Copyright (2022), with permission from Elsevier.

Thuy et al. developed a new fluorescent probe for detecting trace amounts of tetracycline (TC) in food samples by preparing highly stable Cs$_4$PbBr$_6$/CsPbBr$_3$ perovskite nanocrystals (PNCs) protected by perfluorooctane triethoxy-silane fluorocarbons on their surface. The high hydrophobic groups formed a waterproof layer, which rendered the PNCs highly stable in water. The fluorescent probes demonstrated high sensitivity and selectivity for TC detection in food samples under optimized conditions, with a limit of detection (LOD) of 76 nM and a linear range of 0.4–10 μM [65].

In Figure 29, Wang et al. developed a novel CsPbBr$_3$@BN fluorescent probe for the highly sensitive detection of TC with an LOD as low as 6.5 μg/L. The green-emitting CsPbBr$_3$ quantum dots (QDs) were successfully adhered to BN nanosheets, and upon the addition of TC, the CsPbBr$_3$@BN fluorescence was sensitively quenched due to electron transfer between them. The fluorescent probe was successfully applied in the determination of TC in honey and milk samples [66]. Furthermore, Wang et al. synthesized a novel fluorescent probe for TC in highly polar ethanol at room temperature. The silicon layer was easily modified by in situ hydrolysis of 3-amino-propyl triethoxysilane (APTES) on the surface of IPQDs at room temperature, producing a new fluorescent probe without water and initiator that could be stably stored in ethanol. The method had high selectivity and sensitivity to TC in ethanol, and the LOD could even reach 76 nM. The fluorescence quenching mechanism was mainly due to electron transfer between TC and IPQDs. The sensor was successfully applied to the detection of trace TC in practical samples. These studies laid a foundation for improving the stability of PNCs and their development in the field of TC detection [67].

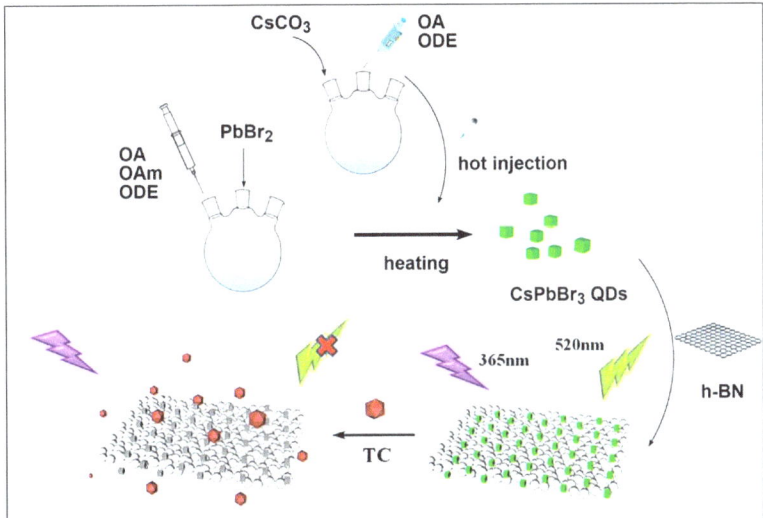

Figure 29. Schematic illustration of the synthesis of CsPbBr$_3$@BN for sensitive detection of TC. Reproduced from ref. [66] Copyright (2021), with permission from Elsevier.

Salari et al. developed a chemiluminescent (CL) probe for the highly selective and sensitive determination of cefazolin (CFZ) antibiotic in food, water, and biological samples. The CL probe consists of CsPbBr$_3$ quantum dots (QDs) in an organic phase and Fe(II) and K$_2$S$_2$O$_8$ in an aqueous medium. The synthesis of CsPbBr$_3$ QDs from CsBr and PbBr$_2$ by a solution-based method at room temperature is illustrated in Figure 30. The designed probe exhibited a linear range of 25–300 nM with a low LOD of 9.6 nM and a recovery rate of 94% to 106% for CFZ, enhancing the CL signal of the probe. This research has the potential to advance the development of CL probes for the detection of antibiotics in various fields [68].

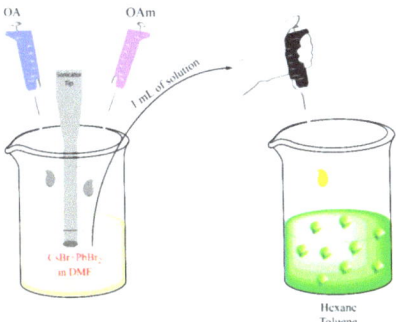

Figure 30. Synthesis process of CsPbBr$_3$ QDs via LARP. Reproduced from ref. [68] Copyright (2023), with permission from Elsevier.

By analyzing the chemical formula of ciprofloxacin hydrochloride ($C_{17}H_{18}FN_3O_3 \cdot HCl \cdot H_2O$), the concentration of its solution can be determined by measuring the Cl^- ionic concentration. $CsPbBr_3$ NCs have superior optical properties and high anion exchange capacity, enabling precise and sensitive detection of Cl^- ionic concentrations. Shi et al. developed colorimetric test strips based on $CsPbBr_{(3-x)}Cl_x$ NCs for rapid and convenient detection of ciprofloxacin hydrochloride in food. As depicted in Figure 31, at room temperature, $CsPbBr_3$ NCs and $C_{17}H_{18}FN_3O_3 \cdot HCl \cdot H_2O$ undergo an anion exchange reaction where some of the Br^- ions are replaced by Cl^- ions. Upon exposure to different concentrations of ciprofloxacin hydrochloride, the emitted light of $CsPbBr_3$ NC changes from 513 nm to 442 nm, and the color of the test strip changes immediately after exposure to different ciprofloxacin solutions [69].

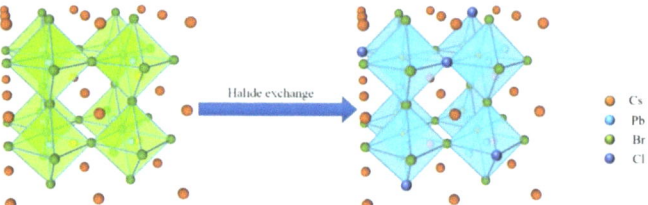

Figure 31. Schematic diagram of the anion exchange reaction of $CsPbBr_3$ to $CsPbBr_{(3-x)}Cl_x$. Reproduced from ref. [69] Copyright (2022), with permission from IoP Publishing.

Table 5 summarizes the PNC-based fluorescent probes developed for detecting antibiotics in food samples. Fluorescent probes using $CsPbBr_3$ and $MAPbBr_3$ NCs were successfully applied to detect antibiotics such as ROX, TC, CFZ, and ciprofloxacin hydrochloride. The results show excellent stability, high selectivity and sensitivity, low LOD, and good recovery rate. Furthermore, the use of color analysis software in smartphones enables qualitative and semi-quantitative visual sensing of antibiotics on various food surfaces, which provides a basis for improving the stability of PQDs and their potential for use in detection applications.

Table 5. Summary of PNC-based fluorescent probes for antibiotic detection in food.

Target Analyte	Molecular Structure of the Target Analyte	Fluorescent Probe	Linearity Range	LOD	Recovery Rate	Ref.
Roxithromycin		MIP-$CsPbBr_3$		1.7×10^5 µg/mL (20.6 pM)		[63]
Tetracycline		$MAPbBr_3$@PbBr(OH)@SiO_2-Cit-Eu	0–25 µM	11.15 nM		[64]
		Cs_4PbBr_6/$CsPbBr_3$	0.4–10 µM	76 nM		[65]
		$CsPbBr_3$@BN	0–0.44 mg/L	6.5 µg/L		[66]
		APTES@IPQDs				[67]
Cefazolin		$CsPbBr_3$ QDs	25–300 nM	9.6 nM	94–106%	[68]
Ciprofloxacin hydrochloride		$CsPbBr_{(3-x)}Cl_x$ NCs				[69]

2.2.2. Detection of Microbial Toxins, Pathogens, and Carcinogens

Aflatoxin B_1 (AFB$_1$) contamination is one of the most serious problems in food safety. Su et al. have developed a photoelectrochemical (PEC) immunosensing platform for sensitive detection of AFB$_1$ in peanut and corn samples using CsPbBr$_3$ NCs and amorphous TiO$_2$ [CsPbBr$_3$/a-TiO$_2$]. AFB$_1$-BSA conjugates labeled with alkaline phosphatases (ALP) were used as competitors in the competitive immunoreaction triggered on anti-AFB$_1$ antibody-coated microplates. In Figure 32, photoelectrochemically active CsPbBr$_3$/a-TiO$_2$ nanocomposites were dropwise added onto a fluorine-doped tin oxide (FTO) electrode. The ALP collected on the microplate hydrolyzed ascorbic acid-2-phosphate (AAP) to produce ascorbic acid (AA), which enhanced the photocurrent of CsPbBr$_3$/a-TiO$_2$ nanocomposites. The concentration of AFB$_1$ in food can be monitored by detecting the photocurrent of the CsPbBr$_3$/a-TiO$_2$-modified electrode. The PEC platform exhibited a working range of 0.01~15 ng/mL and a low detection limit of 2.8 pg/mL. The accuracy of this method was found to be acceptable compared with the reference aflatoxin enzyme-linked immunosorbent assay (ELISA). The developed PEC immunosensing platform shows promise in the sensitive and accurate detection of AFB$_1$ in foodstuffs [70]. In addition, Li et al. synthesized CH$_3$NH$_3$PbBr$_3$ (MAPB) QDs@SiO$_2$ nanospheres via the encapsulation of MAPB QDs with a network structure of APTES hydrolyzed and condensed with SiO$_2$. They constructed an electrochemical luminescence (ECL) platform using MAPB QDs@SiO$_2$ for the highly selective and ultra-sensitive detection of AFB$_1$, with a LOD of 8.5 fg/mL. The sensor exhibited good recovery rates for AFB$_1$ in real corn oil samples [71].

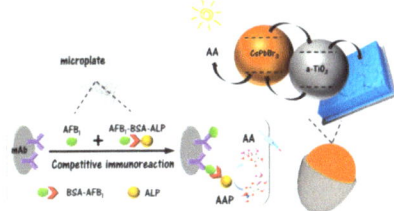

Figure 32. Schematic illustration of the PEC immunosensing platform for detecting AFB$_1$ using CsPbBr$_3$/a-TiO$_2$ nanocomposites and enzyme immunoassay. Reproduced from ref. [70] Copyright (2019), with permission from Royal Society of Chemistry.

Staphylococcus aureus contamination of food at 20–37 °C can result in the production of enterotoxin, causing acute gastroenteritis symptoms such as nausea, vomiting, and diarrhea within 1–5 h of consumption. Food poisoning is mainly caused by enterotoxins rather than bacteria. Staphylococcal enterotoxin (SEs) consists mainly of S. aureus enterotoxins A, B, C, D, and E (SEA, SEB, SEC, SED, and SEE), and the minimum dose of enterotoxin required to cause food poisoning is 1–7.2 g/kg body weight. Xu et al. developed CsPbBr$_3$@mesoporous silica nanomaterials (MSN) and converted it into CsPb$_2$Br$_5$@MSN in an aqueous phase for immunoassay of Staphylococcal enterotoxin (SEs). This additive-free surface-enhanced Raman scattering (SERS) immunoassay can be used for the simple, sensitive, and reproducible detection of SEC. The research can be extended to the development of various perovskite composites, providing the possibility of exploring more SERS detection probes for food safety monitoring [72].

In food processing, waxing can maintain the fresh flavor of fruit. However, artificial wax applied to the surface of fruit often contains morpholine, which is easily converted into nitroso-morpholine in the human body and can cause liver or kidney cancer. In addition, artificial wax also contains protein substances that can cause allergic reactions. Ye et al. [73] prepared a lead-free Cs$_2$PdBr$_6$ perovskite humidity sensor, which was used for the first time to detect artificial wax on fruit. The manufactured humidity sensor has a response time of 0.7 s and a recovery time of 1.7 s. The impedance varies by 10^5 Ω when the relative humidity increases from 11% to 95%. The oleic acid-modified Cs$_2$PdBr$_6$ sensor has the advantages of high sensitivity and fast response/recovery. It can be used to distinguish the waxy/wax-free apples and oranges from the surface moisture.

Table 6 summarizes PNC-based fluorescent probes for the detection of microbial toxins, pathogens, and carcinogens in food. PNCs of sPbBr$_3$, CH$_3$NH$_3$PbBr$_3$, and lead-free Cs$_2$PdBr$_6$ were used to detect B$_1$ (AFB$_1$), SEs, nitrite, and artificial wax on fruit and showed high sensitivity and selectivity and has a wide range of application potential.

Table 6. Summary of PNC-based fluorescent probes for the detection of microbial toxins, pathogens, and carcinogens in food.

Target Analyte	Molecular Structure	Fluorescent Probe	Linearity Range	LOD	Ref.
B$_1$ (AFB$_1$)		CsPbBr$_3$/a-TiO$_2$	0.01~15 ng/mL	2.8 pg/mL	[70]
		MAPB QDs@SiO$_2$ and MAPB		8.5 fg/mL	[71]
SEs	Staphylococcal enterotoxin	CsPb$_2$Br$_5$ @MSN			[72]
Artificial wax on fruit		Cs$_2$PdBr$_6$			[73]

2.2.3. Detection of Food Spoilage Gas

The odor of ammonia is a result of protein breakdown by microorganisms in spoiled food. To detect ammonia, Huang et al. developed a highly sensitive and selective CsPbBr$_3$ QDs film sensor that is fully reversible. Real-time dynamic passivation of PQDs was studied by monitoring changes in PL intensity, as shown in Figure 33. The sensing device consists of a silicon detector and a PQDs film, excited at 365 nm under the flow of ammonia gas. Exposure to ammonia gas led to a remarkable increase in fluorescence of the QDs film, which could even be observed by the naked eye, and returned to its original state upon removal of the gas. The response–recovery cycles of this process are shown in the schematic diagram in Figure 33. The sensor can detect ammonia in a wide range of 25–350 ppm, with an LOD as low as 8.85 ppm, and a fast response time of ≈10 s and a recovery time of ≈30 s are achieved, respectively. This novel sensor has potential applications in food safety and environmental monitoring [74].

Figure 33. Schematic diagram of the response–recovery cycles of the dynamic passivation of PQDs. Reproduced from ref. [74] Copyright (2020), with permission from Wiley-VCH.

Hydrogen sulfide (H$_2$S) is a toxic gas that emits a rotten egg smell, and exposure to high levels of H$_2$S in food can be detrimental to human health. Li et al. developed water-soluble inorganic PQDs (CsPbBr$_3$@PEG-PCL nanoparticles) by encapsulating CsPbBr$_3$ QDs with a block copolymer of polyethylene glycol polycaprolactone (PEG-PCL), as illustrated in Figure 34a. The fluorescence of CsPbBr$_3$@PEG-PCL nanoparticles remains stable in aqueous solution for at least 15 days under light and room temperature conditions, making it ideal for H$_2$S detection. The fluorescence intensity of CsPbBr$_3$@PEG-PCL nanoparticles is quenched by H$_2$S, and thus, a highly sensitive H$_2$S fluorescent probe can be constructed based on fluorescence intensity quenching. The fluorescence signal has a linear relationship with H$_2$S concentration in the range of 0–32.00 μM, with a limit of detection (LOD) of

37.65 nM. CsPbBr$_3$-based sensors are also capable of accurately detecting H$_2$S in wine samples, cells, and zebrafish [75]. Another type of water-soluble inorganic PQDs (CsPbBr$_3$@SBE-β-CD nanoparticles) was obtained by passivating the surface ligands with sulfobutyl ether-β-cyclodextrin (SBE-β-CD). This nanoparticle can be used as a photothermal probe to react with H$_2$S and produce a substance with high photothermal effect. Under 808 nm laser irradiation, the temperature change of the system was measured using a thermometer as a signal reading device to quantitatively analyze H$_2$S, as shown in Figure 34b. There is a linear relationship between temperature and H$_2$S concentration in the range of 0.5–6000.0 μM, with a LOD of 0.19 μM. This sensor can achieve the quantitative detection of H$_2$S in water medium [76].

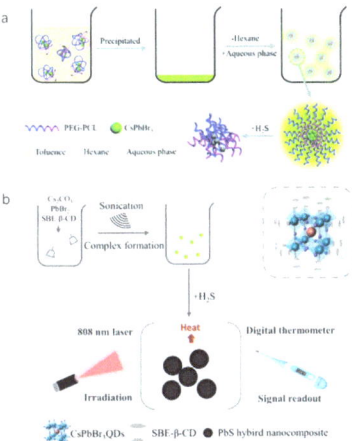

Figure 34. (a) Preparation process of CsPbBr$_3$@PEG-PCL nanoparticle and the detection of H$_2$S. (b) Preparation process of CsPbBr$_3$@SBE-β-CD nanoparticles and their H$_2$S photothermal sensor. Reproduced from ref. [75] Copyright (2021), with permission from Royal Society of Chemistry. Reproduced from ref. [76] Copyright (2022), with permission from Elsevier.

Zu et al. proposed a synthesis method for all-inorganic CsPbBr$_3$ QDs that can be directly stored in aqueous solution, maintaining excellent fluorescence stability and dispersion, as illustrated in Figure 35. By adding CTAB aqueous solution, CsPbBr$_3$ QDs and mineral oil can be rapidly processed by ultrasonic to form CsPbBr$_3$@CO complex. In aqueous solution, when the concentration of CTAB as a surfactant exceeds the critical micellar concentration (CMC), spherical or rodlike micelles may be formed, with hydrophobic nuclei surrounded by hydration shells. These micelles can encapsulate hydrophobic substances, enhancing their solubility and dispersion in water. As fluorescent probes, these micelles exhibit good selectivity for the detection of hydrogen sulfide (H$_2$S) in food and have a short detection time. The fluorescent probe demonstrates a linear relationship between the H$_2$S concentration and the fluorescence intensity in the range of 0.15–105.0 μM, with an LOD of 53.0 nM, satisfying the H$_2$S detection requirements for most food or biological samples [77].

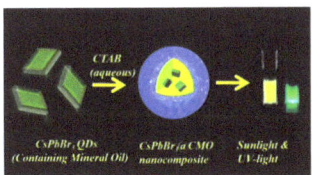

Figure 35. The encapsulation process of the CsPbBr$_3$@CO complex.

Table 7 summarizes fluorescent probes based on PNCs for detecting food spoilage. Modified CsPbBr$_3$ QDs exhibit high sensitivity and selectivity in detecting gases such as ammonia and H$_2$S in wine, beer, and biological samples (zebrafish), with rapid response and recovery times.

Table 7. Summary of PNC-based fluorescent probes for the detection of food spoilage gas in food.

Target Analyte	Fluorescent Probe	Linearity Range	LOD	Ref.
NH$_3$	CsPbBr$_3$ QDs	25–350 ppm	8.85 ppm	[74]
H$_2$S	CsPbBr$_3$@PEG-PCL			[75]
H$_2$S	CsPbBr$_3$@SBE-β-CD	0.5–6000.0 μM	0.19 μM	[76]
H$_2$S	CsPbBr$_3$@CO	0.15–105.0 μM	53.0 nM	[77]

2.2.4. Detection of Harmful Additives in Food

Sudan dyes, classified as carcinogens, are commonly used as phenylazo derivatives in industry but are strictly prohibited from use in food. He et al. utilized CsPbX$_3$ PQDs of varying colors within molecularly imprinted polymers as fluorescent probes for the sensitive and selective detection of Sudan Red I in food. The fluorescence intensity of MIPs-CsPbX$_3$ microspheres was significantly quenched after loading Sudan Red I. The linear response was good within the range of 0.5–150 μg/L, with an LOD of 0.3 μg/L and recoveries between 95.27–105.96% [2]. Wu et al. developed a simple and effective fluorescence detection platform for Sudan I-IV utilizing CsPbBr$_3$ PQDs, as shown in Figure 36. Under UV excitation, the CsPbBr$_3$ QDs exhibited bright green fluorescence, which was efficiently quenched by Sudan I-IV due to significant overlap between the absorption band of Sudan I-IV and the fluorescence spectrum of CsPbBr$_3$ QDs. After optimization, the quantification of fluorescence quenching efficiency of CsPbBr$_3$ QDs was found to be correlated with the logarithmic concentrations of Sudan I-IV (100–10000, 0.1–1000, 0.1–2000, and 0.4–1000 ng). The LODs were 3.33, 0.03, 0.03, and 0.04 ng/mL, respectively [78].

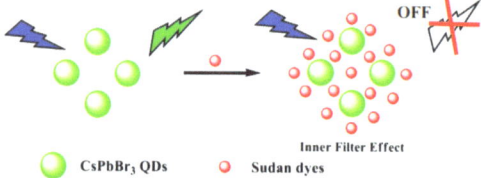

Figure 36. Schematic illustration of the Sudan dye detection mechanism using the CsPbBr$_3$ QDs. Reproduced from ref. [78] Copyright (2018), with permission from Springer Nature.

Rhodamine 6G (Rh6G) is an anthraquinone dye that has found wide application in the textile and food industries. However, due to its teratogenic, carcinogenic, and other biological toxicities to both humans and animals, it is crucial to detect it effectively in low concentration ranges. Chan et al. synthesized monodispersed PEGylated CsPbBr$_3$/SiO$_2$ QDs using the LARP synthesis method at an ambient temperature of 23 °C. In Figure 37, the CsPbBr$_3$/SiO$_2$ QDs were encapsulated in mPEG-DSPE phospholipid micelles to provide excellent aqueous solubility and stability. This nanosensor exhibits good sensitivity and selectivity for Rh6G and can also detect it in biological samples containing complex interference backgrounds. The operation range is 0–10 mg/mL, and the LOD is 0.01 mg/mL [79]. Similarly, Wang et al. prepared a CsPbBr$_3$ PQDs/polystyrene fiber film (CPB QDs/PS FM) using a one-step electrospinning method. Due to the excellent optical properties of CPB QDs, the ultra-low LOD for Rh6G detection is 0.01 ppm, and FRET efficiency is 18.80% in 1 ppm Rh6G aqueous solution [80].

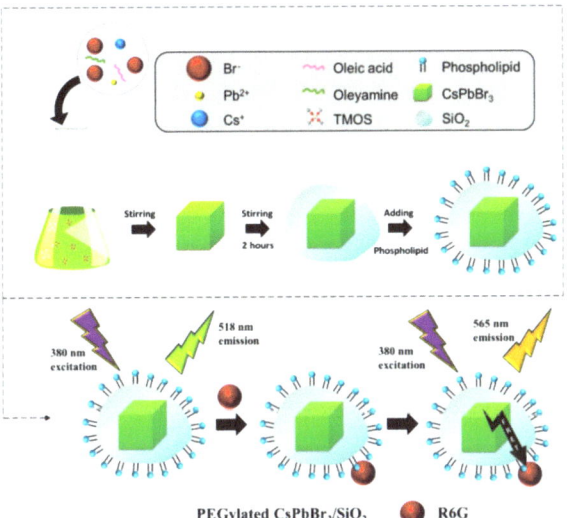

Figure 37. Synthesis and encapsulation of PEGylated CsPbBr$_3$/SiO$_2$ QDs, and FRET-based sensing mechanism of R6G using the proposed nanosensor. Reproduced from ref. [79] Copyright (2022), with permission from Elsevier.

Furthermore, RhoB is a harmful substance that often appears in printing and dye wastewater or as an additive in food. Traditional detection systems are not sensitive enough and often require a large number of sample solutions (>1 mL) to concentrate to a higher concentration. Ni et al. fabricated a reusable perovskite nanocomposite fiber paper using microwave and electrospinning methods, which consisted of CsPbBr$_3$ QDs grown in situ in a high concentration solid polymer fiber. This perovskite fiber paper can be easily recycled and has the advantages of ultra-sensitive detection (0.01 ppm), rapid detection (<3 min), and minimal dose (<25 µL), which are obviously superior to traditional detection systems [81].

Shi et al. synthesized oil-soluble all-inorganic CsPbBrI$_2$ QDs using the thermal injection method for the detection of basic yellow dyes that are illegally added to food and beverages. Their results showed a linear range of 1–500 µg/mL, a LOD of 0.78 µg/mL, and recoveries of 95.27–98.84% at levels of 8, 40, 80, and 400 µg/mL, with relative standard deviations of less than 3.52% [82].

Li et al. developed an ultra-sensitive fluorescent nanosensor for the detection of melamine using PNCs (CsPbBr$_3$NCs@BaSO$_4$). Figure 38 illustrates the nanosensor's principle. The negatively charged citrate-stabilized AuNPs were mixed with positively charged CsPbBr$_3$ NCs@BaSO$_4$, causing them to assemble together through electrostatic interaction. This aggregation led to strong fluorescence quenching of the CsPbBr$_3$ NCs@BaSO$_4$ due to the inner filter effect of the AuNPs. However, upon the addition of melamine, the color of the solution changed from wine red to blue–grey, indicating the agglomeration of the AuNPs. This aggregation decreased the inner filter effect of the AuNPs, resulting in a recovery of the fluorescence of the CsPbBr$_3$ NCs@BaSO$_4$. The nanosensor was used for the determination of trace melamine in dairy products, achieving a LOD of 0.42 nmol/L and a relative standard deviation of 4.0% by repeated determination of 500.0 nmol/L (n = 11) [83].

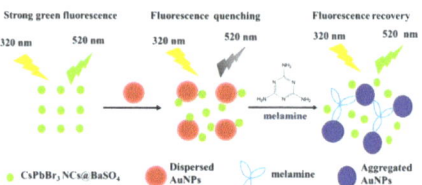

Figure 38. Schematic illustration of turn-on fluorescent melamine nanosensor based on the inner filter effect of the AuNPs on CsPbBr$_3$ NCs@BaSO$_4$. Reproduced from ref. [80] Copyright (2016), with permission from American Chemical Society.

To achieve efficient electrochemical luminescence (ECL) in PQDs, Sun et al. proposed the synthesis of Cs$_4$PbBr$_6$@CsPbBr$_3$ PQD nanoacanthospheres (PNAs) with optimized conditions resulting in ECL activity four times higher than 3D (CsPbBr$_3$) PQDs, which was then applied to Bisphenol A (BPA) ECL sensing [84].

Table 8 summarizes the use of modified CsPbBr$_3$ and CsPbBrI$_2$ QDs as PNC-based fluorescent probes for detecting harmful additives in food. These probes showed good selectivity and sensitivity in detecting Sudan I–IV, Rh6G, RhoB, basic yellow dye, melamine, and bisphenol A, with the advantages of rapid determination and minimal dose. This demonstrates their potential practical applications and superiority over traditional detection systems.

Table 8. Summary of PNC-based fluorescent probes for the detection of harmful additives in food.

Target Analyte	Molecular Structure	Fluorescent Probe	Linearity Range	LOD	Recovery Rate	Ref.
Sudan red I		CsPbX$_3$	0.5–150 µg/L	0.3 µg/L	95.27–105.96%	[2]
Sudan red I–IV		CsPbBr$_3$		3.33, 0.03, 0.03, 0.04 ng/mL		[78]
Rhodamine 6G		CsPbBr$_3$/SiO$_2$ QDs	0–10 mg/mL	0.01 mg/mL		[79]
		CPBQDs/PSFM		0.01 ppm		[80]
RhoB		CsPbBr$_3$-PVDF		0.01 ppm		[81]
Basic yellow dye		CsPbBrI$_2$ QDs	1–500 µg/mL	0.78 µg/mL	95.27–98.84%	[82]
Melamine		CsPbBr$_3$NCs@BaSO$_4$		0.42 nmol/L		[83]
Bisphenol A		Cs$_4$PbBr$_6$@CsPbBr$_3$ PQD				[84]

2.2.5. Edible oil Quality Inspection

Edible oil undergoes complex changes during storage, with the most important being the decomposition of oil into free fatty acids and other products under the influence of light, heat, air, and other factors, commonly referred to as rancidity. Peroxide is an intermediate product formed during the process of oil oxidation and rancidity, which is unstable and can decompose into aldehydes, ketones, and other oxides. This not only affects the flavor and nutritional value of food but also poses risks to human health. China's National Testing Standard GB/T 2716-2018 specifies that the peroxide value of edible oil should be <0.25 g/100 g. Currently, the peroxide value of cooking oil is usually determined by titration or potential methods, which are not reproducible, sensitive, and environmentally friendly, and the methods applied are cumbersome to operate. Although spectrophotometry and high-performance liquid chromatography have been developed to overcome these shortcomings, there is still a need for real-time and visual detection methods that are simple and rapid.

The peroxide number of edible oil is related to its quality, and classical determination methods for the peroxide number are unsatisfactory due to their complexity and poor reproducibility in the analytical process and their incapability of field rapid detection. To overcome this, a colorimetric sensing method based on the wavelength shift of $CsPbBr_3$ NCs with the addition of oleylammonium iodide (OLAM-I) was developed for the determination of peroxide number in edible oil. As shown in Figure 39, by performing halide exchange with OLAM-I, the fluorescence emission wavelength of $CsPbBr_3$ NCs gradually red-shifts, with the degree of red-shift proportional to the amount of OLAM-I added. This results in a color change from green to yellow and finally to red, reflecting the halogen-exchange characteristics of $CsPbBr_3$ NCs and the redox reaction between OLAM-I and peroxides in edible oil. This method enables the visual detection of the peroxidation value of edible oil samples, with a detection process that only takes about 15 min and has been proven to be convenient and accurate [85].

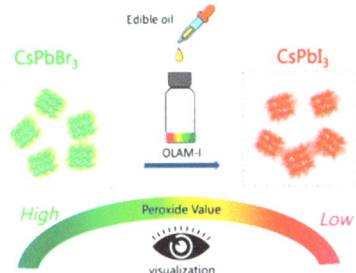

Figure 39. Mechanism for determination of peroxide in edible oils by using $CsPbBr_3$ NCs. Reproduced from ref. [85] Copyright (2019), with permission from American Chemical Society.

The quality of edible oil has a significant impact on human health, with excessive acid number (AN), 3-chloro-1,2-propylene glycol (3-MCPD), and moisture content (MC) being the key factors to monitor. To address this issue, Zhao et al. developed orange luminescent oil-soluble $CsPbBr_{1.5}I_{1.5}$ QDs and applied them to detect AN, 3-MCPD, and MC in edible oil by employing fluorescence quenching and wavelength shift. The mechanism diagram of the detection method for hazardous substances in edible oil is shown in Figure 40. By adding varying concentrations of benzoic acid into the oil, the orange-emitting fluorescence of oil-soluble $CsPbBr_{1.5}I_{1.5}$ QDs was quenched by edible oil with different ANs, and a good linear relationship was observed between the fluorescence intensity of $CsPbBr_{1.5}I_{1.5}$ QDs and AN concentration, enabling quantitative detection of AN in edible oil. Additionally, the unique halogen exchange reaction between $CsPbBr_{1.5}I_{1.5}$ QDs and 3-MCPD allowed for the sensitive and specific detection of 3-MCPD, as the emission peak of $CsPbBr_{1.5}I_{1.5}$ QDs blue-shifted with increasing 3-MCPD concentration. To monitor the MC of edible oil, water-sensitive PQDs were prepared to establish ratiometric fluorescent probes. The orange-emitting CsPbBr$_{1.5}$I$_{1.5}$@MSNs were used as detection probes, which were generally quenched when exposed to water, while green-emitting quasi-two-dimensional (2D) CsPbBr$_3$ NSs served as reference probes due to their stable fluorescence properties in aqueous solution. The ratiometric fluorescent probes based on these two materials were established for MC detection in edible oil, with limits of detection (LODs) of 0.71 mg KOH/g, 39.8 µg/mL 3-MCPD, and 0.45% MC. By leveraging this detection principle, the researchers achieved quantitative detection of AN, 3-MCPD, and MC in edible oil, demonstrating great potential for monitoring the quality of edible oil [86].

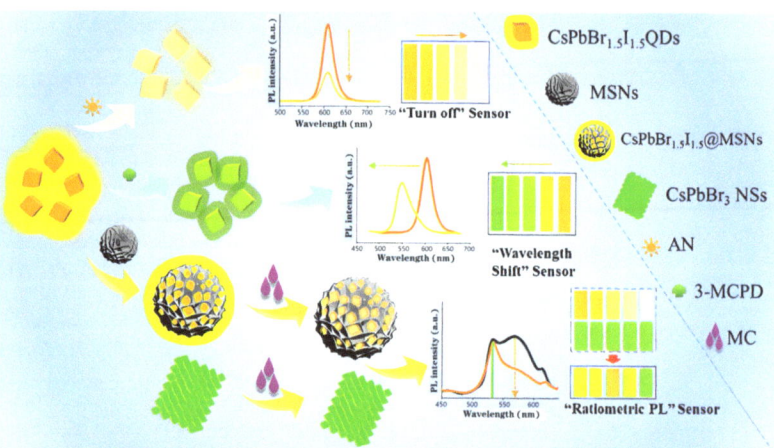

Figure 40. Schematic illustration of CsPbBr$_{1.5}$I$_{1.5}$ QDs -engineered multiplex-mode fluorescence sensing of AN, 3-MCPD, and MC in edible oil. Reproduced from ref. [86] Copyright (2021), with permission from American Chemical Society.

Accurately detecting trace moisture in edible oil is crucial to ensure its quality and safety. Although all inorganic PQDs are potential candidates for optoelectronic applications, their stability is often restricted by environmental factors, especially high polarity materials. In contrast, perovskite-structured QDs are highly suitable as polarity sensors due to their instability to polar materials. To address this, CsPbBr$_3$ PQDs were synthesized at low temperature, which were modified with a functional ligand dimethyl aminoterephthalate (CsPbBr$_3$@DMT-NH$_2$ QDs). The resulting dual-emission QD displayed highly sensitive fluorescence turn-on/off dual response and a distinct fluorescence color change from green to blue in the presence of water. This enabled the establishment of ratiometric fluorescence sensing and visual ratiometric chromaticity detection methods for water assay, with ultralow LODs of 0.006% (v/v) and 0.01% (v/v), respectively. These methods accurately detected trace water in edible oils (with a recovery rate of 93.0% to 108.0%, and a relative error of ≤5.33%). The study also revealed that the high polarity levels of protic solvents can disintegrate the QDs, while the relatively low polarity levels promote their aggregation, ultimately leading to green fluorescence quenching. Furthermore, the deprotonation and polarity of water enhance the blue fluorescence of DMT-NH$_2$ by promoting excited-state intramolecular proton transfer and intramolecular charge transfer, as shown in Figure 41. This is the first report on dual-response ratiometric fluorescence PQDs for ultrasensitive water detection in edible oils [87].

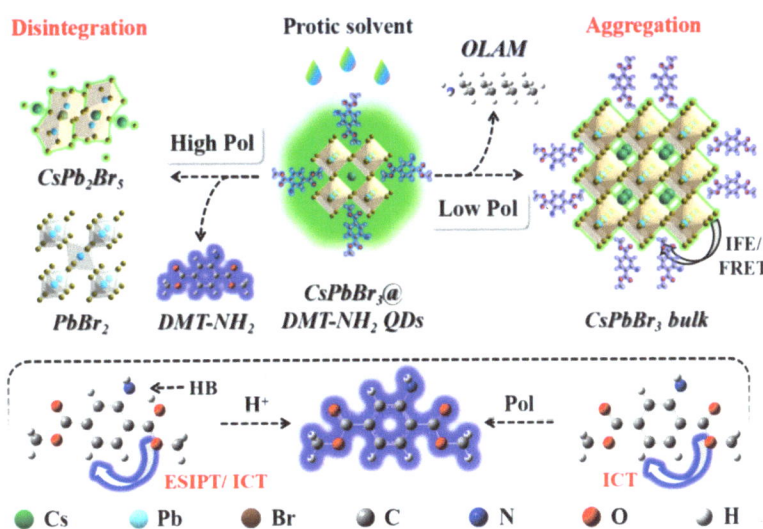

Figure 41. Response mechanism of CsPbBr$_3$@DMT-NH$_2$ QDs to water and polar protic solvents. Reproduced from ref. [87] Copyright (2022), with permission from Elsevier.

To ensure the quality and safety of edible oil, the analysis of total polar substance (TPM) is crucial. However, the development of appropriate TPM analysis methods has been challenging in the food safety field. A paper-based fluorescent probe using CsPbBr$_3$ QDs was synthesized and utilized for rapid detection of TPM in different types of edible oils. The results showed a linear relationship for olive oil (17–31.5%), soybean oil (25–31.5%), and sunflower oil (21.5–33%). Real-time monitoring of TPM content was achieved using the probe, and the potential for a more rapid detection method was explored. These findings indicate that CsPbBr$_3$ QDs can serve as a new fluorescent probe for the rapid detection of TPM in edible oil, with practical application value. Furthermore, Table 9 summarizes the use of PNC-based fluorescent probes for edible oil quality inspection, including modified CsPbBr$_3$ NCs and CsPbBr$_{1.5}$I$_{1.5}$ QDs, which demonstrate the convenience and accuracy of detecting peroxide value, AN, 3-MCPD, MC, and TPM. These perovskite nanomaterials not only expand the application of bioanalysis but also provide a new material and method for monitoring the safety of oil-phase foods in multiple ways [88].

Table 9. Summary of PNC-based fluorescent probes for edible oil quality inspection.

Target Analyte	Molecular Structure	Fluorescent Probe	Linearity Range	LOD	Recovery Rate	Ref.
Peroxide value	CH$_3$(CH$_2$)$_7$CH=CH(CH$_2$)$_7$CH$_2$NH$_3$I	CsPbBr$_3$ NCs				[85]
AN, 3-MCPD, MC	Excessive acid number, H$_2$O	CsPbBr$_{1.5}$I$_{1.5}$@MSNs		0.71 mg KOH/g, 39.8 μg/mL 3-MCPD, 0.45% MC		[86]
H$_2$O	H$_2$O	CsPbBr$_3$@DMT-NH$_2$ QDs		0.006% (v/v), 0.01% (v/v)	93.0~108.0%	[87]
TPM		CsPbBr$_3$ QD	17–31.5%, 25–31.5%, 21.5–33%			[88]

3. Summary and Prospects

In summary, the fluorescent probe based on PNCs is expected to replace the traditional inorganic QDs because of its high fluorescence quantum yield and sensitivity based on

PNCs to detect hazardous substances in agriculture and food. Nowadays, PNCs have been used in agro-ecological environment to detect pesticide residue, environmental pollutant, and harmful ions. At the same time, PNCs have also been applied in food to detect antibiotics, microorganisms, food spoilage gases, additives, and edible oil quality. Compared with the fluorescence analysis method constructed by traditional QDs, PNCs as fluorescent probes are just in their infancy. Therefore, there are still many problems in the practical application of PNCs, and more research is needed.

(1) The external environment (light, humidity, temperature, etc.) is easy to affect photoelectric properties of PNCs. Although, many modification methods have been proved to be effective to achieve PNCs in aqueous solutions, some of these methods are complex or cannot be used as fluorescent probes in the field of agriculture and food. Simple and novel water-stabilized method that is suitable for detection in agriculture and food should be further developed in the future. For example, the exploration of organic semiconductor ligands to prepare PNCs materials with good optoelectrical properties and heterostructures.

(2) Lead is an important component of PNCs structure, and it is a huge obstacle, which hinders PNCs in practice in agriculture and food. Non-contact detection (gas or separated liquid samples) may be more suitable for the lead-based PNCs fluorescent probe. The development of novel and efficient lead-free PNCs fluorescent probe is more preferred and should be the future trend in this area.

(3) The specificity of the fluorescent probe based on PNCs can be further developed to selectively detect specific analyte. Combined with molecular imprinting techniques, it has been proved to be an effective way to improve its selectivity. Especially, the novel characterizing methods should be paid more attention, such as MIECL platform, PEC immunosensing platform, etc.

(4) The integration of fluorescent probe based on PNCs, excitation light source, and fluorescence detector into a portable instrument is still a big challenge. The future practice of fluorescent probe based on PNCs in the field of agriculture and food should tend to portable, real-time, and visualized detection. Fluorescent probe based on PNCs prefer to be developed into the form of test strip. Ratio fluorescence is a good method to get visual sensing with high accuracy. In addition, with the help of the color analysis software in smart phones, qualitative and semi-quantitative visual sensing can be easily realized, which has a good application prospect.

In short, with the deepening of research, fluorescent probe based on PNCs should be constantly improved and innovated. The development of highly sensitive, highly selective, portable, and smart fluorescent probe based on PNCs in the field of agriculture and food is expected.

Author Contributions: Funding acquisition, W.Z. and T.Y.; writing-original draft preparation, W.Z., J.Z. and F.K.; writing—review and editing, W.Z. and T.Y.; validation and supervision, T.Y.; formal analysis and investigation, W.Z. All authors have read and agreed to the published version of the manuscript.

Funding: This work was supported by the Open Fund of the State Key Laboratory of Luminescent Materials and Devices, South China University of Technology (Grant No. 2022-skllmd-08), the National Key Research and Development Program of China (No. 2022YFD1200704--2), the National Key Research and Development Program of China (No. 2021YFD1201000), and the Heilongjiang Provincial Postdoctoral Science Foundation (Grant No. LBH-TZ0604).

Institutional Review Board Statement: Not applicable.

Data Availability Statement: Not applicable.

Conflicts of Interest: The authors declare no conflict of interest.

References

1. Vajubhai, G.N.; Chetti, P.; Kailasa, S.K. Perovskite quantum dots for fluorescence turn-off detection of the clodinafop pesticide in food samples via liquid–liquid microextraction. *ACS Appl. Nano Mater.* **2022**, *5*, 18220–18228. [CrossRef]
2. He, J.; Yu, L.; Jiang, Y.; Lü, L.; Han, Z.; Zhao, X.; Xu, Z. Encoding $CsPbX_3$ perovskite quantum dots with different colors in molecularly imprinted polymers as fluorescent probes for the quantitative detection of Sudan I in food matrices. *Food Chem.* **2023**, *402*, 134499. [CrossRef] [PubMed]
3. Gupta, A.; Mehta, S.K.; Kunal, K.; Mukhopadhyay, K.; Singh, S. Quantum dots as promising nanomaterials in agriculture. In *Agricultural Nanobiotechnology*; Woodhead Publishing: Southton, UK, 2022; pp. 243–296.
4. Huang, S.; Guo, M.; Tan, J.; Geng, Y.; Wu, J.; Tang, Y.; Liang, Y. Novel fluorescent probe based on all-inorganic perovskite quantum dots coated with molecularly imprinted polymers for highly selective and sensitive detection of omethoate. *ACS Appl. Mater. Interfaces* **2018**, *10*, 39056–39063. [CrossRef] [PubMed]
5. Shangguan, Z.; Zheng, X.; Zhang, J.; Lin, W.; Guo, W.; Li, C.; Wu, T.; Lin, Y.; Chen, Z. The stability of metal halide perovskite nanocrystals—A key issue for the application on quantum-dot-based micro light-emitting diodes display. *Nanomaterials* **2020**, *10*, 1375. [CrossRef]
6. Jia, D.; Xu, M.; Mu, S.; Ren, W.; Liu, C. Recent progress of perovskite nanocrystals in chem/bio sensing. *Biosensors* **2022**, *12*, 754. [CrossRef] [PubMed]
7. Wei, Y.; Cheng, Z.; Lin, J. An overview on enhancing the stability of lead halide perovskite quantum dots and their applications in phosphor-converted LEDs. *Chem. Soc. Rev.* **2019**, *48*, 310–350. [CrossRef] [PubMed]
8. Stoumpos, C.C.; Malliakas, C.D.; Kanatzidis, M.G. Semiconducting tin and lead iodide perovskites with organic cations: Phase transitions, high mobilities, and near-infrared photoluminescent properties. *Inorg. Chem.* **2013**, *52*, 9019–9038. [CrossRef]
9. Liu, Y.; Pan, G.; Wang, R.; Shao, H.; Wang, H.; Xu, W.; Song, H. Considerably enhanced exciton emission of $CsPbCl_3$ perovskite quantum dots by the introduction of potassium and lanthanide ions. *Nanoscale* **2018**, *10*, 14067–14072. [CrossRef]
10. Huang, S.; Wang, B.; Zhang, Q.; Li, Z.; Shan, A.; Li, L. Postsynthesis potassium modification method to improve stability of $CsPbBr_3$ perovskite nanocrystals. *Adv. Opt. Mater.* **2018**, *6*, 1701106. [CrossRef]
11. Protesescu, L.; Yakunin, S.; Bodnarchuk, M.I.; Krieg, F.; Caputo, R.; Hendon, C.H.; Kovalenko, M.V. Nanocrystals of cesium lead halide perovskites ($CsPbX_3$, X = Cl, Br, and I): Novel optoelectronic materials showing bright emission with wide color gamut. *Nano Lett.* **2015**, *15*, 3692–3696. [CrossRef]
12. Di Stasio, F.; Christodoulou, S.; Huo, N.; Konstantatos, G. Near-unity photoluminescence quantum yield in $cspbbr_3$ nanocrystal solid-state films via postsynthesis treatment with lead bromide. *Chem. Mater.* **2017**, *29*, 7663–7667. [CrossRef]
13. Chen, M.; Zou, Y.; Wu, L.; Pan, Q.; Yang, D.; Hu, H.; Zhang, Q. Solvothermal synthesis of high-quality all-inorganic cesium lead halide perovskite nanocrystals: From nanocube to ultrathin nanowire. *Adv. Funct. Mater.* **2017**, *27*, 1701121. [CrossRef]
14. Pan, Q.; Hu, H.; Zou, Y.; Chen, M.; Wu, L.; Yang, D.; Zhang, Q. Microwave-assisted synthesis of high-quality "all-inorganic" $CsPbX_3$ (X = Cl, Br, I) perovskite nanocrystals and their application in light emitting diodes. *J. Mater. Chem. C* **2017**, *5*, 10947–10954. [CrossRef]
15. Tong, Y.; Bladt, E.; Aygüler, M.F.; Manzi, A.; Milowska, K.Z.; Hintermayr, V.A.; Feldmann, J. Highly luminescent cesium lead halide perovskite nanocrystals with tunable composition and thickness by ultrasonication. *Angew. Chem. Int. Ed.* **2016**, *55*, 13887–13892. [CrossRef] [PubMed]
16. Zhu, M.; Bai, J.; Chen, R.; Li, H. Synthesis and properties of B-site doped all-inorganic perovskite quantum dots. *Chin. J. Appl. Chem.* **2021**, *38*, 1541.
17. Wei, Y.; Deng, X.; Xie, Z. Enhancing the stability of perovskite quantum dots by encapsulation in crosslinked polystyrene beads via a swelling–shrinking strategy toward superior water resistance. *Adv. Funct. Mater.* **2017**, *27*, 1703535. [CrossRef]
18. An, H.; Kim, W.K.; Wu, C.; Kim, T.W. Highly-stable memristive devices based on poly (methylmethacrylate): $CsPbCl_3$ perovskite quantum dot hybrid nanocomposites. *Org. Electron.* **2018**, *56*, 41–45. [CrossRef]
19. Xuan, T.; Huang, J.; Liu, H.; Lou, S.; Cao, L.; Gan, W.; Wang, J. Super-hydrophobic cesium lead halide perovskite quantum dot-polymer composites with high stability and luminescent efficiency for wide color gamut white light-emitting diodes. *Chem. Mater.* **2019**, *31*, 1042–1047. [CrossRef]
20. Avugadda, S.K.; Castelli, A.; Dhanabalan, B.; Fernandez, T.; Silvestri, N.; Collantes, C.; Arciniegas, M.P. Highly emitting perovskite nanocrystals with 2-year stability in water through an automated polymer encapsulation for bioimaging. *ACS Nano* **2022**, *16*, 13657–13666. [CrossRef]
21. Li, S.; Lei, D.; Ren, W.; Guo, X.; Wu, S.; Zhu, Y.; Jen, A.K.Y. Water-resistant perovskite nanodots enable robust two-photon lasing in aqueous environment. *Nat. Commun.* **2020**, *11*, 1192. [CrossRef]
22. You, C.Y.; Li, F.M.; Lin, L.H.; Lin, J.S.; Chen, Q.Q.; Radjenovic, P.M.; Li, J.F. Ultrastable monodispersed lead halide perovskite nanocrystals derived from interfacial compatibility. *Nano Energy* **2020**, *71*, 104554. [CrossRef]
23. Zhang, Q.; Wang, B.; Zheng, W.; Kong, L.; Wan, Q.; Zhang, C.; Li, L. Ceramic-like stable $CsPbBr_3$ nanocrystals encapsulated in silica derived from molecular sieve templates. *Nat. Commun.* **2020**, *11*, 31. [CrossRef] [PubMed]
24. Qiao, G.Y.; Guan, D.; Yuan, S.; Rao, H.; Chen, X.; Wang, J.A.; Yu, J. Perovskite quantum dots encapsulated in a mesoporous metal–organic framework as synergistic photocathode materials. *J. Am. Chem. Soc.* **2021**, *143*, 14253–14260. [CrossRef] [PubMed]

25. Wu, L.Y.; Mu, Y.F.; Guo, X.X.; Zhang, W.; Zhang, Z.M.; Zhang, M.; Lu, T.B. Encapsulating perovskite quantum dots in iron-based metal-organic frameworks (MOFs) for efficient photocatalytic CO_2 reduction. *Angew. Chem. Int. Ed.* **2019**, *58*, 9491–9495. [CrossRef]
26. Liu, B.; Guo, X.; Gao, D. Research progress on the improvement of the stability of perovskite quantum dots. *Chem. Ind. Eng. Prog.* **2021**, *40*, 247–258. [CrossRef]
27. Pering, S.R.; Deng, W.; Troughton, J.R.; Kubiak, P.S.; Ghosh, D.; Niemann, R.G.; Cameron, P.J. Azetidinium lead iodide for perovskite solar cells. *J. Mater. Chem. A* **2017**, *5*, 20658–20665. [CrossRef]
28. Filip, M.R.; Eperon, G.E.; Snaith, H.J.; Giustino, F. Steric engineering of metal-halide perovskites with tunable optical band gaps. *Nat. Commun.* **2014**, *5*, 5757. [CrossRef]
29. Protesescu, L.; Yakunin, S.; Kumar, S.; Bär, J.; Bertolotti, F.; Masciocchi, N.; Kovalenko, M.V. Dismantling the "red wall" of colloidal perovskites: Highly luminescent formamidinium and formamidinium–cesium lead iodide nanocrystals. *ACS Nano* **2017**, *11*, 3119–3134. [CrossRef]
30. Wu, X.; Jiang, Y.; Chen, C.; Guo, J.; Kong, X.; Feng, Y.; Gao, J. Stable triple cation perovskite precursor for highly efficient perovskite solar cells enabled by interaction with 18C6 stabilizer. *Adv. Funct. Mater.* **2020**, *30*, 1908613. [CrossRef]
31. Xu, L.; Yuan, S.; Zeng, H.; Song, J.J. A comprehensive review of doping in perovskite nanocrystals/quantum dots: Evolution of structure, electronics, optics, and light-emitting diodes. *Mater. Today Nano* **2019**, *6*, 100036. [CrossRef]
32. Wang, J.; Liu, X.; Zhou, L.; Shen, W.; Li, M.; He, R. Highly luminescent and stable quasi-2D perovskite quantum dots by introducing large organic cations. *Nanoscale Adv.* **2021**, *3*, 5393–5398. [CrossRef]
33. Zhou, Y.; Chen, J.; Bakr, O.M.; Sun, H.T. Metal-doped lead halide perovskites: Synthesis, properties, and optoelectronic applications. *Chem. Mater.* **2018**, *30*, 6589–6613. [CrossRef]
34. Zhuang, X.; Sun, R.; Zhou, D.; Liu, S.; Wu, Y.; Shi, Z.; Song, H. Synergistic effects of multifunctional lanthanides doped $CsPbBrCl_2$ quantum dots for efficient and stable $MAPbI_3$ perovskite solar cells. *Adv. Funct. Mater.* **2022**, *32*, 2110346. [CrossRef]
35. Shen, X.; Zhang, Y.; Kershaw, S.V.; Li, T.; Wang, C.; Zhang, X.; Rogach, A.L. Zn-alloyed $CsPbI_3$ nanocrystals for highly efficient perovskite light-emitting devices. *Nano Lett.* **2019**, *19*, 1552–1559. [CrossRef]
36. Song, P.; Qiao, B.; Song, D.; Cao, J.; Shen, Z.; Xu, Z.; Al-Ghamdi, A. Modifying the crystal field of $CsPbCl_3$: Mn^{2+} nanocrystals by co-doping to enhance its red emission by a hundredfold. *ACS Appl. Mater. Interfaces* **2020**, *12*, 30711–30719. [CrossRef]
37. Kim, H.; Bae, S.R.; Lee, T.H.; Lee, H.; Kang, H.; Park, S.; Kim, S.Y. Enhanced optical properties and stability of $CsPbBr_3$ nanocrystals through nickel doping. *Adv. Funct. Mater.* **2021**, *31*, 2102770. [CrossRef]
38. Zeng, Q.; Zhang, X.; Bing, Q.; Xiong, Y.; Yang, F.; Liu, H.; Yang, B. Surface Stabilization of Colloidal Perovskite Nanocrystals via Multi-amine Chelating Ligands. *ACS Energy Lett.* **2022**, *7*, 1963–1970. [CrossRef]
39. Bi, C.; Kershaw, S.V.; Rogach, A.L.; Tian, J. Improved stability and photodetector performance of $CsPbI_3$ perovskite quantum dots by ligand exchange with aminoethanethiol. *Adv. Funct. Mater.* **2019**, *29*, 1902446. [CrossRef]
40. Krieg, F.; Ochsenbein, S.T.; Yakunin, S.; Ten Brinck, S.; Aellen, P.; Süess, A.; Kovalenko, M.V. Colloidal $CsPbX_3$ (X = Cl, Br, I) nanocrystals 2.0: Zwitterionic capping ligands for improved durability and stability. *ACS Energy Lett.* **2018**, *3*, 641–646. [CrossRef]
41. Krieg, F.; Ong, Q.K.; Burian, M.; Rainò, G.; Naumenko, D.; Amenitsch, H.; Kovalenko, M.V. Stable ultraconcentrated and ultradilute colloids of $CsPbX_3$ (X= Cl, Br) nanocrystals using natural lecithin as a capping ligand. *J. Am. Chem. Soc.* **2019**, *141*, 19839–19849. [CrossRef]
42. Wang, S.; Du, L.; Jin, Z.; Xin, Y.; Mattoussi, H. Enhanced stabilization and easy phase transfer of $CsPbBr_3$ perovskite quantum dots promoted by high-affinity polyzwitterionic ligands. *J. Am. Chem. Soc.* **2020**, *142*, 12669–12680. [CrossRef] [PubMed]
43. Pan, Q.; Hu, J.; Fu, J.; Lin, Y.; Zou, C.; Di, D.; Cao, M. Ultrahigh stability of perovskite nanocrystals by using semiconducting molecular species for displays. *ACS Nano* **2022**, *16*, 12253–12261. [CrossRef] [PubMed]
44. Rajput, S.; Sharma, R.; Kumari, A.; Kaur, R.; Sharma, G.; Arora, S.; Kaur, R. Pesticide residues in various environmental and biological matrices: Distribution, extraction, and analytical procedures. *Environ. Dev. Sustain.* **2022**, *24*, 6032–6052. [CrossRef]
45. Damalas, C.A.; Eleftherohorinos, I.G. Pesticide exposure, safety issues, and risk assessment indicators. *Int. J. Environ. Res. Public Health* **2011**, *8*, 1402–1419. [CrossRef]
46. Huang, S.; Tan, L.; Zhang, L.; Wu, J.; Zhang, L.; Tang, Y.; Liang, Y. Molecularly imprinted mesoporous silica embedded with perovskite $CsPbBr_3$ quantum dots for the fluorescence sensing of 2, 2-dichlorovinyl dimethyl phosphate. *Sens. Actuators B Chem.* **2020**, *325*, 128751. [CrossRef]
47. Tan, L.; Guo, M.; Tan, J.; Geng, Y.; Huang, S.; Tang, Y.; Liang, Y. Development of high-luminescence perovskite quantum dots coated with molecularly imprinted polymers for pesticide detection by slowly hydrolysing the organosilicon monomers in situ. *Sens. Actuators B Chem.* **2019**, *291*, 226–234. [CrossRef]
48. Chen, D.; Guan, J.; Zhang, G. Research status of microbial degradation of pesticides. *J. Henan Inst. Sci. Technol. (Nat. Sci. Ed.)* **2020**, *48*, 39–46.
49. Zhang, R.R.; Gan, X.T.; Xu, J.J.; Pan, Q.F.; Liu, H.; Sun, A.L.; Zhang, Z.M. Ultrasensitive electrochemiluminescence sensor based on perovskite quantum dots coated with molecularly imprinted polymer for prometryn determination. *Food Chem.* **2022**, *370*, 131353. [CrossRef]
50. Pan, Q.F.; Jiao, H.F.; Liu, H.; You, J.J.; Sun, A.L.; Zhang, Z.M.; Shi, X.Z. Highly selective molecularly imprinted-electrochemiluminescence sensor based on perovskite/Ru $(bpy)_3^{2+}$ for simazine detection in aquatic products. *Sci. Total Environ.* **2022**, *843*, 156925. [CrossRef]

51. Yang, Y.; Han, A.; Hao, S.; Li, X.; Luo, X.; Fang, G.; Liu, J.; Wang, S. Fluorescent methylammonium lead halide perovskite quantum dots as a sensing material for the detection of polar organochlorine pesticide residues. *Analyst* **2020**, *145*, 6683–6690. [CrossRef]
52. Liu, C.X.; Zhao, J.; Zhang, R.R.; Zhang, Z.M.; Xu, J.J.; Sun, A.L.; Shi, X.Z. Development and application of fluorescent probe and test strip based on molecularly imprinted quantum dots for the selective and sensitive detection of propanil in fish and seawater samples. *J. Hazard. Mater.* **2020**, *389*, 121884. [CrossRef] [PubMed]
53. Halali, V.V.; Rani, R.S.; Balakrishna, R.G.; Budagumpi, S. Ultra-trace level chemosensing of uranyl ions; scuffle between electron and energy transfer from perovskite quantum dots to adsorbed uranyl ions. *Microchem. J.* **2020**, *156*, 104808. [CrossRef]
54. Deng, P.; Wang, W.; Liu, X.; Wang, L.; Yan, Y. A hydrophobic polymer stabilized $CsPbBr_3$ sensor for environmental pollutant detection. *New J. Chem.* **2021**, *45*, 930–938. [CrossRef]
55. Liu, Y.; Tang, X.; Zhu, T.; Deng, M.; Ikechukwu, I.P.; Huang, W.; Qiu, F. All-inorganic $CsPbBr_3$ perovskite quantum dots as a photoluminescent probe for ultrasensitive Cu^{2+} detection. *J. Mater. Chem. C* **2018**, *6*, 4793–4799. [CrossRef]
56. Li, Q.; Zhou, W.; Yu, L.; Lian, S.; Xie, Q. Perovskite quantum dots as a fluorescent probe for metal ion detection in aqueous solution via phase transfer. *Mater. Lett.* **2021**, *282*, 128654. [CrossRef]
57. Lu, L.Q.; Tan, T.; Tian, X.K.; Li, Y.; Deng, P. Visual and sensitive fluorescent sensing for ultratrace mercury ions by perovskite quantum dots. *Anal. Chim. Acta* **2017**, *986*, 109–114. [CrossRef]
58. Chen, M.; An, J.; Hu, Y.; Chen, R.; Lyu, Y.; Hu, N.; Liu, Y. Swelling-shrinking modified hyperstatic hydrophilic perovskite polymer fluorescent beads for Fe (III) detection. *Sens. Actuators B Chem.* **2020**, *325*, 128809. [CrossRef]
59. Wang, J.; Li, M.; Shen, W.; Su, W.; He, R. Ultrastable carbon quantum dots-doped $MAPbBr_3$ perovskite with silica encapsulation. *ACS Appl. Mater. Interfaces* **2019**, *11*, 34348–34354. [CrossRef]
60. Yan, J.; He, Y.; Chen, Y.; Zhang, Y.; Yan, H. CH_3NH_3Br solution as a novel platform for the selective fluorescence detection of Pb^{2+} ions. *Sci. Rep.* **2019**, *9*, 15840. [CrossRef]
61. Wang, S.; Huang, Y.; Zhang, L.; Li, F.; Lin, F.; Wang, Y.; Chen, X. Highly selective fluorescence turn-on determination of Pb (II) in Water by in-situ enrichment of Pb (II) and $MAPbBr_3$ perovskite growth in sulfydryl functionalized mesoporous alumina film. *Sens. Actuators B Chem.* **2021**, *326*, 128975. [CrossRef]
62. Park, B.; Kang, S.M.; Lee, G.W.; Kwak, C.H.; Rethinasabapathy, M.; Huh, Y.S. Fabrication of $CsPbBr_3$ perovskite quantum dots/cellulose-based colorimetric sensor: Dual-responsive on-site detection of chloride and iodide ions. *Ind. Eng. Chem. Res.* **2019**, *59*, 793–801. [CrossRef]
63. Han, J.; Sharipov, M.; Hwang, S.; Lee, Y.; Huy, B.T.; Lee, Y.I. Water-stable perovskite-loaded nanogels containing antioxidant property for highly sensitive and selective detection of roxithromycin in animal-derived food products. *Sci. Rep.* **2022**, *12*, 3147. [CrossRef]
64. Jia, L.; Xu, Z.; Zhang, L.; Li, Y.; Zhao, T.; Xu, J. The fabrication of water-stable perovskite-europium hybrid polychromatic fluorescence nanosensor for fast visual sensing of tetracycline. *Appl. Surf. Sci.* **2022**, *592*, 153170. [CrossRef]
65. Thuy, T.T.; Huy, B.T.; Kumar, A.P.; Lee, Y.I. Highly stable $Cs_4PbBr_6/CsPbBr_3$ perovskite nanoparticles as a new fluorescence nanosensor for selective detection of trace tetracycline in food samples. *J. Ind. Eng. Chem.* **2021**, *104*, 437–444. [CrossRef]
66. Wang, W.; Deng, P.; Liu, X.; Ma, Y.; Yan, Y. A $CsPbBr_3$ quantum dots/ultra-thin BN fluorescent probe for stability and highly sensitive detection of tetracycline. *Microchem. J.* **2021**, *162*, 105876. [CrossRef]
67. Wang, T.; Wei, X.; Zong, Y.; Zhang, S.; Guan, W. An efficient and stable fluorescent sensor based on APTES-functionalized $CsPbBr_3$ perovskite quantum dots for ultrasensitive tetracycline detection in ethanol. *J. Mater. Chem. C* **2020**, *8*, 12196–12203. [CrossRef]
68. Salari, R.; Amjadi, M.; Hallaj, T. Perovskite quantum dots as a chemiluminescence platform for highly sensitive assay of cefazolin. *Spectrochim. Acta Part A Mol. Biomol. Spectrosc.* **2023**, *285*, 121845. [CrossRef] [PubMed]
69. Shi, X.; Kralj, M.; Zhang, Y. Colorimetric test strips based on cesium lead bromide perovskite nanocrystals for rapid detection of ciprofloxacin hydrochloride. *J. Phys. Condens. Matter* **2022**, *34*, 304002. [CrossRef]
70. Su, L.; Tong, P.; Zhang, L.; Luo, Z.; Fu, C.; Tang, D.; Zhang, Y. Photoelectrochemical immunoassay of aflatoxin B_1 in foodstuff based on amorphous TiO_2 and $CsPbBr_3$ perovskite nanocrystals. *Analyst* **2019**, *144*, 4880–4886. [CrossRef] [PubMed]
71. Li, J.; Wang, Q.; Xiong, C.; Deng, Q.; Zhang, X.; Wang, S.; Chen, M.M. An ultrasensitive $CH_3NH_3PbBr_3$ quantum dots@SiO_2-based electrochemiluminescence sensing platform using an organic electrolyte for aflatoxin B_1 detection in corn oil. *Food Chem.* **2022**, *390*, 133200. [CrossRef] [PubMed]
72. Xu, Y.; Shi, L.; Jing, X.; Miao, H.; Zhao, Y. SERS-active composites with Au–Ag Janus nanoparticles/perovskite in immunoassays for Staphylococcus aureus enterotoxins. *ACS Appl. Mater. Interfaces* **2022**, *14*, 3293–3301. [CrossRef]
73. Ye, W.; Cao, Q.; Cheng, X.F.; Yu, C.; He, J.H.; Lu, J.M. Lead-free Cs_2PdBr_6 perovskite-based humidity sensor for artificial fruit waxing detection. *J. Mater. Chem. A* **2020**, *8*, 17675–17682. [CrossRef]
74. Huang, H.; Hao, M.; Song, Y.; Dang, S.; Liu, X.; Dong, Q. Dynamic passivation in perovskite quantum dots for specific ammonia detection at room temperature. *Small* **2020**, *16*, 1904462. [CrossRef] [PubMed]
75. Luo, F.; Li, S.; Cui, L.; Zu, Y.; Chen, Y.; Huang, D.; Lin, Z. Biocompatible perovskite quantum dots with superior water resistance enable long-term monitoring of the H_2S level in vivo. *Nanoscale* **2021**, *13*, 14297–14303. [CrossRef]
76. Li, S.; Zhang, Y.; Zu, Y.; Chen, Y.; Luo, F.; Huang, D.; Lin, Z. Photothermal sensor based on water-soluble $CsPbBr_3$@ sulfobutylether-β-cyclodextrins nanocomposite using a thermometer as readout. *Sens. Actuators B Chem.* **2022**, *355*, 131301. [CrossRef]
77. Luo, F.; Zhang, Y.; Zu, Y.; Li, S.; Chen, Y.; Chen, Z.; Lin, Z. Quick preparation of water-soluble perovskite nanocomposite via cetyltrimethylammonium bromide and its application. *Microchim. Acta* **2022**, *189*, 68. [CrossRef] [PubMed]

78. Wu, C.; Lu, Q.; Miu, X.; Fang, A.; Li, H.; Zhang, Y. A simple assay platform for sensitive detection of Sudan I–IV in chilli powder based on CsPbBr$_3$ quantum dots. *J. Food Sci. Technol.* **2018**, *55*, 2497–2503. [CrossRef]
79. Chan, K.K.; Yap, S.H.K.; Giovanni, D.; Sum, T.C.; Yong, K.T. Water-stable perovskite quantum dots-based FRET nanosensor for the detection of rhodamine 6G in water, food, and biological samples. *Microchem. J.* **2022**, *180*, 107624. [CrossRef]
80. Wang, Y.; Zhu, Y.; Huang, J.; Cai, J.; Zhu, J.; Yang, X.; Li, C. CsPbBr$_3$ perovskite quantum dots-based monolithic electrospun fiber membrane as an ultrastable and ultrasensitive fluorescent sensor in aqueous medium. *J. Phys. Chem. Lett.* **2016**, *7*, 4253–4258. [CrossRef]
81. Ni, D.J.; Zhang, J.; Cao, Z.K.; Li, R.; Xu, T.F.; Sang, H.W.; Long, Y.Z. Supersensitive and reusable perovskite nanocomposite fiber paper for time-resolved single-droplet detection. *J. Hazard. Mater.* **2021**, *403*, 123959. [CrossRef]
82. Shi, L.; Xu, C.; Fan, Y. Determination of basic yellow in food by fluorescence colorimetry based on perovskite nanomaterials. *J. Food Saf. Qual.* **2021**, *12*, 3658–3664.
83. Li, Q.; Wang, H.; Yue, X.; Du, J. Perovskite nanocrystals fluorescence nanosensor for ultrasensitive detection of trace melamine in dairy products by the manipulation of inner filter effect of gold nanoparticles. *Talanta* **2020**, *211*, 120705. [CrossRef] [PubMed]
84. Sun, R.; Yu, X.; Chen, J.; Zhang, W.; Huang, Y.; Zheng, J.; Chi, Y. Highly Electrochemiluminescent Cs$_4$PbBr$_6$@CsPbBr$_3$ perovskite nanoacanthospheres and their application for sensing bisphenol A. *Anal. Chem.* **2022**, *94*, 17142–17150. [CrossRef] [PubMed]
85. Zhu, Y.; Li, F.; Huang, Y.; Lin, F.; Chen, X. Wavelength-shift-based colorimetric sensing for peroxide number of edible oil using CsPbBr$_3$ perovskite nanocrystals. *Anal. Chem.* **2019**, *91*, 14183–14187. [CrossRef] [PubMed]
86. Zhao, Y.; Xu, Y.; Shi, L.; Fan, Y. Perovskite nanomaterial-engineered multiplex-mode fluorescence sensing of edible oil quality. *Anal. Chem.* **2021**, *93*, 11033–11042. [CrossRef]
87. Wang, S.; Yu, L.; Wei, Z.; Xu, Q.; Zhou, W.; Xiao, Y. Dual-response ratiometric fluorescence based ligand-functionalized CsPbBr$_3$ perovskite quantum dots for sensitive detection of trace water in edible oils. *Sens. Actuators B Chem.* **2022**, *366*, 132010. [CrossRef]
88. Huangfu, C.; Feng, L. High-performance fluorescent sensor based on CsPbBr$_3$ quantum dots for rapid analysis of total polar materials in edible oils. *Sens. Actuators B Chem.* **2021**, *344*, 130193. [CrossRef]

Disclaimer/Publisher's Note: The statements, opinions and data contained in all publications are solely those of the individual author(s) and contributor(s) and not of MDPI and/or the editor(s). MDPI and/or the editor(s) disclaim responsibility for any injury to people or property resulting from any ideas, methods, instructions or products referred to in the content.

Article

A Light/Pressure Bifunctional Electronic Skin Based on a Bilayer Structure of PEDOT:PSS-Coated Cellulose Paper/CsPbBr₃ QDs Film

Wenhao Li [2], Jingyu Jia [2], Xiaochen Sun [2], Sue Hao [2,*] and Tengling Ye [1,2,*]

1. CAS Key Laboratory of Renewable Energy, Guangzhou Institute of Energy Conversion, Guangzhou 510640, China
2. Department of Applied Chemistry, School of Chemistry and Chemical Engineering, Harbin Institute of Technology, Harbin 150001, China; liwenhao0824@163.com (W.L.); jjy271828@163.com (J.J.); 15684172397@163.com (X.S.)
* Correspondence: haosue@hit.edu.cn (S.H.); ytl@hit.edu.cn (T.Y.)

Citation: Li, W.; Jia, J.; Sun, X.; Hao, S.; Ye, T. A Light/Pressure Bifunctional Electronic Skin Based on a Bilayer Structure of PEDOT:PSS-Coated Cellulose Paper/CsPbBr₃ QDs Film. *Polymers* **2023**, *15*, 2136. https://doi.org/10.3390/polym15092136

Academic Editor: Zhaoling Li

Received: 6 April 2023
Revised: 27 April 2023
Accepted: 28 April 2023
Published: 29 April 2023

Copyright: © 2023 by the authors. Licensee MDPI, Basel, Switzerland. This article is an open access article distributed under the terms and conditions of the Creative Commons Attribution (CC BY) license (https://creativecommons.org/licenses/by/4.0/).

Abstract: With the continuous development of electronic skin (e-skin), multifunctional e-skin is approaching, and in some cases even surpassing, the capabilities of real human skin, which has garnered increasing attention. Especially, if e-skin processes eye's function, it will endow e-skins more powerful advantages, such as the vision reparation, enhanced security, improved adaptability and enhanced interactivity. Here, we first study the photodetector based on CsPbBr₃ quantum dots film and the pressure sensor based on PEDOT: PSS-coated cellulose paper, respectively. On the base of these two kinds of sensors, a light/pressure bifunctional sensor was successfully fabricated. Finally, flexible bifunctional sensors were obtained by using a flexible interdigital electrode. They can simultaneously detect light and pressure stimulation. As e-skin, a high photosensitivity with a switching ratio of 168 under 405 nm light at a power of 40 mW/cm² was obtained and they can also monitor human motions in the meantime. Our work showed that the strategy to introduce perovskite photodetectors into e-skins is feasible and may open a new way for the development of flexible multi-functional e-skin.

Keywords: e-skin photodetector; pressure sensor; CsPbBr₃; PEDOT: PSS; e-skin

1. Introduction

Electronic skin (e-skin) is an electronic system to simulate the function of human skin. It is a kind of important flexible wearable electronic device that has a promising prospect in the field of health monitoring, human-machine interaction, software robots, and the Internet of Things [1–4]. Recently, multifunctional e-skins have become a hot direction to fulfill the requirements of specific wearable and portable applications. In the many exploration processes to realize multifunctional e-skins, it is the main work to simulate the function of human skin, such as the sensory abilities to stress, strain, temperature, humidity and chemicals, etc. [5–8]. However, in addition to the function of human skin, there are many ways for human beings to obtain external information. The eyes, ears, nose, and mouth are indispensable channels for obtaining external information. Especially, for realizing the eye's function, the retina is a key film-shaped tissue, which directly detects light and outputting/transmitting a bioelectricity signal to the nervous system [9]. If e-skins have the function of the retina, they may be endowed with some very interesting advantages [10]. For example, e-skins with visual perception can perceive light and images such as eyes, thus the user's safety in dangerous environments can be improved, avoiding potential dangers, such as high-energy radiation damage, and improving environmental adaptability [11]. Through the visual perception ability of e-skins, the robot with these e-skins can better interact with humans and enhance the ability of human–machine interactions. In addition, it may be more powerful in future medical care and may play an important role in vision reparation for people who are blind [10]. Anyway, it will bring broader application prospects and more functional advantages.

Perovskite photodetectors have recently become popular due to their exceptional photoelectric properties and easy fabrication process. [12]. Inorganic halide $CsPbBr_3$ perovskite quantum dots (PQDs) are considered as a new generation of photoelectric material for the photodetector due to their merits including a high light absorption coefficient, a long charge carrier life, a tunable bandgap, relatively high stability, inexpensiveness, and flexible compositional control [13,14]. With the deepening of research, more and more multifunctional photodetectors have been developed [15,16]. What is more, the absorption of $CsPbBr_3$ QDs is in the high energy region from UV to blue-green, and it is suitable for the high-energy radiation detection. On the other hand, piezoresistive flexible pressure sensors (PFPSs) can transduce pressure into corresponding resistance signals and have been extensively investigated as e-skin owing to their low cost, simple assembly process and excellent sensing performances [17,18]. Generally, PFPSs usually consist of an elastic, porous, conductive interlayer sandwiched between two flexible electrodes. Porous cellulose paper is considered as a good candidate for the interlayer framework materials owing to its low cost, light weight, high surface area, outstanding deformability, excellent flexibility, and air permeability. When combining the porous cellulose paper with common conductive materials, such as carbon nanomaterials, metal nanomaterials or conductive polymers, highly sensitive FPSs with a low cost can be easily obtained. However, the conductive materials in these FPSs are usually opaque, which cannot meet the requirement of light sensing e-skins [19,20].

In this paper, we propose to combine a perovskite photodetector with a piezoresistive pressure sensor to design a bifunctional sensor with both light and pressure response. The light/pressure bifunctional e-skins were prepared based on PEDOT: PSS-coated cellulose paper/$CsPbBr_3$ QDs film. Firstly, we studied the photodetectors based on $CsPbBr_3$ PQDs and optimized the fabrication process on the rigid conductive glass. Then, the transparent conductive polymer PEDOT: PSS and cellulose papers were selected to fabricate the piezoresistive pressure sensor. After that the two sensors are integrated together to prepare the light/pressure bifunctional sensors. Finally, the flexible bifunctional sensors are successfully obtained by replacing the rigid conductive glass with a flexible interdigital electrode. In addition, flexible bifunctional sensors are successfully applied to human physiological signal monitoring.

2. Materials, Methods, and Characterization

2.1. Materials

$PbBr_2$ (99%) were supplied by Shanghai Energy Chemical Co., Ltd., Shanghai, China. and Cs_2CO_3 (99.9%) were purchased from ZhongNuo Advanced Material (Beijing) Technology Co., Ltd., Beijing, China; Oleamine (OLA with 90%) and Octadecylene (90%) were purchased from Shanghai Macklin Biochemical Technology Co., Ltd., Shanghai, China; Oleic acid (OA, analytical reagent) was supplied by DaMao Chemical Reagent Factory, Tianjin, China; n-Hexane, Isopropyl alcohol and acetone (analytical pure) were purchased from Tianjin Fuyu Chemical Reagent Co., Ltd., Tianjin, China; PEDOT:PSS, 4083, was purchased from Xi'an Bao Laite Optoelectronics Technology Co., Ltd., Xi'an, China; Cellulose paper (plain weave, 60 g, 27.5 cm × 57.5 cm) was supplied from Sichuan Sanheng Yishu Technology Co., Ltd., Chengdu, China.

2.2. Methods

2.2.1. Synthesis of $CsPbBr_3$ PQDs

The synthesis follows the modified method reported by Wang et al. [21]. 0.65 g Cs_2CO_3, 2.5 mL of OA, and 18 mL of ODE were mixed and kept in an N_2 atmosphere. The mixture was heated up to 130 °C for 1 h. After that, the mixture was further heated up to 150 °C for reaction and lasted for 0.5 h. Then, the solution was cooled down to room temperature.

On the other side, 2 mL ODE, 0.3 mL OA, 0.3 mL OLA and 0.2 mmol of $PbBr_2$, were put together and kept in an N_2 atmosphere for 1 h at temperature of 120 °C. Then, the solution was further heated up to 160 °C for reaction for 10 min. Finally, the Cs precursor (0.2 mL) was quickly added into the solution and cooled down with an ice bath. After centrifugation, the resulting pellet was dispersed with n-hexane.

2.2.2. Preparation of Photodetectors Based on $CsPbBr_3$ QDs Film

ITO glass with a slit was firstly washed with alkaline dish soap to remove dust from the surface. After that, three alternating ultrasounds were used with acetone and isopropanol for 30 min each time. Next, a certain amount of washing solvent (isopropanol or acetone) is added into the obtained QDs solution in n-hexane according to the n-hexane/solvent volume ratio of 1:3. Then, we repeat the process of centrifugation and dispersion, and the washed $CsPbBr_3$ was redispersed in n-hexane. Finally, the indium tin oxide glass (ITO) is soaked in the washed $CsPbBr_3$ solution, and a $CsPbBr_3$ film was obtained by the centrifugation method (5000 r/min, 20 mg/mL). The $CsPbBr_3$ film on ITO was placed on a hot plate at 120 °C for 10 min to evaporate the excess solvent. ITO is 1.5 cm × 1.0 cm in size, and there is an etched area with a width of 105 μm.

2.2.3. Preparation of Pressure Sensors Based on PEDOT: PSS-Coated Cellulose Paper

The spunlace and needled non-woven paper are made of polyethylene terephthalate (PET) and cellulose paper was made of cellulose. All of them were cut into the size of 0.6 cm × 1 cm. Next, different volumes of PEDOT: PSS solution were dropwise added to the paper, and then they were dried in an oven at 80 °C for 20 min. The PEDOT: PSS-coated cellulose paper was placed on top of the ITO glass. Finally, a polydimethylsiloxane (PDMS) film is used to package and fix the pressure sensor with the structure of PEDOT: PSS-coated cellulose paper/ITO.

2.2.4. Preparation of Light/Pressure Bifunctional Sensors

Firstly, the $CsPbBr_3$ QDs film on ITO was fabricated using the centrifugation method as described in 2.2.2. After that, 20 μL of PEDOT: PSS aqueous solution was added dropwise to the cellulose paper (0.6 cm × 1 cm) which was dried in an oven at 80 °C for 20 min. Next, the conductive paper was placed on top of the ITO glass. Finally, the PDMS film was used to package and fix the bifunctional sensor. For the flexible bifunctional sensors, we just simply replaced the ITO glass with a flexible interdigital electrode (Au/polyimide, 10 mm × 20 mm × 13 μm, Au width: 100 μm, Au distance: 100 μm, 20 pairs of electrodes and the electrode length is 6.3 mm).

2.3. Characterization

The morphology of $CsPbBr_3$ QDs were performed on a transmission electron microscope (TEM, JEM 1400 plus, 100 kV). The Fourier transform infrared (FTIR) spectrum of $CsPbBr_3$ QDs was recorded by a Fourier infrared spectrometer (FTIR-7600, Lambda). X-ray diffraction (XRD) patterns in the range of 5–85° were recorded on an PANalytical X'Pert Powder diffractometer using Cu-Kα radiation (λ = 1.5406 Å) (PANalytical, Almelo, The Netherlands). Optical microscope photographs were obtained by using a TXS06-02H Fluorescence Microscope (Srate Optical Instrument Manufactory, Nanyang, China). The absorption spectrum of $CsPbBr_3$ QDs were measured by a Ultraviolet-visible spectrophotometer (Cary 100, Agilent Technologies Inc., Santa Clara, CA, USA) and the fluorescence spectrums were measured using an optic fiber associated with a spectrometer (Ocean Optics USB 4000). The samples were dissolved in n-hexane with a concentration of 1×10^{-4} mol/L. Furthermore, the pressure sensing performances of the pressure sensor were implemented on a homemade test system. A 50 g weight was connected to a dip coater to apply reciprocating pressure on pressure sensors and a KEITHLEY 2400 source meter connected to a computer was used for the in-situ electrical signal collection during the expansive and compressive process. The current–voltage curves of the photodetector and the pressure sensors were also recorded using the KEITHLEY 2400 source meter; the applied bias is 1 V or 10 V. For photoelectric measurement, the excitation light source was an ultra-violet source with a 365 filter (EXECURE 4000, HOYA, Tokyo, Japan) and a LED light source (405 nm, 532 nm, 660 nm), Guangzhou Jingyi Optoelectronic Technology Co., Ltd.). When the LED light source of JY18102301 is used, the distance between the source and photodetector was determined by the power densities which were estimated using a power meter (Newport, Model 2936-C). The switching ratio of the photo detector was calculated using the switching ratio = $(I_p/I_d) = ((I_{on} - I_d)/(I_d))$.

3. Results and Discussion

3.1. Preparation and Properties of PQDs Photodetector

$CsPbBr_3$ QDs were successfully synthesized by the typical hot injection method [22,23] and the morphology and structure can be confirmed by the TEM and XRD. The TEM image of $CsPbBr_3$ QDs is shown in Figure 1a and the average size of $CsPbBr_3$ QDs is 10~15 nm. The XRD pattern of $CsPbBr_3$ QDs in Figure 1b shows that its structure belongs to cubic phase (JC//PDS No. 54-0752). As described in the introduction, the $CsPbBr_3$ QDs may be well applicable for high energy photodetectors because of their range of light absorptivity and high stability. As shown in Figure 1c and Figure S1, The UV–vis absorption spectrum indicates that the $CsPbBr_3$ QDs have a direct bandgap of about 2.49 eV and the fluorescence spectrum in Figure 1d shows an emission peak at 521 nm. Compared to the bulk crystal, the decrease in size of the $CsPbBr_3$ QDs results in a blue-shift of the absorption onset due to quantum size effects [24,25]. In addition, the absorbance range indicates that the $CsPbBr_3$ QDs can be used as the active layer for UV-blue-green photodetectors.

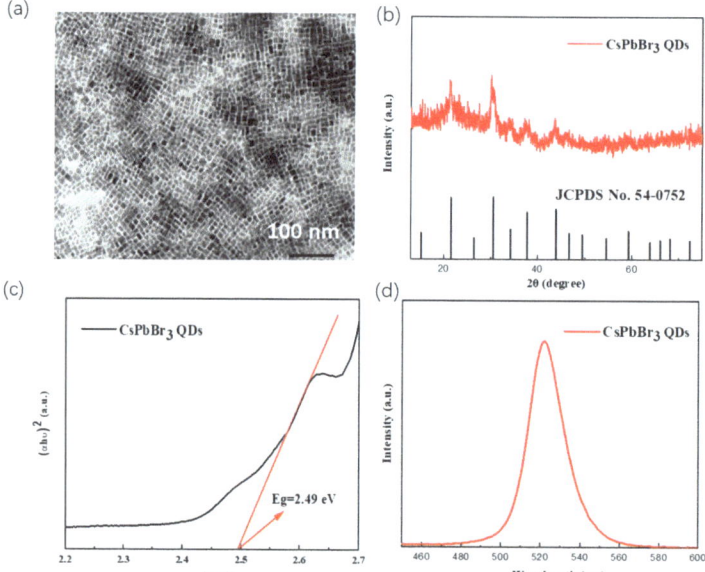

Figure 1. $CsPbBr_3$ QDs: (**a**) TEM; (**b**) XRD pattern; (**c**) Optical bandgap, red lines indicate optical bandgaps determined from Tauc plots; (**d**) Fluorescence spectrum.

The planar photodetector device based on $CsPbBr_3$ QDs is illustrated in Figure 2a. Here, we fabricated $CsPbBr_3$ QDs films on rigid conductive ITO substrates for conveniently optimizing the detector performance. The $CsPbBr_3$ QDs films formed via centrifugal casting served as the photoactive layer and the ITO glass with a slit was used as the two electrodes. In the process of the synthesis of $CsPbBr_3$ QDs, oleic acid (OA), Octadecene and oleyamine (OAM) were used as capping ligands. Figure 2b shows the FTIR spectrum of CsPbBr$_3$ QDs. There are rotational or vibration absorption peaks at 2920 cm^{-1}, 2850 cm^{-1}, 1730 cm^{-1} and 1460 cm^{-1}, among which 2920 cm^{-1} and 2850 cm^{-1} can be regarded as vibration absorption peaks of C-H single bond. While 1730 cm^{-1} can be regarded as the stretching vibration absorption peak of C=O, 1460 cm^{-1} can be inferred as the deformation vibration absorption peak of NH$_2$. The existence of organic ligands can effectively stabilize QDs and improve film-forming property. On the other hand, the excessive capping ligand also makes an insulating layer outside perovskite QDs, which inevitably restricts the charge carrier transfer efficiency for perovskite QDs film. Ligand engineering is needed to balance surface passivation and carrier transport of CsPbBr$_3$ QDs [21,26].

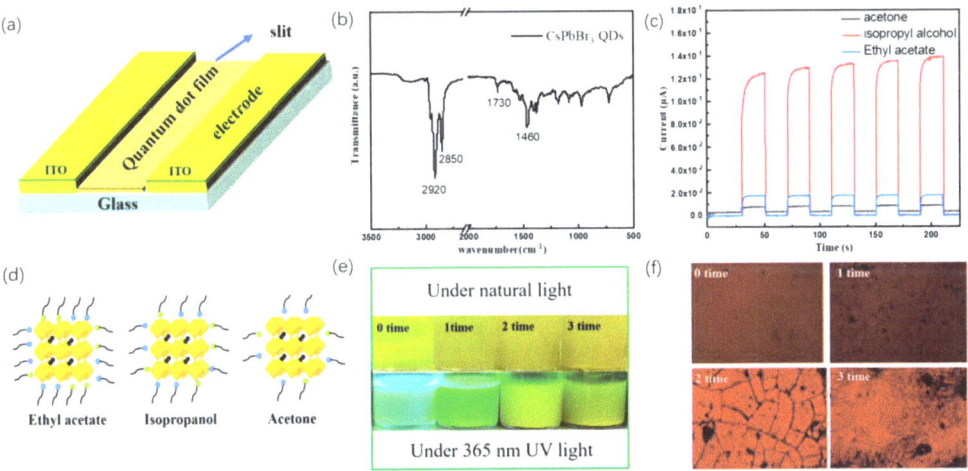

Figure 2. (**a**) Schematic diagram of the PQDs photodetector; (**b**) Infrared spectrum of CsPbBr$_3$ QDs; (**c**) I-t curves of photodetectors washed by different solvents under UV lamp (10 v); (**d**) Schematic diagram of CsPbBr$_3$ QDs washed by different solvents; (**e**) Micromorphology of devices with different washing times (isopropanol + n-hexane); (**f**) Micromorphology of the device with different washing times.

Here, we demonstrated a solvent-washed method to control the surface ligand density of CsPbBr$_3$ QDs and a balance surface passivation, and carrier transport is obtained via solvent treatment. Washing solvents with different polarities and washing times were optimized. Different solvents with different polarities will result in different elutive power to the ligand on CsPbBr$_3$ QDs which affects the morphology and carrier transport of CsPbBr$_3$ QDs. Firstly, ethyl acetate, isopropanol and acetone were used as solvents to wash CsPbBr$_3$ QDs, and the photoelectric performance of the corresponding detectors was tested as shown in Figure 2c. The switching ratio of the photodetector washed by isopropanol was as high as 384, which was much higher than that of detectors washed by ethyl acetate and acetone (78 and 1.35, respectively). To illustrate the effect of different solvent treatments, the photographs of CsPbBr$_3$ QDs dispersed in n-hexane solution under daylight and 365 nm UV light are shown in Figure S2, respectively. The order of Figure S2a–d are the cases without treatment and washed with ethyl acetate, isopropanol, and acetone, respectively. It can be seen that CsPbBr$_3$ QDs solutions change from green (clarification) to yellow (turbid). When excited under the UV lamp, we can observe the fluorescent to gradually get dark and the solution washed by acetone is very weak. These phenomena can be explained by the polarity order of the washing solvents (ethyl acetate < isopropanol < acetone). The surface ligand density of CsPbBr$_3$ QDs can be controlled by the polarity of solvents. As shown in Figure 2d, acetone with the largest polarity washed away most of the ligands on the surface of QDs. As the organic ligand was washed away, the stability of the QDs gradually weakened, and the aggregation of QDs occurred. Therefore, the solution treated by acetone was turbid and dark; ethyl acetate is the mild one and isopropanol is in the middle. The microscopic morphology of photodetectors treated by different solvents further confirms this conclusion. Figure S3 showed the film morphology of the device washed by ethyl acetate and isopropanol is continuous, while the film morphology of the device washed by acetone is poor and crackle. In addition, the washing times of isopropanol were also optimized as shown in Figure 2e,f, and one time provides the best switching ratio. With the increasing number of washes, the solution gradually became turbid from clarification, and the fluorescence intensity gradually disappeared, the film forming morphology of the device also gradually deteriorates. Therefore, detectors washed by isopropanol for one time gets the balance between good morphology and carrier transport of $CsPbBr_3$ QDs film. It can also be seen from Figure S4 that the optimal switching ratio can be obtained with the device washed once with isopropyl alcohol.

To evaluate the detectivity of $CsPbBr_3$ QDs to different incident wavelengths, the I-t and I-V curves of the corresponding photodetectors under different irradiation wavelengths at 40 mW/cm^2 were recorded in Figure S4 (Supplementary Materials) and Figure 3a. As expected, the $CsPbBr_3$ photodetector is the most sensitive to the light at 405 nm owing to the strong absorption shown in Figure S5, followed by 532 nm, and there is no photoelectric response to light at 660 nm. These results fully demonstrated that the $CsPbBr_3$ photodetector is well sensitive to high energy wavelengths. What is more, we also measured I-t and I-V curves at 405 nm as a function of incident light intensity. As shown in Figure S6, the current linearly increased with the incident light intensity. It is because the number of photogenerated carriers is proportional to the absorbed photon flux, which is related to the high light absorptivity of the perovskites. Similarly, it can be found in Figure 3b that the photocurrent gradually increases with the incident light intensity. Figure S7 shows that with the increase in voltage, the corresponding photocurrent also increases linearly, which reflects the good contact between $CsPbBr_3$ QDs film and ITO glass. In addition, $CsPbBr_3$ photodetectors show good stability as shown in Figure 3c. It shows that the switching ratio nearly remained unchanged when $CsPbBr_3$ photodetectors were measured for 180 cycles at 40 mW/cm^2, 10 V. The primary cycle and response time of the photodetector are shown in Figure S8, the rise time is 0.176 s after triggering and the decay time is 0.09 s after termination of irradiation. These response times are as good as other $CsPbBr_3$ photodetectors with a similar structure.

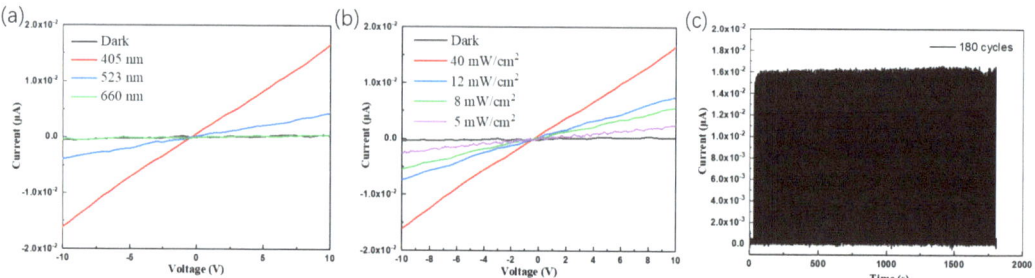

Figure 3. (**a**) Current-voltage curves of photodetectors under different wavelengths (40 mW/cm^2); (**b**) Current-voltage curves of photodetectors under different optical power density (405 nm); (**c**) Cycle diagram of a photodetector (405 nm, 10 v).

3.2. Study on the Performance of Paper Based Pressure Sensor

The device structure of the pressure sensor based on PEDOT: PSS-coated cellulose paper is shown in Figure 4a. The porous conductive structure of the PEDOT: PSS-coated cellulose paper is on the top of ITO glass with a slit, and a transparent PDMS film is used to pack the whole device. Generally, when the bias is applied on the two ITO electrodes, the total resistance (R_{total}) of the pressure sensor is composed of the bulk resistance of the conductive paper (R_{bulk}) and the contact resistance ($R_{contact}$) between the conductive paper and the electrodes. When pressure is applied to the sensor, the bulk is compressed and the contact area increases, and then all these lead to a significant decrease in R_{bulk} and $R_{contact}$. As the deformed part of the pressure sensor, the quality of porous materials greatly affects the device performance. There are several reported ways to conformally grow PEDOT onto porous substrate, such as the oxidative chemical vapor deposition (oCVD), spin-coating and drop coating [27,28]. Here, we select drop coating because it is free oxidant, the preparing process is simple and the content of PEDOT: PSS in the cellulose paper is well controllable. Therefore, we first studied the influence of different papers on the sensing performance of the pressure sensor. The test results are shown in Figure 4b. Compared with spunlace and needled non-woven paper, the sensitivity of the pressure based on PEDOT: PSS-coated cellulose paper is the highest. It is possible that the hydrophobic nature of the spunlace and needled non-woven PET paper may hinder the infiltration of the PEDOT:PSS aqueous solution during the wetting-drying process. In contrast, PEDOT:PSS can readily infiltrate cellulose paper and adhere to the fiber surface. The continuous and uniform conductive network of the cellulose paper exhibits a relative improvement in the sensitivity of pressure sensors.

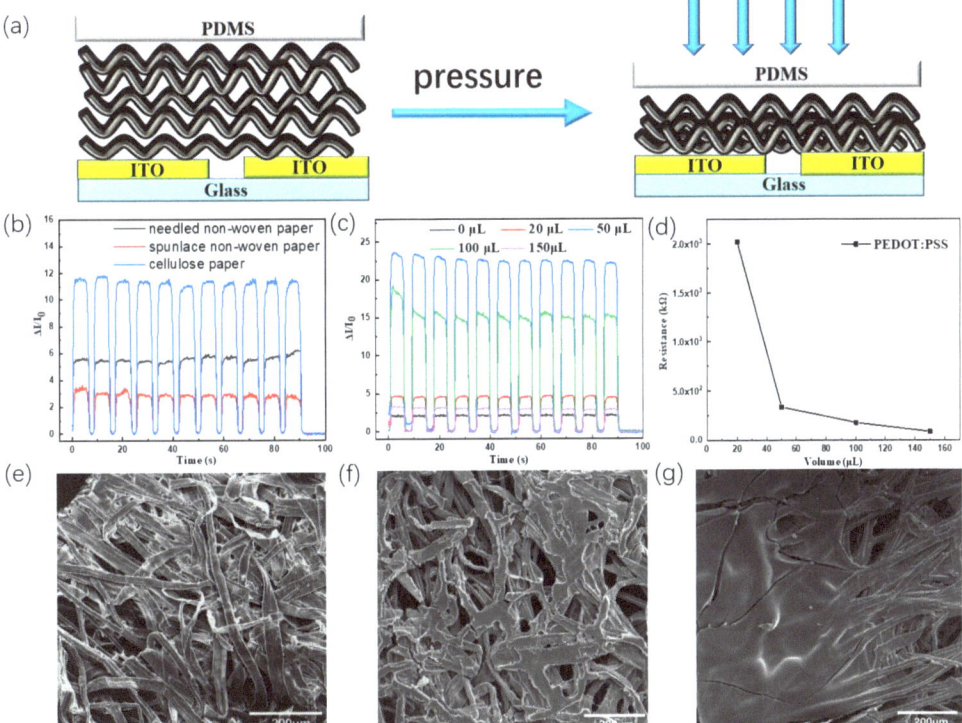

Figure 4. (**a**) Schematic diagram of the pressure sensor and its sensing mechanism; (**b**) The sensing performance of the pressure sensor based on different conductive papers; (**c**) $\Delta I/I_0$ vs. t curves of the pressure sensor at different PEDOT:PSS concentrations; (**d**) The resistance of the conductive paper with different PEDOT:PSS concentrations; SEM images of (**e**) Cellulose paper; (**f**) PEDOT: PSS-coated 50 uL cellulose paper (**g**) PEDOT: PSS-coated 150 uL cellulose paper.

To further improve the sensitivity of pressure sensors, we optimize the content of PEDOT: PSS in the cellulose paper. Figure 4c shows the influence of different PEDOT:

PSS content on the performance of pressure sensors. With the increase in the PEDOT: PSS volume, the optimum volume is obtained at 50 uL. The corresponding device sensitivity reaches 2.64 kPa^{-1} and the response time is 0.406 s (Figure S9). Above 50 uL, the sensitivity of the paper-based pressure sensor began to gradually decrease. From Figure 4d, we can see that the resistance of the conductive paper greatly decreased from 20 uL to 50 uL. After that, the resistance of the conductive paper changes a little. 50 uL is the percolation threshold for the conductive network. Excess PEDOT: PSS may lead to a hard paper and the deformation of the conductive paper will be limited and the detector performance decrease. This result can be further proved by the SEM measurement. Figure 4e–g show the SEM images for cellulose paper, PEDOT: PSS-coated 50 uL cellulose paper, and PEDOT: PSS-coated 150 uL cellulose paper, respectively. It can be seen in Figure 4f that PEDOT:PSS is obviously absorbed on cellulose paper and there is some holes for deformation. However, the cellulose paper is almost fulfilled by 150 uL PEDOT: PSS and the deformation ability of the conductive paper have been greatly limited. Therefore, the optimal PEDOT: PSS content is 50 uL. In addition, the stability of the optimal pressure sensor was measured and shown in Figure S10. During the initial cycles, the value of $\Delta I/I_0$ gradually decreases, which is due to irreversible deformation of the conductive paper and the loose contact at the interface between conductive paper and ITO glass. After about 300 cycles, the deformation of the conductive paper and the contact at the interface become stable, and then the value of $\Delta I/I_0$ of the pressure sensor remains stable even after 1000 cycles, indicating that this press sensor based on PEDOT: PSS-coated cellulose paper can provide a continuous and stable sensitivity after the initial transition.

3.3. Study on the Performance of Perovskite QDs Dual-Function Sensor

The light/pressure bifunctional sensor is fabricated by a simple superposition combination of the above two kinds of sensors. Its structure diagram is shown in Figure 5a, the photodetector based on CsPbBr$_3$ QDs film is at the bottom, and then the PEDOT: PSS-coated cellulose paper was simply stacked up on top of the photodetector. Finally, a transparent PDMS film is used to pack the whole device. In order to evaluate the light/pressure bifunctional sensor, we conduct both optical sensing and pressure sensing tests, respectively. Figure 5b shows the I-t curves when light is from the paper side or the ITO glass side. It shows that the photocurrent from the paper side is lower than that from the ITO side, which is due to the light absorption of PDMS film and conductive paper. The photocurrent from the ITO side is similar with that from the pure photodetector, which indicates that the introduction of conductive paper into the bifunctional sensor nearly has no effect on the performance of the photodetector from the ITO side. At the same time, the pressure sensing performance of the bifunctional sensor was compared with that of the pure pressure sensor. As shown in Figure 5c, the former can also detect the pressure easily. However, the sensitivity has dropped to almost 25% of the pure pressure sensor. We can attribute decrement to the large resistance introduced by the CsPbBr$_3$ QDs film, resulting in a smaller current on the pressure sensor and then a smaller $\Delta I/I_0$ value. Although the pressure sensitivity has declined after compounding, the combined sensor can achieve separate responses to both light and pressure. Therefore, simultaneous responses to both light and pressure are well expected. Figure 5d shows the sensing situation when pressure is applied during a certain illumination. When a UV light was on the bifunctional sensor, the value of $\Delta I/I_0$ was stabilized at once, and then a periodic pressure was applied. The periodic $\Delta I/I_0$ responses were observed on the base of the stable photocurrent. It demonstrates that bifunctional sensor can respond to pressure when illuminated and realizes the light/pressure bifunctional sensors were successfully obtained.

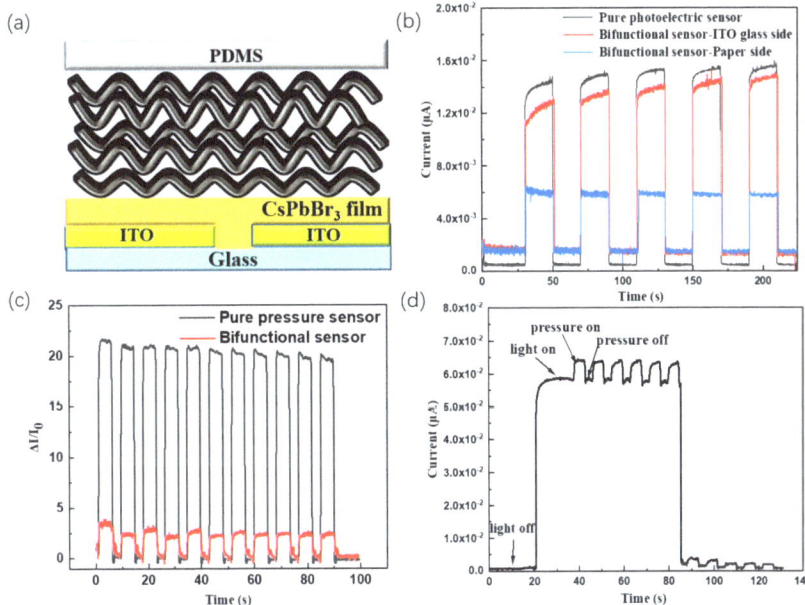

Figure 5. (**a**) Schematic diagram of the bifunctional sensors; (**b**) I-t curves of pure photodetector and the photodetectors lighted at 405 nm from the paper side or the ITO glass side; (**c**) ΔI/I$_0$ vs. t curves of the pure pressure sensor and bifunctional sensor; (**d**) I-t curves of light/pressure bifunctional sensor (365 nm); all applied bias is 10 v.

3.4. Application Research of Flexible Dual-Function Sensor

To explore these bifunctional sensors in the application of e-skin, flexible bifunctional sensors were fabricated. The rigid ITO glass was replaced by the flexible Au interdigital electrodes. The flexible photodetector device based on our CsPbBr$_3$ QDs film is illustrated in Figure 6(ai). The continuous and uniform CsPbBr$_3$ QDs films can also be formed on flexible Au interdigital electrodes by centrifugal casting. A photograph of the curved CsPbBr$_3$ photodetector is shown in Figure 6(aii), demonstrating its high flexibility. We measured the I-V curves using a laser diode at 405 nm, as a function of incident light intensity. It can be seen in Figure 6b that the flexible photoconductive detector based on all-inorganic perovskite CsPbBr$_3$ QDs exhibited a high performance. Upon illumination with a 405 nm laser, the typical linear and symmetrical photocurrent versus voltage (I–V) plots indicate that the CsPbBr$_3$ QDs were well dispersed on the Au interdigital electrode with good ohmic contact. The demonstrated ohmic contact is due to the energy alignment between the CsPbBr$_3$ QDs and the Au interdigital electrode. The current under dark and light conditions for the CsPbBr$_3$ QDs photodetector device were 3.32×10^{-9} A and 5.6×10^{-7} A, respectively, under an applied voltage of 10 V. Therefore, the switching ratio was 168, which indicated that the flexible device showed better light-switching behavior than the rigid device. The main reason is the interdigital electrode structure, which provides more electrode fingers to extract electrons and holes separately.

Similarly, bifunctional sensors based on Au interdigital electrodes are well flexible and shown in Figure 6(aiii,aiv). Benefiting from their bifunctional and flexible properties, our bifunctional sensors as e-skin exhibit promising potential to monitor human motions (Figure 6c). By monitoring the ΔI/I$_0$ as shown in Figure 6d, it was observed that the bilayer sensor could accurately detect the flexion of the finger knuckle at different angles (30°, 60° and 90°) and the shape of peaks in curves is almost the same while consecutive bending at the same angles, indicating the excellent repeatability of this e-skin sensor.

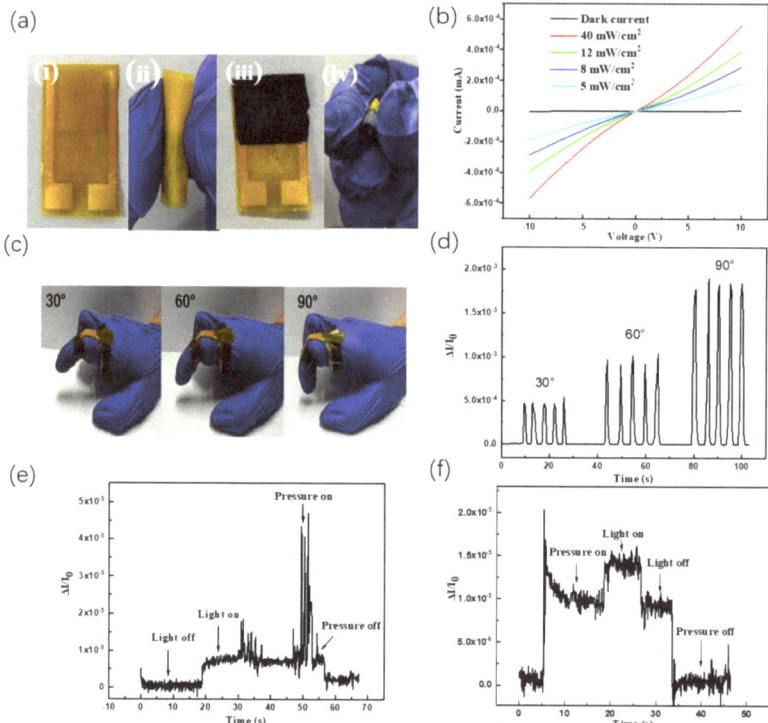

Figure 6. (**a**) Flexible optical sensor (**i,ii**) and flexible bifunctional sensor (**iii,iv**); (**b**) Current-voltage curves of photodetector under different optical power densities (405 nm); (**c**) Photos and (**d**)$\Delta I/I_0$ vs. t curves of finger bending at different angles based on flexible bifunctional sensor; $\Delta I/I_0$ vs. t curves of the flexible bifunctional sensor: (**e**) the force is applied when the light (532 nm) is on and (**f**) the light (532 nm) is on when is the force applied; all applied bias is 1 v.

To illustrate the light/pressure bifunctional property, we further studied the effects of simultaneous pressure and light stimulation on flexible bifunctional sensors. Figure 6e,f show the $\Delta I/I_0$ vs. time curves of the bifunctional sensor. Figure 6e shows the sensing situation when pressure is applied during a certain illumination. When a 532 nm laser was lit on the bifunctional sensor, the value of $\Delta I/I_0$ was stabilized at once. Then, different levels of pressure were applied. The value of $\Delta I/I_0$ increased on the base of the stable photocurrent. On the other hand, Figure 6f shows the sensing situation when illumination is applied during a certain of pressure. When a certain pressure was applied on the bifunctional sensor, the value of $\Delta I/I_0$ can also stabilized, and then the light was on, and the value of $\Delta I/I_0$ increased on the base of the stable photocurrent. These results demonstrate that our flexible bifunctional sensor can respond to both light and pressure and further illustrate the great application potential of flexible bifunctional sensors in the flexible multifunctional e-skin for next-generation wearable electronics.

4. Conclusions

In conclusion, a UV-blue-green photodetector based on $CsPbBr_3$ QDs films was prepared via a simple centrifugal casting method. The n-hexane/isopropanol mixed solvent treatment was conducted to adjust the surface ligand density of $CsPbBr_3$ PQDs, which got the balance between good morphology and carrier transport of $CsPbBr_3$ QDs film and consequently results in an optimal switching ratio for the photodetector. A novel pressure sensor is prepared by using PEDOT: PSS-coated cellulose paper. It was found that the optimal pressure sensitivity was 2.64 kPa^{-1} when the PEDOT: PSS content was 50 μL. Based on the photodetector and the pressure sensor, we successfully realize light/pressure bifunctional sensors, which can simultaneously respond to light and pressure stimulation. Furthermore, flexible bifunctional sensors were successfully applied as e-skin to monitor light and human motions. They exhibited a high photosensitivity with a switching ratio of 168 under 405 nm light at a power of 40 mW/cm^2. At the same time, the flexion of the finger knuckle at different angles (30°, 60° and 90°) could be accurately detected. Our work showed that the strategy to introduce perovskite photodetectors into e-skins is feasible and may open a new way for the development of flexible multi-functional e-skin.

Supplementary Materials: The following supporting information can be downloaded at: https://www.mdpi.com/article/10.3390/polym15092136/s1, Figure S1. The absorption spectrum of CsPbBr$_3$ QDs in n-hexane solution; Figure S2. Solutions of CsPbBr$_3$ QDs washed by different solvents under fluorescent lamps (upper photos) and under 365 nm UV lamps (lower photos); Figure S3. Microscopic morphologies of photodetectors washed by different solvents; Figure S4. I-t curves of photodetectors under different washing times (405 nm, 10v); Figure S5. I-t curves of photodetectors under different wavelengths (40 mW/cm^2,10v); Figure S6. I-t curves of photodetectors under different optical power density (405 nm, 40 mW/cm^2); Figure S7. I-t curves of photodetectors at different voltages (405 nm, 10v); Figure S8. One cycle of photodetector and its response time (b is light on and c is light off); Figure S9. One cycle of pressure sensor and its response time (b is pressure on and c is pressure off); Figure S10. $\Delta I/I_0$ vs. t curves of the pressure sensor (1000 cycles); all applied bias is 1 v.

Author Contributions: Conceptualization, T.Y.; methodology, T.Y. and S.H.; validation, X.S.; data curation, J.J. and W.L.; writing—original draft preparation, W.L. and J.J.; writing—review and editing, T.Y.; supervision, T.Y. and S.H.; funding acquisition, T.Y. All authors have read and agreed to the published version of the manuscript.

Funding: This work was supported by CAS Key Laboratory of Renewable Energy, Guangzhou Institute of Energy Conversion (No. E229kf0901) and Heilongjiang Provincial Postdoctoral Science Foundation (Grant No. LBH-TZ0604).

Institutional Review Board Statement: Not applicable.

Informed Consent Statement: Not applicable.

Data Availability Statement: The data presented in this study are available on request from the corresponding author.

Conflicts of Interest: The authors declare no conflict of interest.

References

1. Schwartz, G.; Tee, B.C.; Mei, J.; Appleton, A.L.; Kim, D.H.; Wang, H.; Bao, Z. Flexible Polymer Transistors with High Pressure Sensitivity for Application in Electronic Skin and Health Monitoring. *Nat. Commun.* **2013**, *4*, 1859. [CrossRef]
2. Cao, J.; Lu, C.; Zhuang, J.; Liu, M.; Zhang, X.; Yu, Y.; Tao, Q. Multiple Hydrogen Bonding Enables the Self-healing of Sensors for Human-Machine Interactions. *Angew. Chem. Int. Ed.* **2017**, *56*, 8795–8800. [CrossRef]
3. Byun, J.; Lee, Y.; Yoon, J.; Lee, B.; Oh, E.; Chung, S.; Lee, T.; Cho, K.-J.; Kim, J.; Hong, Y. Electronic Skins for Soft, Compact, Reversible Assembly of Wirelessly Activated Fully Soft Robots. *Sci. Robot.* **2018**, *3*, eaas9020. [CrossRef]
4. Liu, G.; Tan, Q.; Kou, H.; Zhang, L.; Wang, J.; Lv, W.; Dong, H.; Xiong, J. A Flexible Temperature Sensor Based on Reduced Graphene Oxide for Robot Skin Used in Internet of Things. *Sensors* **2018**, *18*, 1400. [CrossRef]
5. Tee, B.C.-K.; Chortos, A.; Berndt, A.; Nguyen, A.K.; Tom, A.; McGuire, A.; Lin, Z.C.; Tien, K.; Bae, W.-G.; Wang, H.; et al. A Skin-inspired Organic Digital Mechanoreceptor. *Science* **2015**, *350d*, 313–316. [CrossRef]
6. Fastier-Wooller, J.W.; Dinh, T.; Tran, C.D.; Tran, C.-D.; Dao, D.V. Pressure and Temperature Sensitive E-skin for in situ robotic applications. *Mater. Des.* **2021**, *208*, 109886. [CrossRef]
7. Takei, K.; Gao, W.; Wang, C.; Javey, A. Physical and Chemical Sensing with Electronic Skin. *Proc. IEEE* **2019**, *107*, 2155–2167. [CrossRef]
8. Li, W.D.; Ke, K.; Jia, J.; Pu, J.; Zhao, X.; Bao, R.; Liu, Z.; Bai, L.; Zhang, K.; Yang, M.; et al. Recent Advances in Multiresponsive Flexible Sensors towards E-skin: A Delicate Design for Versatile Sensing. *Small* **2022**, *18*, 2103734. [CrossRef] [PubMed]
9. Kim, J.N.; Shadlen, M.N. Neural Correlates of a Decision in the Dorsolateral Prefrontal Cortex of the Macaque. *Nat. Neurosci.* **1999**, *2*, 176–185. [CrossRef] [PubMed]

10. Dai, Y.; Fu, Y.; Zeng, H.; Xing, L.; Zhang, Y.; Zhan, Y.; Xue, X. A Self-Powered Brain-Linked Vision Electronic-skin Based on Triboelectric-Photodetecing Pixel-Addressable Matrix for Visual-image Recognition and Behavior Intervention. *Adv. Funct. Mater.* **2018**, *28*, 1800275. [CrossRef]
11. Zhang, W.; Liu, Y.; Pei, X.; Yuan, Z.; Zhang, Y.; Zhao, Z.; Hao, H.; Long, R.; Nan, L. Stretchable MoS$_2$ Artificial Photoreceptors for E-Skin. *Adv. Funct. Mater.* **2021**, *32*, 2107524. [CrossRef]
12. Vuong, V.-H.; Pammi, S.V.N.; Ippili, S.; Jella, V.; Thi, T.N.; Pasupuleti, K.S.; Kim, M.-D.; Jeong, M.J.; Jeong, J.-R.; Chang, H.S.; et al. Flexible, Stable, and Self-Powered Photodetectors Embedded with Chemical Vapor Deposited Lead-Free Bismuth Mixed Halide Perovskite Films. *Chem. Eng. J.* **2023**, *458*, 141473. [CrossRef]
13. Bu, H.; He, C.; Xu, Y.; Xing, L.; Liu, X.; Ren, S.; Yi, S.; Chen, L.; Wu, H.; Zhang, G.; et al. Emerging New-Generation Detecting and Sensing of Metal Halide Perovskites. *Adv. Electron. Mater.* **2022**, *8*, 2101204. [CrossRef]
14. Yan, S.; Li, Q.; Zhang, X.; Tang, S.; Lei, W.; Chen, J. A Vertical Structure Photodetector Based on All-Inorganic PQDs. *J. Soc. Inf. Disp.* **2020**, *28*, 9–15. [CrossRef]
15. Ren, B.; Yuen, G.; Deng, S.; Jiang, L.; Zhou, D.; Gu, L.; Xu, P.; Zhang, M.; Fan, Z.; Yueng, F.S.Y.; et al. Multifunctional Optoelectronic Device Based on an Asymmetric Active Layer Structure. *Adv. Funct. Mater.* **2019**, *29*, 1807894. [CrossRef]
16. Shi, Q.; Liu, D.; Hao, D.; Zhang, J.; Tian, L.; Xiong, L.; Huang, J. Printable, ULtralow-Power Ternary Synaptic Transistors for Multifunctional Information Processing System. *Nano Energy* **2021**, *87*, 106197. [CrossRef]
17. Chen, W.; Yan, X. Progress in Achieving High-Performance Piezoresistive and Capacitive Flexible Pressure Sensors: A Review. *J. Mater. Sci. Technol.* **2020**, *43*, 175–188. [CrossRef]
18. Zheng, Y.; Yin, R.; Zhao, Y.; Liu, H.; Zhang, D.; Shi, X.; Zhang, B.; Liu, C.; Shen, C. Conductive MXene/cotton Fabric Based Pressure Sensor with Both High Sensitivity and Wide Sensing Range for Human Motion Detection and E-skin. *Chem. Eng. J.* **2021**, *420*, 127720. [CrossRef]
19. Bu, Y.; Shen, T.; Yang, W.; Yang, S.; Zhao, Y.; Liu, H.; Zheng, Y.; Liu, C.; Shen, C. ULtrasensitive Strain Sensor Based on Superhydrophobic Microcracked Conductive Ti$_3$C$_2$T$_x$ MXene/paper for Human-Motion Monitoring and E-skin. *Sci. Bull.* **2021**, *66*, 1849–1857. [CrossRef]
20. Guo, Y.; Wei, X.; Gao, S.; Yue, W.; Li, Y.; Shen, G. Recent Advances in Carbon Material-based Multifunctional Sensors and Their Applications in Electronic skin Systems. *Adv. Funct. Mater.* **2021**, *31*, 2104288. [CrossRef]
21. Wang, Y.; Li, X.; Nalla, V.; Zeng, H.; Sun, H. Solution-Processed Low Threshold Vertical Cavity Surface Emitting Lasers from All-Inorganic Perovskite Nanocrystals. *Adv. Funct. Mater.* **2017**, *27*, 1605088. [CrossRef]
22. Li, S.; Lei, D.; Ren, W.; Guo, X.; Wu, S.; Zhu, Y.; Rogach, A.L.; Chhowalla, M.; Jen, A.K.-Y. Water-Resistant Perovskite Nanodots Enable Robust Two-Photon Lasing in Aqueous Environment. *Nat. Commun.* **2020**, *11*, 1192. [CrossRef] [PubMed]
23. Oh, J.M.; Venters, C.C.; Di, C.; Pinto, A.M.; Wan, L.; Younis, I.; Cai, Z.; Arai, R.; So, B.R.; Duan, J.; et al. U1 Snrnp Regulates Cancer Cell Migration and Invasion in Vitro. *Nat. Commun.* **2020**, *11*, 1. [CrossRef] [PubMed]
24. Tyagi, P.; Arveson, S.M.; Tisdale, W.A. Colloidal Organohalide Perovskite Nanoplatelets Exhibiting Quantum Confinement. *J. Phys. Chem. Lett.* **2015**, *6*, 1911–1916. [CrossRef]
25. Song, J.; Xu, L.; Li, J.; Xue, J.; Dong, Y.; Li, X.; Zeng, H. Monolayer and Few-Layer All-Inorganic Perovskites as a New Family of Two-Dimensional Semiconductors for Printable Optoelectronic Devices. *Adv. Mater.* **2016**, *28*, 4861–4869. [CrossRef]
26. Wu, H.; Kang, Z.; Zhang, Z.; Si, H.; Zhang, S.; Zhang, Z.; Liao, Q.; Zhang, Y. Ligand Engineering for Improved All-inorganic Perovskite Quantum Dot-MoS$_2$ Monolayer Mixed Dimensional Van Der Waals Phototransistor. *Small Methods* **2019**, *3*, 1900117. [CrossRef]
27. Heydari Gharahcheshmeh, M.; Gleason, K.K. Recent Progress in Conjugated Conducting and Semiconducting Polymers for Energy Devices. *Energies* **2022**, *15*, 3661. [CrossRef]
28. Smolin, Y.Y.; Soroush, M.; Lau, K.K.S. Influence of Ocvd Polyaniline Film Chemistry in Carbon-Based Supercapacitors. *Ind. Eng. Chem. Res.* **2017**, *56*, 6221–6228. [CrossRef]

Disclaimer/Publisher's Note: The statements, opinions and data contained in all publications are solely those of the individual author(s) and contributor(s) and not of MDPI and/or the editor(s). MDPI and/or the editor(s) disclaim responsibility for any injury to people or property resulting from any ideas, methods, instructions or products referred to in the content.

Environmentally Friendly Photoluminescent Coatings for Corrosion Sensing

Carmen R. Tubio [1,*], Laura Garea [1], Bárbara D. D. Cruz [2,3], Daniela M. Correia [2], Verónica de Zea Bermudez [4] and Senentxu Lanceros-Mendez [1,5]

1. BCMaterials, Basque Center for Materials, Applications and Nanostructures, UPV/EHU Science Park, 48940 Leioa, Spain; laura.garea.mayordomo@gmail.com (L.G.); senentxu.lanceros@bcmaterials.net (S.L.-M.)
2. Centre of Chemistry, University of Minho, 4710-057 Braga, Portugal; barbaraddcruz@gmail.com (B.D.D.C.); dcorreia@quimica.uminho.pt (D.M.C.)
3. Physics Centre of Minho and Porto Universities (CF-UM-UP) and Laboratory of Physics for Materials and Emergent Technologies (LapMET), University of Minho, 4710-057 Braga, Portugal
4. Chemistry Department and CQ-VR, University of Trás-os-Montes e Alto Douro, 5000-801 Vila Real, Portugal; vbermude@utad.pt
5. IKERBASQUE, Basque Foundation for Science, 48009 Bilbao, Spain
* Correspondence: carmen.rial@bcmaterials.net

Abstract: Although an increasing number of studies are being devoted to the field of corrosion, with topics from protection to sensing strategies, there is still a lack of research based on environmentally eco-friendly materials, which is essential in the transition to sustainable technologies. Herein, environmentally friendly composites, based on photoluminescent salts dispersed in vegetable oil-based resins, are prepared and investigated as corrosion sensing coatings. Two salts NaA, where A- is a lanthanide complex anion (with Ln = Nd^{3+}, and Yb^{3+}), are incorporated into the resins as active functional fillers and different coatings are prepared on carbon steel substrates to assess their functional properties. The influence exerted by a corrosive saline solution on the morphology, structural, and photoluminescent properties of the coatings is evaluated, and their suitability for the practical detection of the early corrosion of coated carbon steel is demonstrated. The photoluminescence of the composite coatings depends on the corrosion time, with the effect being more important for the coatings doped with Nd^{3+}. The present work shows that the composites obtained are suitable candidates for corrosion sensing coating applications, offering improved sustainability.

Keywords: corrosion; photoluminescence salts; lanthanide complexes; sensing

1. Introduction

Undesired corrosion is a relevant problem in several sectors and industries, including the energy, transport, and automotive sectors. In addition, it is relevant to infrastructure, such as bridges, pipelines, and highways, among other areas [1]. Consequently, in the past decades, extensive research has been performed on the development of efficient strategies to improve the protection, continuous monitoring, and early detection of corrosion processes. The great majority of studies related to corrosion protection have focused on organic/inorganic coatings, the removal of corrosive constituents, the use of inhibitors, and anodic/cathodic protection [2–4]. In parallel, during the past decade, extensive self-repairing strategies have been introduced in protective coatings [5,6].

In addition to the well-established approaches, the implementation of sensing capabilities is recognized as one of the most effective ways to early detect, monitor, and control corrosion processes. The implementation of sensing technologies essentially relies on the use of optical fibers [7,8], electrochemical noise detection [9,10], pH-indicating agents [11,12], and smart materials which can provide qualities of self-healing, damage sensing, and stimulus responsiveness [13,14]. In particular, fluorescence-based sensors have been proposed as a suitable sensing solution due to their high sensitivity and real-time detection [15]. Their working principle is based on the use of fluorescent indicators, which can be sensitive to pH or metal ion variations and can be added to solutions or coatings. Several reports have demonstrated that these fluorescent indicators can act as sensors, with promising results in the corrosion field [13,16]. However, one of the most critical issues is the design and synthesis of composites with environmentally friendly materials.

In this context, many luminescent salts are environmentally friendly compounds which properties can be tailored through the judicious choice of the nature of the cations and anions. This advantage, plus the fact those materials may be incorporated into specific polymer matrices, has boosted their use in a wide range of applications [17]. Concerning luminescence characteristics, the introduction of lanthanide metal ions, such as lanthanide (Ln^{3+}) ions, can yield highly luminescent compounds with narrow emission bands, photostability, long decay times, and high quantum efficiency [17,18]. The potential of luminescent salts has been exploited in various applications, such as anticounterfeiting [19,20] and biomedical imaging [21,22], among others [23–25]. Yet, in the field of corrosion sensing, such luminescent compounds have scarcely been explored. Therefore, taking into consideration the current pressing need for more sustainable approaches and materials, it is suitable to evaluate the use of luminescent materials as active fillers to develop practical sensors in the area of corrosion. Another important aspect is the selection of the host polymeric matrix, which should also be environmentally sustainable to reduce the use of synthetic materials and address the requirement of eco-friendly formulations towards sustainable materials, processes, and technologies. In this regard, vegetable oil-based resins have emerged as promising candidates on account of being abundantly availabl, and environmentally friendly. At present, vegetable resins are being increasingly employed in the coating industry, demonstrating potential applicability in corrosive environments [26].

In this work, we propose a novel approach to develop sensing materials for corrosive environments. This implies the use of a soybean-based vegetable UV-curable resin doped with Na[Ln(tta)4], where Ln = Nd or Yb and tta– is tetrakis(thenoyltrifluoroacetonate. The latter sodium salts are the precursor compounds of luminescent salts, the applicability of which has been already demonstrated in anticounterfeiting [19,20]. Herein, we demonstrate the applicability of such composite materials in corrosion sensing through the evaluation of the sensing performance in coated carbon steel substrates in saline environments. Accordingly, immersion experiments are performed in 5 wt. % NaCl solution for several days. In the present work, the coating is directly placed on top of the corroding surface, whereas in specific applications, such as very harsh environments, the sensing layer can be placed on top of another more protective layer. This work represents a basis for a new sensing approach in the area of steel corrosion, with improved performance and sustainability.

2. Materials and Methods

2.1. Materials

A transparent soybean-based and biodegradable photopolymer resin (ECO UV resin, Anycubic, Shenzhen, China) was used as received.

2-thenoyltrifluoroacetone (Htta) (99%, Sigma-Aldrich, St. Louis, MO, USA), ethanol (EtOH) (99.8%, Fisher Chemical, Loughborough, UK), neodymium(III)chloride hexahydrate $NdCl_3 \cdot 6H_2O$, 99.9%, Sigma-Aldrich, USA), ytterbium (III)chloride hexahydrate

(YbCl$_3$·6H$_2$O, 99.9%, Sigma-Aldrich), and sodium hydroxide (NaOH, Merck, Rahway, NJ, USA) were used as received.

2.2. Preparation of Ln-Based Salts

Two Ln ternary complexes were prepared according to the procedure reported in [19,20]. Briefly, the synthesis involved the dissolution of Htta in EtOH under magnetic stirring, followed by deprotonation with NaOH at 50–60 °C for 2 h. Then, NdCl$_3$·6H$_2$O, was dissolved in EtOH and was added dropwise into the previous solution. The mixture was kept at 50–60 °C for 1 h, and then the EtOH was evaporated. Afterwards, the resultant salt was dried at 50 °C for 3 days. Similarly, Na[Yb(tta)$_4$] synthesis was achieved using YbCl$_3$·6H$_2$O as a lanthanide metal precursor. Figure 1 shows the chemical structure of Na[Ln(tta)$_4$], where Ln = Nd or Yb, and tta− is tetrakis(thenoyltrifluoroacetonate.

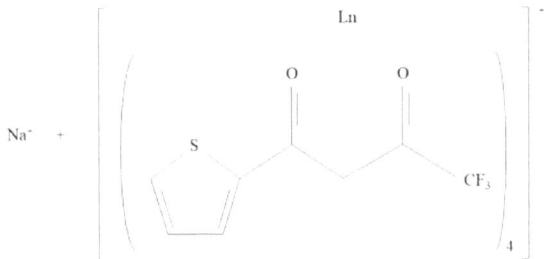

Figure 1. Chemical structure of Na[Ln(tta)$_4$].

2.3. Sample Preparation and Corrosion Experiments

The coating solutions were based on the vegetable UV-curable resin mixed with the sodium salts in a planetary mixer (Thinky ARE-250, Tokyo, Japan) at 2000 rpm for 4 min several times. The salt loading was controlled at 5 and 10 wt. %. Carbon steel substrates (DC01, Laser Norte S.A., Biscay, Spain) with dimensions of 5 mm × 2 mm × 0.5 mm were coated with the Ln-doped solutions using the doctor blade technique and the as-coated films were cured in a UV curing chamber (UVACUBE 400, Hönle AG, Gilching, Germany) for 3 min at 400 W.

The influence of saline solutions on the coated steel substrates was evaluated by immersing the samples in a 5 wt. % NaCl solution. This study was carried out for 4 days, and the results were evaluated on days 1 and 4. The nomenclature adopted to represent the prepared samples was resin-xLn (with x%wt.).

2.4. Characterization of the Samples

The Fourier Transform Infrared (FTIR) spectra were recorded on a Jasco FT/IR-6100 equipped with an Attenuated Total Reflection (ATR) accessory (Easton, MD, USA). The samples were scanned in the 600–4000 cm^{-1} spectral range with a 4 cm^{-1} resolution.

The morphology and elemental compositions of the samples were observed by a scanning electron microscope (SEM, Carl Zeiss EVO-40, Berlin, Germany) equipped with an energy-dispersive mode (EDX, Oxford Instruments, Abingdon, UK). Before the analysis, the samples were sputtered with a 10 nm thin gold layer.

Energy-dispersive X-ray fluorescence (ED-XRF) analysis was performed using the Spectro Midex SD system (Ametek, Inc., Leicester, UK).

Photoluminescence (PL) spectra were recorded using a fluorescence spectrometer (FLS980, Edinburgh Instruments Ltd., Livingston, UK) with a Xenon lamp 450 W. The excitation wavelength was 300 nm.

3. Results

3.1. Active Coating Characterization

FTIR analysis was carried out to evaluate the nature of the interactions between the UV-curable resin matrix and the Na[Ln(tta)$_4$] salts. The FTIR spectra of the neat resin and resin-10Ln (Ln = Nd and Yb) samples in the 3000–600 cm^{-1} wavenumber range are shown in Figure 2.

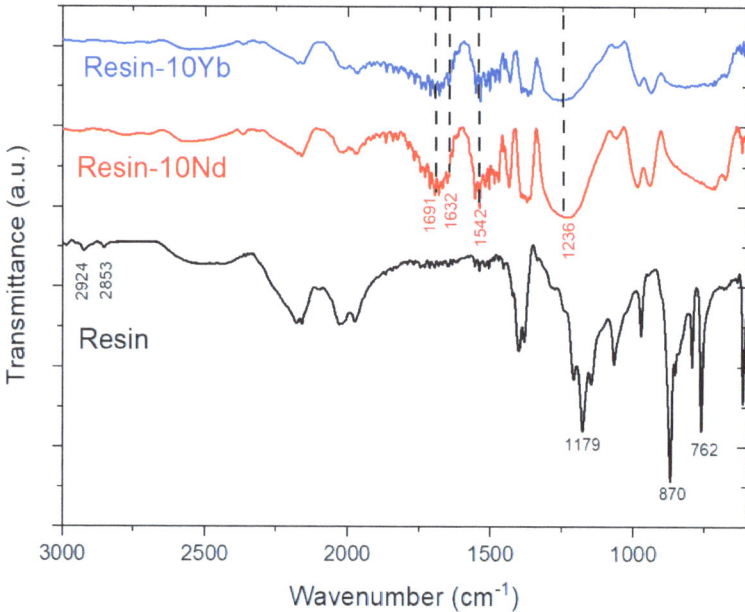

Figure 2. FTIR spectra of the neat resin and resin-10Ln samples.

The FTIR signature of the doped samples was markedly different from that of the neat resin. The FTIR spectrum of the pristine resin showed the typical bands of soybean oil [27–29]. In particular, the characteristic bands of the C-O bond were observed at 1179 cm^{-1}, as were the bands at 2924 and 2853 cm^{-1}. This was related to the presence of the weak asymmetric stretching of the -C-H bond. In addition, the bands of higher intensity at 870 and 762 cm^{-1} allowed us to identify the =C-H bond bending mode. On the other hand, with the presence of the Ln-based salts, new peaks were identified in the composites. These were related to the presence of the filler. In particular, bands that were specific to the β-diketonate ligand were identified at 1542 cm^{-1} due to the presence of the C=C bond, at 1691 and 1632 cm^{-1}, which was ascribed to the carbonyl (C=O) group, and at 1236 cm^{-1}, which was associated with the C-CF3 group [30]. No new bands or energy shifts of the resin and Ln-based salts bands were observed in the composites, indicating a lack of strong interactions between the different composite components.

To study the morphology and microstructure of the samples, SEM and EDX measurements were performed (Figure 3). Figure 3a shows a representative SEM image taken from the cross-sectional area of the neat resin, and the corresponding EDX mapping of the C and O elements. The SEM image reveals that the as-prepared sample is characterized by a smooth and compact morphology, where C and O are homogeneously distributed all along the polymer. The relative distribution of these elements was in agreement with the composition of the vegetable resin made from soybean oil. Furthermore, as observed in Figure 3b,c, the dense morphology was maintained after salt addition and the corresponding EDS map demonstrates the uniform distribution of C, O, Nd, and Yb within the resin matrix, i.e., the excellent distribution of the filler within the polymer matrix, with no large aggregates or voids.

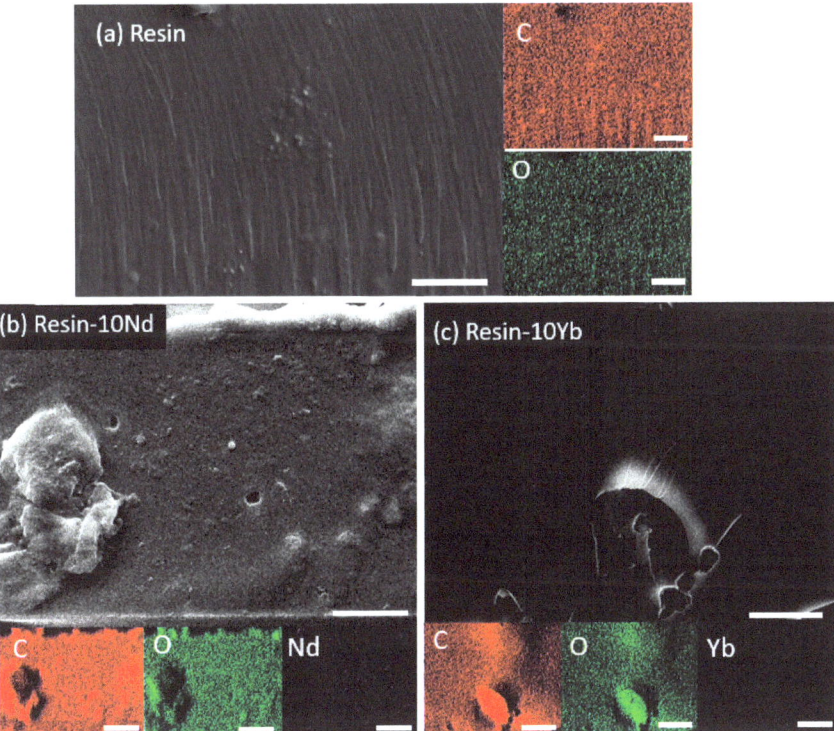

Figure 3. Cross-sectional SEM image and corresponding EDS elemental mapping of the resin-xLn as-prepared samples: (**a**) neat resin; (**b**) resin-10Nd; (**c**) resin-10Yb. Scale bar: 10 μm.

3.2. Corrosion Tests

Figure 4 shows the SEM images of the samples doped with 10 wt. % of Na[Ln(tta)$_4$] before and after immersion in 5 wt. % NaCl solution. A roughened morphology with defects was visible for the samples treated for 1 and 4 days, indicating that the saline solution caused significant corrosive attack on the samples. Additionally, the EDX results represented in Table 1 evidence the high content of C and O elements in all the samples. As expected, the samples contained elements of Na[Ln(tta)$_4$] before and after immersion in the saline solution. Importantly, we noticed that small quantities of Fe were found in the samples exposed to the saline solution for 4 days. This was associated with the corrosion process of the carbon steel substrate. The findings are presented in Figure 5, with the EDX mapping images showing the homogeneous distribution of Fe elements.

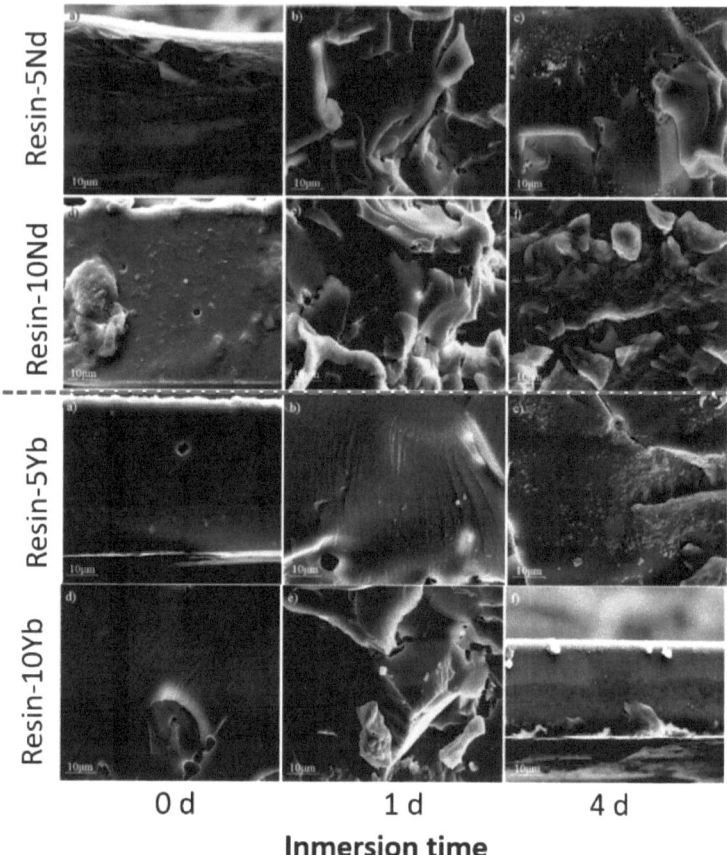

Figure 4. SEM images of the resin-xLn samples before and after immersion in 5 wt. % NaCl for 1 and 4 days. The scale is 10 μm.

Table 1. Elemental compositions of the resin-xLn as a function of immersion time (t_i) in the NaCl solution as calculated by EDX.

Ln	x	t_i (day)	Elemental Composition (wt. %)								
			C	O	Cl	Ln	F	Na	K	Ca	Fe
Nd	5	0	70.02	26.19	0.99	0.49	0.84	0.46	0.74	0.25	0
		1	67.27	29.49	0.45	0.69	1.76	0.34	0	0	0
		4	65.85	31.04	0.51	0.38	1.09	0.5	0	0	0.62
	10	0	63.36	30.24	1.07	1.02	3.03	0.73	0.37	0.19	0
		1	60.17	34.1	0.74	0.99	3.04	0.45	0.41	0.1	0
		4	64.95	29.82	0.37	1.12	2.8	0.32	0	0	0.62
Yb	5	0	62.48	34.25	0.17	0.53	1.82	0.19	0.07	0.08	0
		1	67.32	29.2	0.23	0.65	1.4	0	0	0	0
		4	60.15	28.19	4.13	0.57	0.88	4.53	0.08	0	0.96
	10	0	66.04	27.62	0.2	1.58	2.67	0.22	0	0	0
		1	57.45	34.54	1.49	1.43	2.87	0.98	0.59	0.65	0
		4	60.67	32.39	0.37	1.39	2.71	0.35	0.27	0.7	0.18

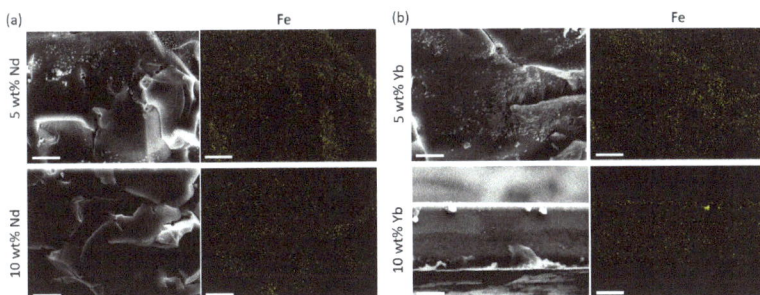

Figure 5. SEM images and corresponding EDS elemental mapping for the distribution of Fe elements in the resin-xLn after immersion in 5 wt. % NaCl for 4 days: (**a**) resin-5Nd (up) and resin-10Nd (down), and (**b**) resin-5Yb (up) and resin-10Yb (down). Scale bar: 10 μm.

The results obtained from the EDX analysis were complemented with an XRF test. Figure 6a and Table 1 depict the Fe contents found in the coatings after 1 and 4 days of immersion in the saline solution, showing that the Fe contents increased with the increasing immersion time, revealing the effect of the Cl$^-$ ions on the carbon steel substrate. Thus, substrate corrosion leads to Fe diffusion within the coating, as illustrated in Figure 6b. In addition, it is important to draw attention to the differences found with respect to the above EDX analysis. Both XRF and EDX are analytical tools for semi-quantitative elemental analysis, and they have different detection limits. While EDX tends to be more effective for lighter elements, XRF shows the opposite trend.

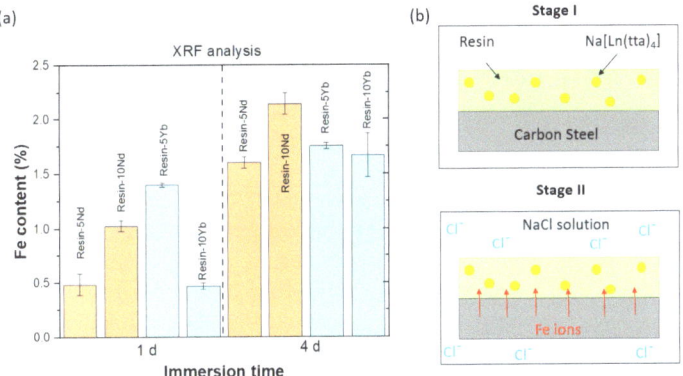

Figure 6. (**a**) A comparison of the Fe content values of the resin-xLn samples after 1 and 4 immersion days in 5 wt. % NaCl solution, as determined by XRF testing. (**b**) A schematic illustration of the system in the initial stage and after immersion in 5 wt. % NaCl solution.

Finally, the functional performance of the systems was evaluated. The PL spectra were measured to determine the sensing efficiency of the coatings against steel corrosion within the saline solution. Figure 7 show the PL changes of the resin with 10 wt. % of Na[Ln(tta)$_4$] for Ln = Nd^{3+} and Yb^{3+}, respectively, as a function of exposure time to a saline medium. In the case of the sample doped with Nd^{3+} (Figure 7a), one emission band, peaking at 1058 nm, is observed. In contrast, in the case of the resin incorporating Yb^{3+} (Figure 7b), two distinct peaks are identified at 976 nm (sharp) and around 1000 nm (broad). These results are in agreement with the characteristic emission wavelengths of these lanthanide ions, with the 1058 and 976 nm peaks corresponding to the characteristic

Nd^{3+} ($^4F_{3/2} \rightarrow {}^4I_{11/2}$) and Yb^{3+} ($^2F_{5/2} \rightarrow {}^2F_{7/2}$) intra-4f transitions in the near-infrared (NIR) spectral range, respectively [31].

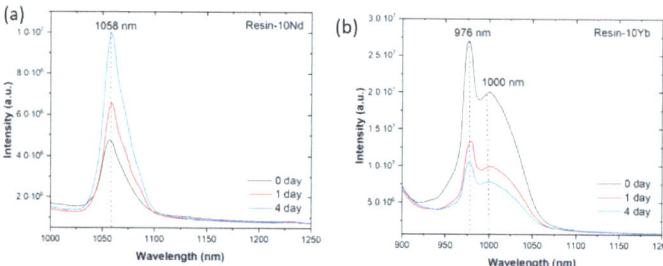

Figure 7. Emission spectra of the resin-10Ln samples, excited at 300 nm, before and after 1 and 4 immersion days in 5 wt. % NaCl solution: (**a**) Ln = Nd and (**b**) Ln = Yb.

With respect to the effect of the immersion time on the PL spectra, the first observation is that the intensity of the PL emission gradually increases in the case of the sample containing Nd^{3+} ions (Figure 7a), whereas a decrease in intensity is observed for the sample doped with Yb^{3+} (Figure 7b), showing the significant differences of both ions upon exposure to a corrosive medium. Those differences are related to the specific characteristics of the ions, including the luminescence lifetime, cross-sectional energy gap, and quantum efficiency, among others, as well as their different interactions with the host matrix. For example, Nd^{3+} has a large energy gap between the most dominantly emissive $^4F_{3/2}$ spin orbit level and the lower $^4I_{5/2}$ level, while Yb^{3+} possesses a large absorption cross section of around 980 nm. In fact, this effect is also reported in related studies, regarding emission variations in lanthanide compounds, for sensing applications [32–34]. The values of the variation in the PL intensity as a function of time have also been determined. The PL intensity of Nd^{3+} increases up to around 27% after 1 day and 52% after 4 days. Meanwhile, for the 976 nm emission transition of Yb^{3+}, this percentage difference is reduced in 50.3% after 1 day and 61.13% after 4 days. This result is in accordance with the observed reduction in the 1000 nm band, with variations of up to 50.13% and 60.95% found after 1 and 4 days, respectively.

The obtained PL emission intensity results suggest that resins containing a functional luminescent lanthanide complex salt can potentially interact with the Fe compound and change the PL properties. They absorb energy and transfer it to the lanthanide metal to enhance the luminescence intensity or quench the luminescent signal. Several works have attempted to develop luminescent Ln-based sensors for Fe detection. For example, Orcutt et al. [35] developed a lanthanide-based chemosensor for bioavailable Fe^{3+}. The studies showed that the displacement of the lanthanide-associated bands with Fe (III) is achieved in organic solvents. Yu et al. [36] demonstrated that several lanthanide metal–organic framework (MOF) compounds present an ultra-sensitive luminescent sensing response towards Fe (III). In addition, Fan et al. [37] studied the application of lanthanide compounds as fluorescent sensors for Fe (III). The results showed that samples are highly selective and sensitive luminescence probes for Fe (III) ions. Focusing on the luminescent sensing mechanism in these previous studies, it is related to variations in the charge transfer and energy transfer transitions. Despite these advances in designing sensors for iron detection, there are still challenges to overcome, such as improving the understanding the energy level modifications and electron–photon coupling parameters, which can provide deeper insights about luminescence sensing mechanisms. Therefore, it is necessary to support experimental analysis with theoretical methods, such as frontier molecular orbital analysis or quantum chemical approaches.

Thus, the variation in the PL emission spectra of the coatings in the presence of Fe ions allows us to monitor the evolution of corrosion over time. Therefore, they work as corrosion sensing coatings. Further, the sensing layer can be placed directly on the corroding surface of on top of another protective layer in a multilayer protective sensing system.

4. Conclusions

This study focused on the development of sustainable coatings with the ability of sensing corrosion. Specifically, the composite coatings fabricated by the combination of luminescent salts and the soybean vegetable matrix were applied to carbon steel substrates, and the system was stored in a saline medium. The morphological and structural features of samples were studied and the results were confirmed by the effective corrosion of carbon steel in the saline solution. Importantly, the functional sensing capability was evaluated though the variation in the PL of the coatings. The PL results show that the type of filler plays a profound role in the sensing response, the Nd-based salt showing a better performance than the Yb-based salt. Our results demonstrate that composites based on eco-friendly and sustainable materials show great potential to be used in sensing-based devices for corrosive environments.

Author Contributions: Conceptualization, C.R.T. and S.L.-M.; methodology, C.R.T. and S.L.-M.; validation, C.R.T.; formal analysis, C.R.T., L.G., B.D.D.C., D.M.C., V.d.Z.B. and S.L.-M.; investigation, C.R.T., L.G., B.D.D.C., D.M.C., V.d.Z.B. and S.L.-M.; resources, S.L.-M.; data curation, C.R.T., L.G., B.D.D.C., D.M.C. and V.d.Z.B.; writing—original draft preparation, C.R.T., D.M.C. and V.d.Z.B.; writing—review and editing, C.R.T. and S.L.-M.; supervision, C.R.T. and S.L.-M.; project administration, S.L.-M.; funding acquisition, S.L.-M. All authors have read and agreed to the published version of the manuscript.

Funding: This study forms part of the Advanced Materials programme and was supported by MCIN with funding from European Union NextGenerationEU (PRTR-C17.I1) as well as by IKUR Strategy under the collaboration agreement between Ikerbasque Foundation and Fundación BCMaterials on behalf of the Department of Education of the Basque Government. Funding from the Basque Government Industry Department under the ELKARTEK program is also acknowledged. The authors thank the Fundação para a Ciência e Tecnologia (FCT) for financial Support under the framework of Strategic Funding UID/QUI/00686/2020 and UID/FIS/04650/2020 and the Investigator FCT Contract 2020.02915.CEECIND (10.54499/2020.02915.CEECIND/CP1600/CT0029) (D.M.C.).

Institutional Review Board Statement: Not applicable.

Informed Consent Statement: Not applicable.

Data Availability Statement: Data are contained within the article.

Acknowledgments: Technical and human support provided by Sgiker (UPV/EHU) and University of Burgos (UBU) are gratefully acknowledged.

Conflicts of Interest: The authors declare no conflict of interest.

References

1. National Research Council; Committee on Research Opportunities in Corrosion Science and Engineering. *Research Opportunities in Corrosion Science and Engineering*; The National Academies Press: Washington, DC, USA, 2011.
2. Leal, D.A.; Kuznetsova, A.; Silva, G.M.; Tedim, J.; Wypych, F.; Marino, C.E.B. Layered materials as nanocontainers for active corrosion protection: A brief review. *Appl. Clay Sci.* **2022**, *225*, 106537. [CrossRef]
3. Alao, A.O.; Popoola, A.P.; Sanni, O. The Influence of Nanoparticle Inhibitors on the Corrosion Protection of Some Industrial Metals: A Review. *J. Bio-Tribo-Corros.* **2022**, *8*, 68. [CrossRef]
4. Refait, P.; Grolleau, A.-M.; Jeannin, M.; Sabot, R. Cathodic Protection of Complex Carbon Steel Structures in Seawater. *Corros. Mater. Degrad.* **2022**, *3*, 439–453. [CrossRef]

5. Poornima Vijayan, P.; Al-Maadeed, M. Self-Repairing Composites for Corrosion Protection: A Review on Recent Strategies and Evaluation Methods. *Materials* **2019**, *12*, 2754. [CrossRef]
6. Liu, C.; Wu, H.; Qiang, Y.; Zhao, H.; Wang, L. Design of smart protective coatings with autonomous self-healing and early corrosion reporting properties. *Corros. Sci.* **2021**, *184*, 109355. [CrossRef]
7. Dong, S.; Liao, Y.; Tian, Q.; Luo, Y.; Qiu, Z.; Song, S. Optical and electrochemical measurements for optical fibre corrosion sensing techniques. *Corros. Sci.* **2006**, *48*, 1746–1756. [CrossRef]
8. Sousa, I.; Pereira, L.; Mesquita, E.; Souza, V.L.; Araújo, W.S.; Cabral, A.; Alberto, N.; Varum, H.; Antunes, P. Sensing System Based on FBG for Corrosion Monitoring in Metallic Structures. *Sensors* **2022**, *22*, 5947. [CrossRef]
9. Tan, Y. Sensing localised corrosion by means of electrochemical noise detection and analysis. *Sens. Actuators B Chem.* **2009**, *139*, 688–698. [CrossRef]
10. Permeh, S.; Lau, K. Electrochemical Behavior of Steel in Alkaline Sulfate Solutions with Low Level Chloride Concentrations. *Meet. Abstr.* **2021**, *MA2021-02*, 598. [CrossRef]
11. Maia, F.; Tedim, J.; Bastos, A.C.; Ferreira, M.G.S.; Zheludkevich, M.L. Nanocontainer-based corrosion sensing coating. *Nanotechnology* **2013**, *24*, 415502. [CrossRef]
12. Dararatana, N.; Seidi, F.; Crespy, D. PH-Sensitive Polymer Conjugates for Anticorrosion and Corrosion Sensing. *ACS Appl. Mater. Interfaces* **2018**, *10*, 20876–20883. [CrossRef] [PubMed]
13. Augustyniak, A.; Tsavalas, J.; Ming, W. Early detection of steel corrosion via "turn-on" fluorescence in smart epoxy coatings. *ACS Appl. Mater. Interfaces* **2009**, *1*, 2618–2623. [CrossRef] [PubMed]
14. Liu, C.; Qian, B.; Hou, P.; Song, Z. Stimulus Responsive Zeolitic Imidazolate Framework to Achieve Corrosion Sensing and Active Protecting in Polymeric Coatings. *ACS Appl. Mater. Interfaces* **2021**, *13*, 4429–4441. [CrossRef]
15. Chen, C.; Yu, M.; Tong, J.; Xiong, L.; Li, Y.; Kong, X.; Liu, J.; Li, S. A review of fluorescence based corrosion detection of metals. *Corros. Commun.* **2022**, *6*, 1–15. [CrossRef]
16. Su, F.; Du, X.; Shen, T.; Qin, A.; Li, W. Aggregation-induced emission luminogens sensors: Sensitive fluorescence 'Turn-On' response for pH and visually chemosensoring on early detection of metal corrosion. *Prog. Org. Coat.* **2021**, *153*, 106122. [CrossRef]
17. Zhang, S.; Zhang, Q.; Zhang, Y.; Chen, Z.; Watanabe, M.; Deng, Y. Beyond solvents and electrolytes: Ionic liquids-based advanced functional materials. *Prog. Mater. Sci.* **2016**, *77*, 80–124. [CrossRef]
18. Tang, S.F.; Mudring, A.V. Highly Luminescent Ionic Liquids Based on Complex Lanthanide Saccharinates. *Inorg. Chem.* **2019**, *58*, 11569–11578. [CrossRef]
19. Correia, D.M.; Polícia, R.; Pereira, N.; Tubio, C.R.; Cardoso, M.; Botelho, G.; Ferreira, R.A.S.; Lanceros-Méndez, S.; de Zea Bermudez, V. Luminescent Poly(vinylidene fluoride)-Based Inks for Anticounterfeiting Applications. *Adv. Photonics Res.* **2022**, *3*, 2100151. [CrossRef]
20. Cruz, B.D.D.; Correia, D.M.; Polícia, R.; Pereira, N.; Nunes, P.; Fernandes, M.; Tubio, C.R.; Botelho, G.; Lanceros-Méndez, S.; de Zea-Bermudez, V. Photoluminescent alginate-based composite inks for anti-counterfeiting security and soft actuator applications. *Chem. Eng. J.* **2023**, *472*, 144813. [CrossRef]
21. Wang, W.; Fu, A.; Lan, J.; Gao, G.; You, J.; Chen, L. Rational Design of Fluorescent Bioimaging Probes by Controlling the Aggregation Behavior of Squaraines: A Special Effect of Ionic Liquid Pendants. *Chem. Eur. J.* **2010**, *16*, 5129–5137. [CrossRef]
22. Bwambok, D.K.; El-Zahab, B.; Challa, S.K.; Li, M.; Chandler, L.; Baker, G.A.; Warner, I.M. Near-Infrared Fluorescent NanoGUMBOS for Biomedical Imaging. *ACS Nano* **2009**, *3*, 3854–3860. [CrossRef] [PubMed]
23. Chen, X.-W.; Liu, J.-W.; Wang, J.-H. A highly fluorescent hydrophilic ionic liquid as a potential probe for the sensing of biomacromolecules. *J. Phys. Chem. B* **2011**, *115*, 1524–1530. [CrossRef] [PubMed]
24. Bwambok, D.K.; Challa, S.K.; Lowry, M.; Warner, I.M. Amino acid-based fluorescent chiral ionic liquid for enantiomeric recognition. *Anal. Chem.* **2010**, *82*, 5028–5037. [CrossRef] [PubMed]
25. Loe-Mie, F.; Marchand, G.; Berthier, J.; Sarrut, N.; Pucheault, M.; Blanchard-Desce, M.; Vinet, F.; Vaultier, M. Towards an efficient microsystem for the real-time detection and quantification of mercury in water based on a specifically designed fluorogenic binary tasl-specific ionic liquid. *Angew. Chem. Int. Ed.* **2010**, *49*, 424–427. [CrossRef]
26. Sharmin, E.; Zafar, F.; Akram, D.; Alam, M.; Ahmad, S. Recent advances in vegetable oils based environment friendly coatings: A review. *Ind. Crops Prod.* **2015**, *76*, 215–229. [CrossRef]
27. Saremi, K.; Tabarsa, T.; Shakeri, A.; Babanalbandi, A. Epoxidation of Soybean Oil. *Ann. Biol. Res.* **2012**, *3*, 4254–4258.
28. Wang, R.; Schuman, T.P. Vegetable Oil-derived Epoxy Monomers and Polymer Blends: A Comparative Study with Review. *Express Polym. Lett.* **2013**, *7*, 272–292. [CrossRef]
29. Zhao, K.; Shi, L.; Liu, Z.; Li, J. Quality Analysis of Reheated Oils by Fourier Transform Infrared Spectroscopy. In Proceedings of the 2015 International Conference on Electromechanical Control Technology and Transportation, Zhuhai, China, 31 October–1 November 2015; pp. 245–249. [CrossRef]
30. Medina-Velazquez, D.Y.; Barraza, M.; Barron, M.; Hileio, I.; Colin, V. Influence of Synthesis Parameters on Luminescence of Thenoyltrifluoroacetone Europium Powders. *J. Mater. Sci. Chem. Eng.* **2018**, *6*, 1–8. [CrossRef]

31. Wang, R.; Zhang, F. CHAPTER 1 Lanthanide-Based Near Infrared Nanomaterials for Bioimaging. In *Near Infrared Nanomaterials: Preparation, Bioimaging, and Therapy Applications*; Nanoscience and Nanotechnology Series; Royal Society of Chemistry: London, UK, 2016; pp. 1–39. [CrossRef]
32. Suta, M.; Antić, Ž.; Đorđević, V.; Kuzman, S.; Dramićanin, M.D.; Meijerink, A. Making Nd^{3+} a Sensitive Luminescent Thermometer for Physiological Temperatures—An Account of Pitfalls in Boltzmann Thermometry. *Nanomaterials* **2020**, *10*, 543. [CrossRef]
33. Pedraza, F.J.; Rightsell, C.; Kumar, G.A.; Giuliani, J.; Monton, C.; Sardar, D.K. Emission Enhancement through Nd^{3+}-Yb^{3+} Energy Transfer in Multifunctional $NaGdF_4$ Nanocrystals. *Appl. Phys. Lett.* **2017**, *110*, 223107. [CrossRef]
34. Aulsebrook, M.L.; Graham, B.; Grace, M.R.; Tuck, K.L. Lanthanide complexes for luminescence-based sensing of low molecular weight analytes. *Coord. Chem. Rev.* **2018**, *375*, 191–220. [CrossRef]
35. Orcutt, K.M.; Jones, W.S.; McDonald, A.; Schrock, D.; Wallace, K.J. A Lanthanide-Based Chemosensor for Bioavailable Fe^{3+} Using a Fluorescent Siderophore: An Assay Displacement Approach. *Sensors* **2010**, *10*, 1326–1337. [CrossRef] [PubMed]
36. Yu, X.; Ryadun, A.A.; Pavlov, D.I.; Guiselnikova, T.Y.; Potapov, A.S.; Fedin, V.P. Highly Luminescent Lanthanide Metal-Organic Frameworks with Tunable Color for Nanomolar Detection of Iron(III), Ofloxacin and Gossypol and Anti-counterfeiting Applications. *Angew. Chem. Int. Ed.* **2023**, *62*, e202306680. [CrossRef] [PubMed]
37. Fan, B.; Wei, J.; Ma, X.; Bu, X.; Xing, N.; Pan, Y.; Zheng, L.; Guan, W. Synthesis of Lanthanide-Based Room Temperature Ionic Liquids with Strong Luminescence and Selective Sensing of Fe(III) over Mixed Metal Ions. *Ind. Eng. Chem. Res.* **2016**, *55*, 2267–2271. [CrossRef]

Disclaimer/Publisher's Note: The statements, opinions and data contained in all publications are solely those of the individual author(s) and contributor(s) and not of MDPI and/or the editor(s). MDPI and/or the editor(s) disclaim responsibility for any injury to people or property resulting from any ideas, methods, instructions or products referred to in the content.

Article

Near-Infrared Responsive Composites of Poly-3,4-Ethylenedioxythiophene with Fullerene Derivatives

Oxana Gribkova, Varvara Kabanova, Ildar Sayarov, Alexander Nekrasov and Alexey Tameev *

A.N. Frumkin Institute of Physical Chemistry and Electrochemistry RAS, Leninskii Prospect 31, Moscow 119071, Russia; gribkova@elchem.ac.ru (O.G.); kabanovavar@phyche.ac.ru (V.K.); sayarovir@phyche.ac.ru (I.S.); secp@elchem.ac.ru (A.N.)
* Correspondence: tameev@elchem.ac.ru

Abstract: Electrochemical polymerization of 3,4-ethylenedioxythiophene in the presence of water-soluble fullerene derivatives was investigated. The electronic structure, morphology, spectroelectrochemical, electrochemical properties and near-IR photoconductivity of composite films of poly(3,4-ethylenedioxythiophene) with fullerenes were studied for the first time. It was shown that fullerene with hydroxyl groups creates favorable conditions for the formation of PEDOT chains and more effectively compensates for the positive charges on the PEDOT chains. The near-IR photoconductivity results from the generation of charge carriers due to electron transfer from the photoexcited PEDOT molecule to the fullerene acceptor.

Keywords: PEDOT; fullerene; electrochemical polymerization; spectroelectrochemistry; photoconductivity

Academic Editors: Tengling Ye and Bixin Li

Received: 15 November 2024
Revised: 19 December 2024
Accepted: 20 December 2024
Published: 25 December 2024

Citation: Gribkova, O.; Kabanova, V.; Sayarov, I.; Nekrasov, A.; Tameev, A. Near-Infrared Responsive Composites of Poly-3,4-Ethylenedioxythiophene with Fullerene Derivatives. *Polymers* **2025**, *17*, 14. https://doi.org/10.3390/polym17010014

Copyright: © 2024 by the authors. Licensee MDPI, Basel, Switzerland. This article is an open access article distributed under the terms and conditions of the Creative Commons Attribution (CC BY) license (https://creativecommons.org/licenses/by/4.0/).

1. Introduction

Electrochemical synthesis of conductive polymers combines the polymerization and deposition of polymer films of the desired thickness onin a single step. The thickness of the polymer films can be easily adjusted by controlling the total amount of charge spent on the electropolymerization. The electrochemically deposited layers usually are more uniform and the electrical contact between the electrode and layer is stronger than that of the layer cast from a solution. By comparison, the very popular poly(3,4-ethylenedioxythiophene):polystyrenesulfonate (PEDOT:PSS) layers of typical thickness 20 to 50 nm obtained by spin-coating tend to form relatively large grains and have a high defect density [1], which results in unfavorable reverse recombination of charge carriers and, consequently, high noise in the current, which is detrimental to the performance of organic photodiodes [2].

Composite layers of conductive polymers can be obtained by introducing various carbon nanomaterials into the polymerization solution. PEDOT has a narrow band gap, which ensures reversible electrochemical p- and n-doping, excellent thermal and electrochemical stability, and the possibility of using environmentally friendly aqueous electrolytes without polymer degradation. The formation of PEDOT composites with fullerenes can lead to products with combined properties inherent in both components. PEDOT can act as an electron donor that absorbs light in a wide range of the spectrum, while fullerene is a well-known electron acceptor that can facilitate the separation of photogenerated electron–hole pairs. The combination of both compounds in one thin film allows for obtaining a material with unique properties. Electrically conductive PEDOT in semi-oxidized and oxidized states has electron levels of polarons and bipolarons that are located inside the band gap and provide absorption of the polymer in the near-IR range. Thus, the introduction of fullerene derivatives into PEDOT can promote photoconductivity in the near-IR range of the spectrum.

There are a small number of works in the literature describing the electrochemical synthesis of composite layers of PEDOT with fullerenes. Thus, to obtain functional layers in organic solar cells (OSC), the electrocodeposition of thiophene and C60 [3] or thiophene derivative and C60 [4] were carried out in a wide range of potential cycling in organic solvents. In another case, monomers of 3,4-ethylenedioxythiophene (EDOT) containing pendant fragments C60 were synthesized and then were electropolymerized on an electrode [5,6]. To obtain electrochromic layers, C60 was used as a doping anion to compensate for the positive charge of the oxidized polymer [7,8]. As a result, C60 was deposited on the film simultaneously with electropolymerized EDOT or its derivatives and was evenly distributed within the layer. In [9], photoactive hybrid PEDOT films with C60 and C60 with a grafted thiophene anchor group were electrochemically synthesized and their photocurrent generation properties in the presence of methyl viologen in the range of 400–700 nm were investigated. In all described cases, electropolymerization was carried out in organic solvents with background electrolytes. In [10] only, PEDOT films were electrochemically deposited on electrodes using aqueous solutions containing EDOT and fullerene sulfonated calixarenes.

In cases of using PEDOT–fullerene as a photogeneration layer in a diode structure, it should be isolated from the anode by a hole-transport (electron-blocking) layer. A commercial PEDOT–PSS composition is most often used as such a layer. In order to create an OSC, the authors of [3,4] performed a copolymerization of thiophene and C60 on the layer of electrodeposited PEDOT. To improve film formation and interfacial charge transfer between the polymer film and the ITO electrode, the authors of [11] used an electrodeposited PEDOT–polystyrene sulfonic acid (PSSA) sublayer. Our previous studies [12,13] demonstrated the efficiency of electrodeposition of composites of PEDOT with Zn and Cu phthalocyaninates and polypyrrole with Zn phthalocyaninate on a thin sublayer of electrode deposited PEDOT complex with poly(2-acrylamido-2-methyl-1-propanesulfonic acid) (PAMPSA). This sublayer was chosen based on our studies of the effect of the hole-transport layer composition on the performance of inverted perovskite solar cells (PSCs) [14]. A PSC with a PEDOT–PAMPSA layer demonstrated the maximum short-circuit current and efficiency. Earlier [15], we prepared electrodeposited PEDOT–fullerenol photosensitive layers and demonstrated their applicability for the development of organic photodiodes. In this case PEDOT-fullerenol layers were electrodeposited without sublayer. When studying the influence of the structure of fullerene derivatives on the electrodeposition of PEDOT, it was found that in the presence of fullerene with 5 carboxyl groups PEDOT can only be electrodeposited onto thin PEDOT sublayer.

In the presented work, we have carried out a comparative investigation of the electrochemical synthesis of PEDOT composite films with water-soluble fullerenes of various structures on the thin sublayer of electrodeposited PEDOT–PAMPSA. The electronic structure and morphology, and the spectroelectrochemical and electrochemical properties of PEDOT composite films with fullerene derivatives were studied. Finally, the performance of the obtained layers in near-IR organic photodiodes was compared.

2. Materials and Methods

2.1. Materials

Na^+-containing fullerene with hydroxyl groups ($Na_4[C_{60}(OH)_x]$, where x~30) (NaFl) [16], synthesized according to the method described in [17] and K^+-containing fullerene with 5 carboxyl groups (KPCF) [16], synthesized according to the method described in [18], were used. Figure 1 shows the structural formulas (a, b) and spectra of aqueous solutions (c) of the fullerenes used. It is evident that they absorb in the UV–visible regions up to 600 nm. By drawing tangents to the decay of the absorption bands of fullerenes, we determined the widths of their optical band gaps (Table 2). The shoulder near 260–280 nm, observed for KPCF, corresponds to absorption of phenyl groups.

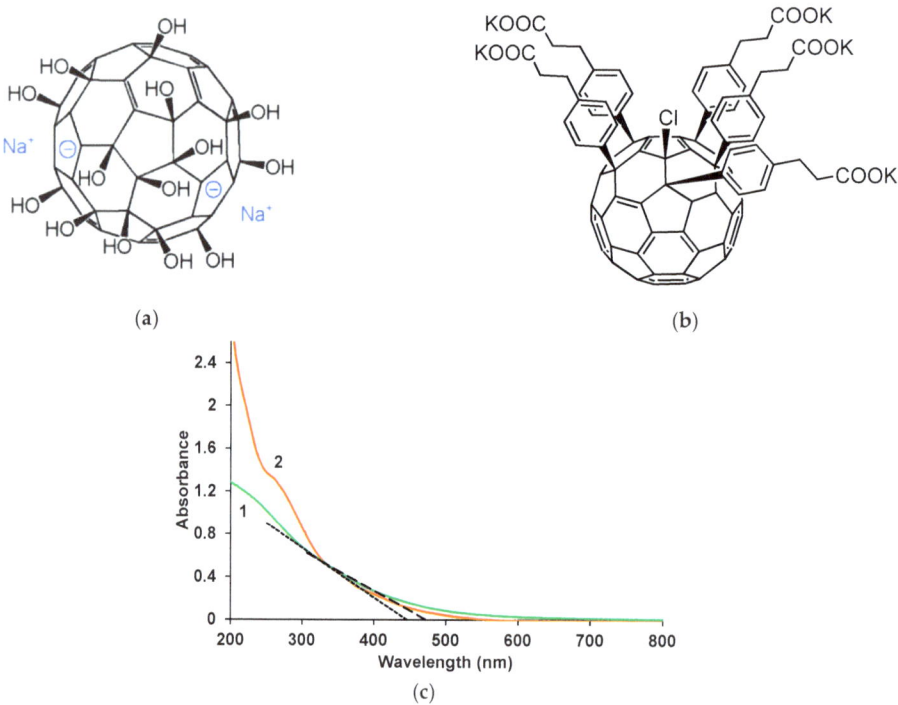

Figure 1. Structure (**a**,**b**) [16] and electronic absorption spectra (**c**) of 0.0001 M aqueous solution of NaFl (1) and KPCF (2).

In NaFl, the most probable localization of the negative charge is on the oxygen atoms with partial localization on the carbon framework [16,19]. Amphiphilic water-soluble fullerene derivative KPCF tends to form various supramolecular structures including vesicles and their clusters in aqueous media. According to dynamic light scattering data, the hydrodynamic radii of the vesicles are 20, 70 nm and more than 10 μm [20].

To reveal the influence of these aggregates on the ability of the substituted fullerenes to act as charge-compensating counter ions during EDOT electropolymerization, we have measured the zeta-potentials in 0.002 M solutions of NaFl or KPCF with the addition of 0.001 M NaCl. The results are given in the Supplementary Materials (SM). From Figure S1a, it is clear that NaFl exhibits one sharp peak, corresponding to the potential values near −24 mV. On the contrary, KPCF (Figure S1b) exhibits one sharp and one wide peak, the first corresponding to the potential values near −12–15 mV and the second one to those near −39–42 mV. The sharp peaks most possibly correspond to non-associated fullerene derivatives. The fact that zeta-potential of single NaFl is higher than that of single KPCF may be explained by higher content of charged moieties on the surface of the former fullerene. The wide peak for KPCF most possibly corresponds to the abovementioned aggregates. By comparing potentials of sharp and wide peaks, one may suppose that the aggregates include approximately three molecules of KPCF on the average. However, the long tail of the wide peak extending to the area with more negative potentials testifies to the existence of aggregates of a higher degree association. Taking into account that the KPCF molecule is asymmetric and has hydrophilic (-COOK groups) and hydrophobic (fullerene body) parts, the most probable structure of the aggregates in aqueous medium is fullerene bodies in the inner sphere surrounded by -COOK groups in the outer sphere.

2.2. PEDOT Synthesis in the Presence of Fullerenes

EDOT (Sigma–Aldrich (St. Louis, MO, USA)) was distilled under argon. The freshly distilled product was used. Fullerenes were dissolved in water to the desired concentration, then EDOT was added and the solution was intensively stirred for 2 h with heating to ~60 °C.

The PEDOT–PAMPSA sublayer was electrodeposited onto glass–FTO electrodes (7 Ohm/sq., Solaronix SA, Aubonne, Switzerland) with an area of ~1.3 CM^2 in galvanostatic mode at a current density of 0.05 mA/cm^2 in an aqueous solution containing 0.01 M EDOT and 0.02 M PAMPSA (Sigma–Aldrich, 15% aqueous solution, $Mw \approx 2 \cdot 10^6$). Electrodeposition of the sublayer was carried out until the electropolymerization charge of 7 mC/cm^2 was reached, which corresponded to a thickness of ~30 nm. The thickness was measured with a stylus profiler Alfa-Step D-100 (KLA Corp., Milpitas, CA, USA).

For the polymerization of EDOT, the following aqueous solutions were prepared: 0.01 M EDOT and 0.0014 M NaFl; 0.01 M EDOT and 0.004 M KPCF. These optimal concentrations were chosen on the base of preliminary experiments. Electrodeposition of PEDOT composite films was carried out in potentiodynamic (PD) in the potential range of −0.6–1.0 V at a scan rate of 50 mV/s, galvanostatic (GS) at a current density of 0.05 mA/cm^2 and potentiostatic (PS) at a potential of 0.9 V on FTO electrodes covered with the PEDOT–PAMPSA sublayer. Composite PEDOT layers were electrodeposited until a charge of 43 mC/cm^2 was reached. The total charge (including sublayer) was 50 mC/cm^2, and the total layer thickness was 200–300 nm. The polymerization was carried out in a three-electrode cell based on a 2 cm spectrophotometric cuvette with a special lid for fixing the electrodes and the salt bridge to a separate volume with a reference electrode. Platinum foil was used as a counter electrode, and a saturated silver–silver chloride electrode (Ag/AgCl) was used as a reference electrode. All potentials in this work are presented relative to this electrode.

To determine the lowest unoccupied molecular orbital (LUMO) and highest occupied molecular orbital (HOMO) energy levels, the composite layers of PEDOT were deposited in the GS mode on a Pt electrode with an area of ~0.5 cm^2 with a PEDOT–PAMPSA sublayer until a charge of 100 mC/cm^2 was reached.

To study the photoelectric properties of the composite films, electrochemical polymerization of EDOT was carried out on optically transparent glass–ITO (indium-tin oxide) electrodes (15 Ohm/sq., Kintec, Hong Kong, China) with an area of 1.12 cm^2 with PEDOT–PAMPSA as a sublayer in the presence of NaFl and KPCF in the GS mode until a charge of 50 mC/cm^2 was reached.

2.3. Preparation and Characterization Techniques

During the EDOT electropolymerization, electrochemical data and, simultaneously, in situ optical absorption spectra in the UV–visible region (350–950 nm) were recorded. Control and recording of the electrochemical parameters during the synthesis and electrochemical studies of the obtained films were carried out using an Autolab PGSTAT302N potentiostat (Metrohm, Utrecht, The Netherlands). Optical absorption spectra during the PEDOT synthesis with a repetition frequency of 2 s, as well as spectroelectrochemical studies of the obtained films at fixed potentials in an aqueous solution of 0.5 M $NaClO_4$, were carried out using a high-speed scanning single-beam spectrophotometer Avantes 2048 (Avantes BV, Apeldoorn, The Netherlands).

The LUMO and HOMO energy levels were determined by cyclic voltammetry (CV) in non-aqueous medium. The detailed procedure is described in the Supplementary Materials. The Fermi level and valence band energy states of the solid films were measured by ultraviolet photoelectron spectroscopy (UPS) using a Thermo Fisher Scientific ESCAlab 250Xi spectrometer (Waltham, MA, USA), with a base pressure of $1 \cdot 10^{-10}$ mbar. For the UPS measurements, an ultraviolet (He I, 21.22 eV) light source was employed and a sample bias of -5 V was applied to obtain the secondary electron cutoff.

The electron absorption spectra of the obtained films in air in the range of 300–1300 nm were recorded using a Shimadzu UV-3101PC spectrophotometer (Shimadzu Deutschland GmbH, Duisburg, Germany).

The ζ-potential of the fullerenes solutions was measured by means of a Zetasizer Nano ZS (Malvern, England) analyzer.

The surface morphology of the composite films obtained in the galvanostatic mode was investigated using an Enviroscope atomic force microscope (AFM) with a Nanoscope V controller (Bruker, Billerica, MA, USA) in the semi-contact mode. Scanning electron microscopy (SEM) of the obtained films was performed using a Tescan Amber GMH scanning microscope (Tescan Orsay Holding, a.s., Brno–Kohoutovice, Czech Republic). The images were obtained using an Everhart–Thornley SE detector at $\times 10{,}000$–$300{,}000$ magnifications and an accelerating voltage of 0.5 kV.

The photoelectric characteristics of the composite layers were investigated in samples of the ITO/sublayer/composite/C_{60}/BCP/Al diode structure, where C_{60} (MST, St. Petersburg, Russia) is a 40 nm fullerene layer as an electron acceptor and transport layer, BCP (Kintec, Hong Kong, China) is an 8 nm layer of 2,9-dimethyl-4,7-diphenyl-1,10-phenanthroline blocking holes, and Al is an electrode (80 nm) All layers, except the composite and sublayer, were deposited by vacuum evaporation.

A standard Keithley 2400 source-meter unit (Solon, OH, USA) was used to perform photoelectric measurements. The samples were illuminated with a xenon UV-IR light source, a model 66477 (Newport Corp., Irvine, CA, USA) through a combination of SZS-26 and KS19 filters (LOMO, St. Petersburg, Russia), which transmit radiation within the 700 nm and 900 nm bandwidth. The radiation power of the light flux incident on the samples was measured using a PE25-SH Ophir energy sensor (Ophir, Jerusalem, Israel). All photoelectric measurements were performed in a sealed glove box in a dry argon atmosphere.

3. Results and Discussion

3.1. Electropolymerization of EDOT in the Presence of Fullerenes

Figure 2 shows the cyclic voltammograms (CVA) obtained during the synthesis of PEDOT in the PD mode in the potential range from -0.6 to 1.0 V. It is evident that the shape of the curve is characteristic of PEDOT films obtained in aqueous solutions [17] (Figure S2). The onset potentials of EDOT oxidation in aqueous solutions of NaFl and KPCF are 0.76 and 0.57 V, respectively. It is evident that in the presence of NaFl, the synthesis of PEDOT proceeds faster with higher currents. The fullerenes used do not exhibit electroactivity in this region of potential cycling (Figure 2c,d).

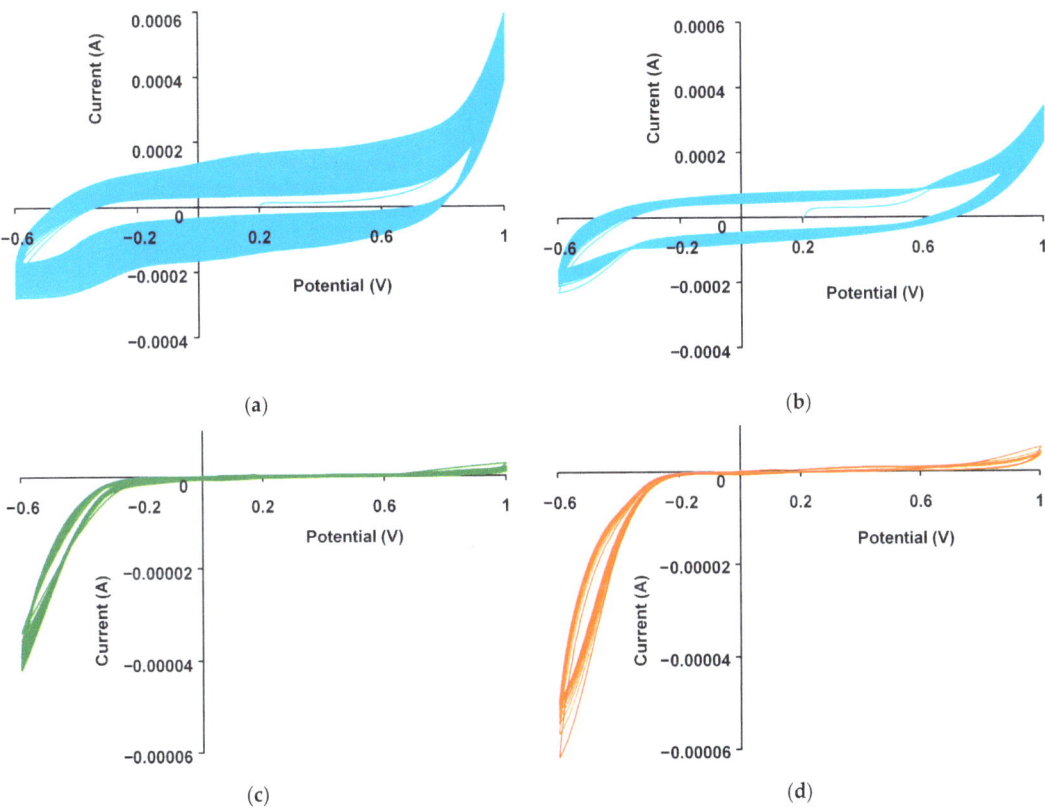

Figure 2. Cyclic voltammograms in aqueous solutions: (**a**) FTO electrode with a PEDOT–PAMPSA sublayer in 0.01 M EDOT and 0.0014 M NaFl; (**b**) FTO electrode with the sublayer in 0.01 M EDOT and 0.004 M KPCF; (**c**) bare FTO electrode in 0.0014 M NaFl and (**d**) bare FTO electrode in 0.004 M KPCF. The potential scan rate was 50 mV/s.

The changes in potential over time obtained during the polymerization of EDOT in the presence of fullerenes in the GS mode are shown in Figure 3a. It is evident that in the presence of NaFl, (curve 1) as PAMPSA (curve 3) the synthesis proceeds at a lower potential than that in the presence of KPCF. In the PS mode, the synthesis charge grows significantly faster in the presence of NaFl (Figure 3b, curve 1) similar in PAMPSA (Figure 3b, curve 3). It can be assumed that NaFl creates favorable conditions for the formation of PEDOT chains and more effectively compensates for the positive charges on the PEDOT chains.

Simultaneously with the electrochemical parameters, electronic absorption spectra were recorded during the synthesis, the evolution of which is shown in Figure 4. It should be noted that the nature and dynamics of the spectral changes in the case of NaFl are characteristic of the synthesis of PEDOT in aqueous solutions [21] including in PAMPSA (Figure S3). A monotonic increase in the optical absorption is observed, most noticeable in the wavelength region above 600 nm and extending into the near-IR region. This indicates the formation of a PEDOT layer in the conducting form. In the presence of KPCF, an increase in the absorption in the 700 nm region is clearly visible, which corresponds to the polaronic form of PEDOT. Strong noise in the 350–550 nm region is associated with the subtraction of the intense absorption of fullerene solutions (Figure 1c), used as a background when recording the spectra.

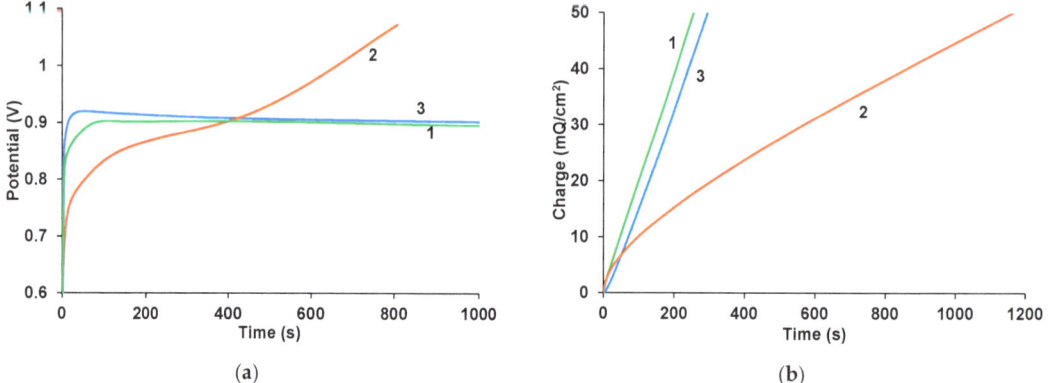

Figure 3. Time dependency of the potential during GS (**a**) and the charge during PS (**b**) electropolymerization of 0.01 M EDOT in the aqueous solutions of 0.0014 M NaFl (1) and 0.004 M KPCF (2) and 0.02 M PAMPSA (3).

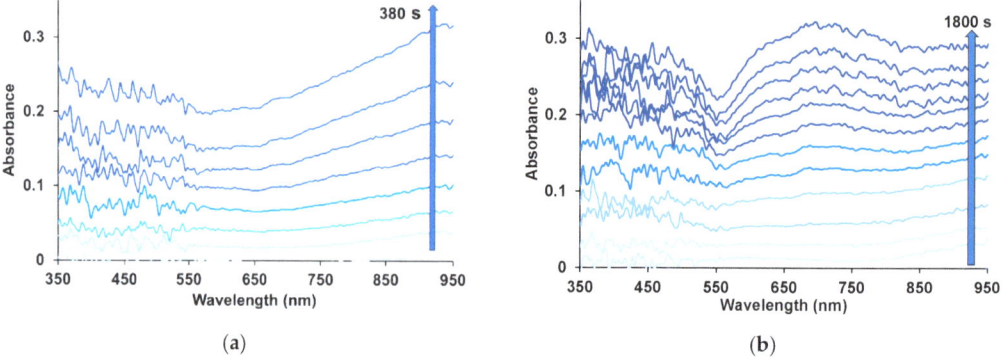

Figure 4. Electron absorption spectra of the PEDOT film formed on the working electrode during the polymerization of EDOT in the PS mode at the potential of 0.9 V in aqueous solutions of 0.0014 M NaFl (**a**) and 0.004 M KPCF (**b**). The optical path in the solution is 1.6 cm.

Figure 5 shows the dynamics of changes in the absorption of growing PEDOT films at characteristic wavelengths. It is evident that the polymerization of EDOT occurs at a higher rate in the presence of NaFl similar PAMPSA. Also, the absorption at 900 nm, characteristic of the bipolaron form of PEDOT, grows more actively in the case of NaFl (PAMPSA). In the presence of KPCF, the growth of the polaron form is more pronounced, and the growth of the bipolaron form begins to slow down toward the end of the synthesis. Such differences in kinetics of the syntheses may be associated with the tendency of KPCF to form aggregates of a large size in aqueous media, which may block the electrode surface and hinder the formation of PEDOT–KPCF films. This is also probably a result of steric hinderances caused by the asymmetric structure (a small part of the hydrophobic fullerene cage is covered with hydrophilic addends with negatively charged carboxyl groups) of KPCF and, accordingly, a worse ability to compensate for charges on the growing PEDOT chains.

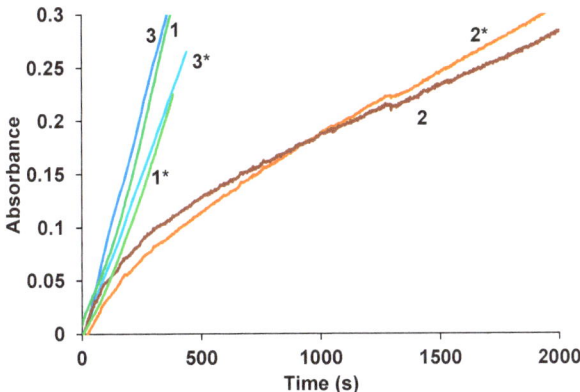

Figure 5. Time dependency of the optical absorption at 900 and 700 (*) nm during PS electrodeposition of PEDOT composite films in the presence of NaFl (1, 1*) and KPCF (2, 2*) and PAMPSA (3, 3*).

3.2. Electron Absorption Spectroscopy of Composite Films in the UV-Visible and Near-IR Regions

Figure 6 shows the electron absorption spectra in the UV–visible and near-IR regions for PEDOT composite films with fullerenes. More intense absorption in the short-wavelength region is clearly visible compared to the PEDOT film obtained in the PAMPSA solution (curve 4), which corresponds to the absorption of both fullerenes (Figure 1c). Moreover, more intense absorption is observed in the case of NaFl, which indicates its higher content in the film. In contrast to the PEDOT–PAMPSA film (curve 3), the spectrum of which indicates that the polymer is in a highly conductive bipolaron state [22], for the composite films a pronounced absorption maximum is observed in the region of 700–800 nm, characteristic of PEDOT in the polaronic state [23]. Lower absorption in the near-IR region in the latter case is associated with a lower content of the bipolaron form. This trend is more pronounced for the PEDOT–KPCF composite (curve 2). In addition, in the spectrum of the PEDOT–KPCF composite, the absorption maximum of the polaron form is shifted to the short-wavelength region, which indicates shorter conjugation length [24].

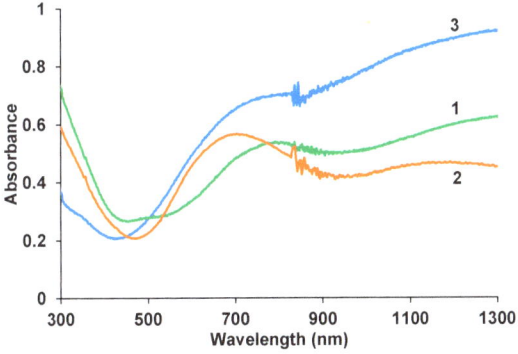

Figure 6. Electron absorption spectra in air of the PEDOT composite films electrodeposited in aqueous solutions of NaFl (1), KPCF (2), PAMPSA (3) at the charge density of 50 mC/cm^2. The spectra were recorded in air.

3.3. Spectroelectrochemical Studies

The evolution of spectra depending on the fixed potentials are similar for both of the composite films. At low potentials (from −0.8 to −0.2 V), the band around 575 nm is observed, caused by π–π* transitions in the reduced form of PEDOT (Figure 7). As the potential increases (oxidation), the intensity of this band decreases, while within the potential range from −0.1 to 0.4 V, the absorption band around 800 nm is formed (the polaron form of oxidized PEDOT). Simultaneously, the absorption in the near IR region of the spectrum increases (bipolaron form) [25], which is most clearly expressed at the potentials from 0.5 to 0.8 V. The observed changes in the spectra during oxidation of the PEDOT composite film are characteristic of the PEDOT films studied in an aqueous medium [22,26]. However, the absorption maxima of the reduced and polaron forms of the composites are slightly shifted to the short-wavelength region compared to the PEDOT–PAMPSA film (Figure S4) (which has conventional spectroelectrochemical behavior) (Table 1), this shift being more pronounced for PEDOT–KPCF. This may be due to the formation of shorter PEDOT chains [24]. The steric hindrances for orientation of KPCF molecules with asymmetrically arranged carboxyl groups during the polymer formation lead to a worse ability to compensate positive charges on growing PEDOT chains. The same influence of steric hindrance on the conjugation length of PEDOT layers were observed in [26,27]. PEDOT layers were prepared in the presence of rigid-chain amid-containing polysulfonic acid and sulfonated poly(β-hydroxyethers) polyester with different content of sulfonic groups. The structure (rigidity of polyacid backbone and distribution of sulfonic groups) of the dopants creates a more strained structure for PEDOT chains and worse compensation of charges which results in the formation of chains with shorter conjugation.

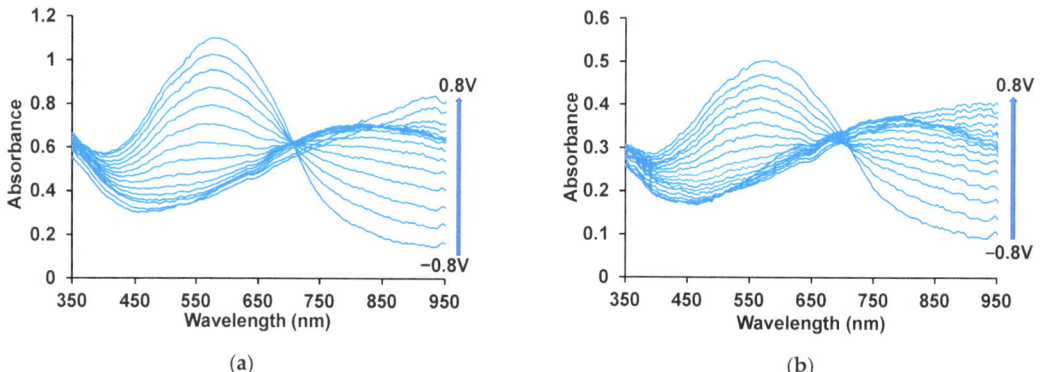

Figure 7. Electron absorption spectra of the composite films PEDOT–NaFl (**a**) and PEDOT–KPCF (**b**) measured at fixed potentials in 0.5 M NaClO4 aqueous solution.

Table 1. Spectral characteristics and doping degrees of PEDOT with different counter ions.

Counterion	Maximum of the Reduced Form, nm	Maximum of the Polaronic Form, nm	DD_{CVA}	DD_{EDX}
NaFl	575	840	0.15	0.11
KPCF	573	776	0.08	0.06
PAMPSA	615	878	0.25	0.84

By drawing tangents to the decay of the bands corresponding to the π-π* transition of the reduced form of PEDOT in composites of PEDOT with fullerenes (Figure 7), we determined the widths of the optical band gap for composites (Table 2).

Table 2. Energy characteristics of the substances determined by different methods: E_g—bandgap; E_f—Fermi energy level; E_{HOMO}—energy level of highest occupied molecular orbital; E_{LUMO}—energy level of lowest unoccupied molecular orbital.

		From UPS Data		From CV Data			
	$E_{g(opt)}$, eV	E_{HOMO}, eV	E_f, eV	E_{LUMO} PEDOT, eV	E_{LUMO} Fullerene, eV	E_{HOMO}, eV	E_g, eV
NaFl	2.78	−6.6	−4.8				
KPCF	2.63	−6.5	−4.9				
PEDOT–PAMPSA	1.57	−4.8	−4.8				
PEDOT–NaFl	1.55			−2.97	−4.05	4.50	1.53
PEDOT–KPCF	1.49			−2.95	−4.05	−4.33	1.38

3.4. Electrochemical Studies

Figure 8 shows the CVA of the composite films measured in 0.5 M NaClO$_4$ solution in the range of potentials from −0.6 to 0.6 V (50 mV/s). Using them, the doping degree DD$_{CVA}$ (Table 1) was calculated according to the relationship DD$_{CVA}$ = 2Q$_{ox}$/(Q$_{poly}$−Q$_{ox}$) [28,29], where Q$_{ox}$ is the oxidation charge calculated by integrating the anodic waves of CVA measured in the same potential cycling range, and Q$_{poly}$ is the polymerization charge spent on the electrosynthesis. In parallel with recording of SEM images we have performed energy dispersive X-ray (EDX) analysis of the samples (Figure S5). We calculated DD$_{EDX}$ of composite films from these data. One can see that these values are in good correspondence with that determined from CVA (Table 1). The doping degree of the PEDOT–KPCF composite is two times lower, which correlates with the lower absorption in the near-IR region of the spectrum. Apparently, in the presence of KPCF with unevenly distributed grafted benzene rings with carboxylate groups, some of these molecules are associated in low-mobile aggregates and there are steric hindrances for efficient electrostatic interactions of these groups with positively charged PEDOT chains. Moreover, the aggregates of KPCF may block some of the active growth sites on the electrode surface and thus hinder film formation. In the case of NaFl, the charge is uniformly distributed over the fullerene molecule. So, NaFl more efficiently compensates for the positive charges on the PEDOT chain, and a more optimal composite structure is formed. While comparing these values with those obtained for PEDOT-PAMPSA composite, one can notice that DD$_{CVA}$ is much higher in the latter case and DD$_{EDX}$ >> DD$_{CVA}$. This situation is quite essential for PEDOT doped with polymeric acids [29] because growing PEDOT film may occlude excessive quantity of long macromolecules and not all sulfonic groups belonging to these macromolecules participate in the charge compensation (doping).

Figure S6 shows the CVA curves recorded in the non-aqueous medium. In the potential range of −0.6–0.2 V (Figure S6a) for the PEDOT–NaFl composite, a front of increasing anodic current of PEDOT oxidation is observed, from which it is possible to determine the HOMO level of PEDOT equal to −4.50 eV. A front of increasing cathodic current is also visible in the potential range of −1.2–1.0 V, which most likely characterizes the LUMO level of fullerene NaFl (E_{LUMO} = −4.05 eV) [30]. In the range of more negative potentials, a front of increasing cathodic current of PEDOT reduction is observed corresponding to $E_{LUMO(EC)}$ = −2.97 eV. From the presented data, it is possible to calculate the width of the bandgap of the PEDOT–NaFl composite: E_g = 4.50 − 2.97 = 1.53 eV. Of note, this value is in good correspondence with that determined from the electron absorption spectra (Figure 7a) $E_{g(opt)} = 1.55$ eV.

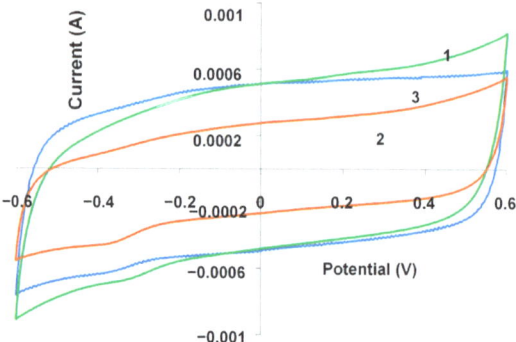

Figure 8. Cyclic voltammograms of PEDOT–NaFl (1), PEDOT–KPCF (2) and PEDOT-PAMPSA (3) films measured in 0.5 M NaClO$_4$ aqueous solution at a scan rate of 100 mV/s. The films have equal electropolymerization charges.

For the PEDOT–KPCF composite, the front of the increasing anodic current of PEDOT oxidation is observed in the potential range of -0.7–0.5 V (Figure S6b), which is more cathodically than for PEDOT–NaFl and corresponds to $E_{HOMO} = -4.33$ eV. Taking into account approximately the same value of $E_{LUMO} = -2.95$ eV, the width of bandgap of the PEDOT–KPCF is $E_g = 4.33 - 2.95 = 1.38$ eV. This value is close to that determined from the electron absorption spectra (Figure 7b), $E_{g(opt)} = 1.49$ eV. Importantly, E_g for PEDOT–KPCF is lower than that for PEDOT–NaFl, demonstrating the same tendency as $E_{g(opt)}$. Inside the bandgap in the range of potentials -1.4–1.0 V, one can see a front of increasing cathodic current, which characterizes the LUMO level of fullerene KPCF ($E_{LUMO} = -4.05$ eV). The energy characteristics of the substances are summarized in Table 2.

Relative positions of energy levels listed in Table 2 are graphically depicted in Figure S8.

Remark: for CV data and UPS data see in the Supplementary Materials in Figure S6 and Figure S7, respectively.

The bandgap reduction for the composite films with shorter chain length may seem a paradox. We can propose the only explanation - loose morphology of KPCF film (Figure 9f). From Table 1 it is clear that maxima of the absorption bands of the reduced forms of PEDOT-fullerene composites are blue-shifted compared to PEDOT-PAMPSA. This is normal situation for shorter chain length. At the same time loose, imperfect structure of the film contributes to increased dispersion of the energy levels, which manifests itself in broadening of absorption band. Since we determine the bandgap value from the intersection of a tangent to the red slope of absorption band with the axis of wavelengths, for broadened absorption band we obtain underestimated bandgap value. Similar situation is for CVA determination: loose structure of the film may facilitate appearing of the current onset at lower values of potential.

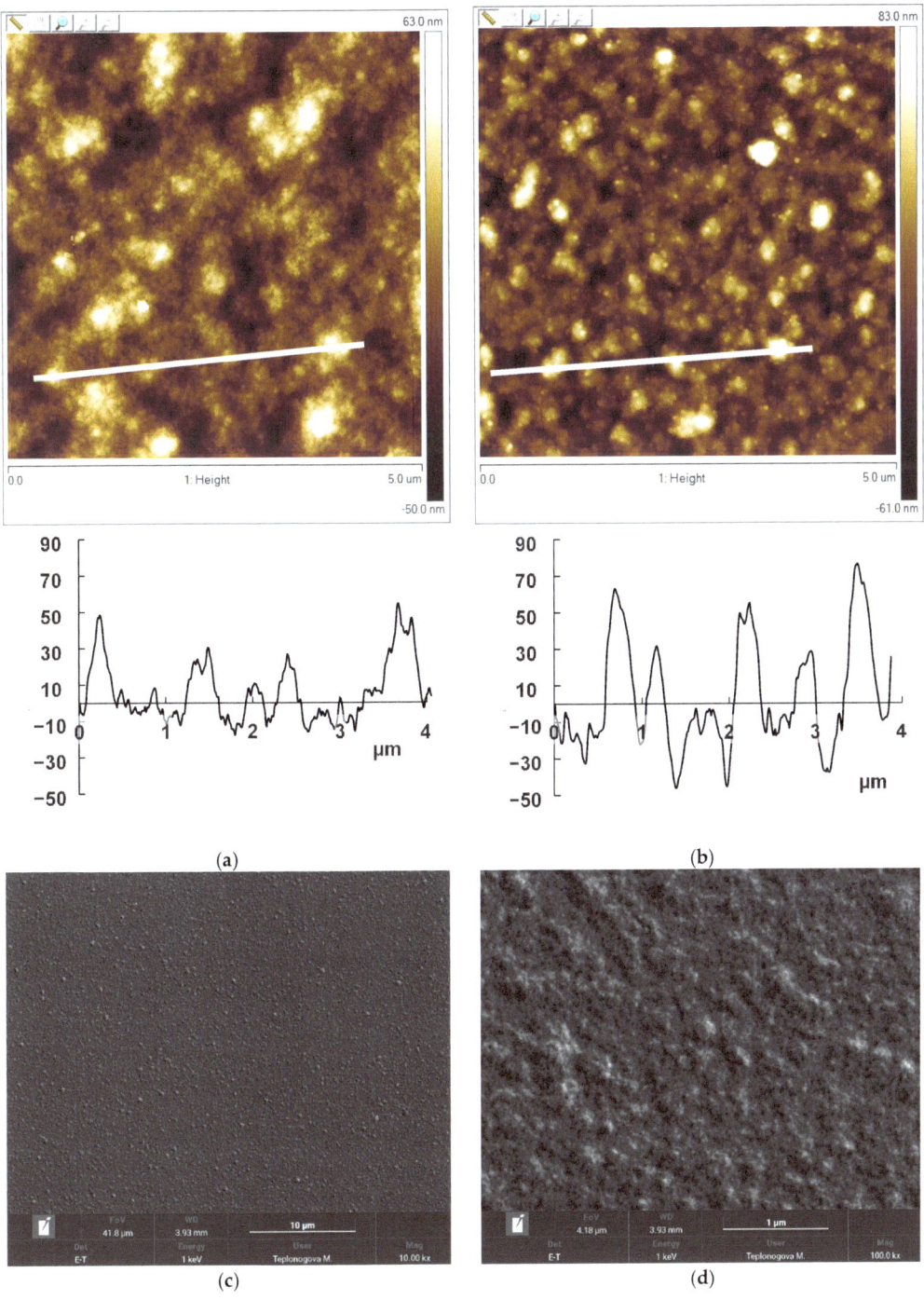

Figure 9. *Cont.*

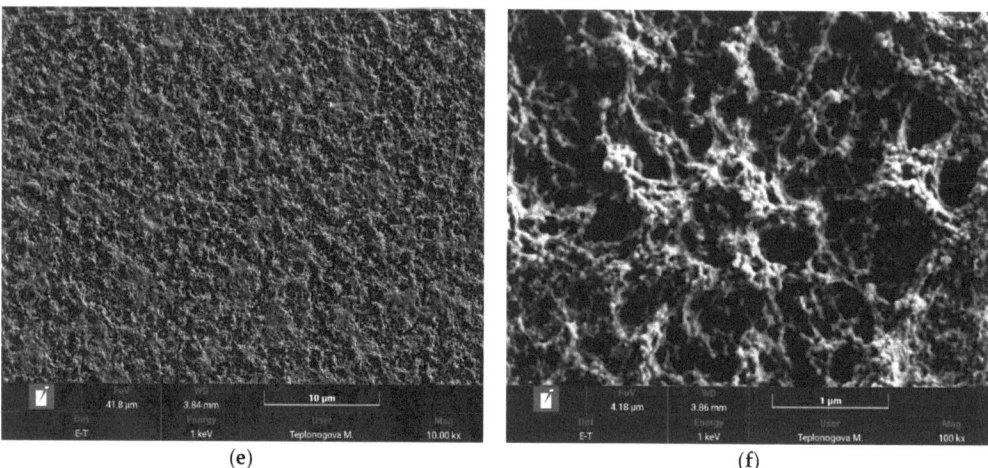

Figure 9. AFM images with profiles along white lines (**a**,**b**) and SEM (**c**–**f**) images of the surfaces of PEDOT composite films deposited onto FTO–electrodes with a PEDOT–PAMPSA sublayer in the presence of NaFl (**a**,**c**,**d**) and KPCF (**b**,**e**,**f**).

3.5. Morphology

The surface morphology of the PEDOT–NaFl and PEDOT–KPCF composite films was studied by AFM and SEM. From the AFM images one can see that PEDOT–NaFl (Figure 9a) has a filament-like structure that forms agglomerates with a lateral size 200–500 nm and height 30–50 nm and roughness 15 nm. PEDOT-PAMPSA film has similar structure with higher roughness 31.5 nm (Figure S9a,b). It should be noted that the roughness of PEDOT-NaFl film deposited on PEDOT-PAMPSA sublayer is lower than without sublayer (40–50 nm) [15]. It is known that the most defective morphology is formed on the stage of nucleation of a polymer phase on naked electrode. In our case the sublayer serves as nucleation centers resulting in formation of more homogeneous PEDOT-NaFl film. The surface of the PEDOT–KPCF film (Figure 9b) has a more globular structure with dense globules of a lateral size 250–350 nm, height 50–80 nm and a depression of depth about 50 nm. The roughness of PEDOT-KPCF film is 20 nm. The SEM overview images (Figure 9c,e) of the composites demonstrated uniform coatings on the electrodes without holes and cracks. The surface of the PEDOT–NaFl composite is more uniform and denser. At a higher magnification of the SEM image (Figure 9d), a granular surface of PEDOT–NaFl formed by compact and intergrown agglomerates was observed. The surface of PEDOT–KPCF is rougher with some hollows and hills (Figure 9f). One can see that their sizes (lateral and height/depth) are similar and correspond to the dimensions of aggregates 20 and 70 nm [20]. Some of them may be covered by PEDOT, which is forming the hills. The hollows may be formed as a result of washing out the substance from the surface during rinsing the films with water after the electropolymerization.

3.6. Near-IR Photoinduced Current

The photoconductivity was measured in devices of an ITO/sublayer/composite/C60/BCP/Al structure. The studied composites absorb radiation in the near-IR range due to PEDOT (Figure 6). The devices exhibited diode behavior. Typical current voltage characteristics are shown in Figure 10. Table 3 lists the near-IR photoinduced current and photosensitivity (photocurrent/incident power) in the PEDOT composites under irradiation in the 700 nm–900 nm bandwidth. In the fullerene-free PEDOT–PAMPSA layer, prepared for comparison, the change in conductivity under the near-IR irradiation was much smaller than that in the composites. In [31,32], the increase in current upon IR irradiation of a PEDOT:PSS film deposited by casting from a solution is explained to be caused by slight heating of the polymer film followed by increasing the charge carrier mobility. In the studied PEDOT–fullerene composites, we attribute the current induced by near-IR photoexcitation to an increase in the concentration of photogenerated charge carriers since fullerene can provide rapid exciton separation [33] by the transfer of photoexcited electrons from PEDOT to fullerene molecules. The transfer is energetically favorable according to the LUMO levels of the PEDOT donor and fullerene acceptor (Table 2). The higher response of PEDOT–NaFl compared to PEDOT–KPCF shows that the photoconductivity is associated with the influence of the fullerene structure on the degree of PEDOT doping and the morphology of the resulting layers. NaFl creates more favorable conditions for the formation of PEDOT chains and, as a consequence, a more uniform and denser layer of the composite with NaFl than with KPCF. The low current induced by near-IR radiation in the fullerene-free PEDOT–PAMPSA film indicates that the increase in current (charge carrier mobility) in the composites due to possible heating of the films makes a negligible contribution.

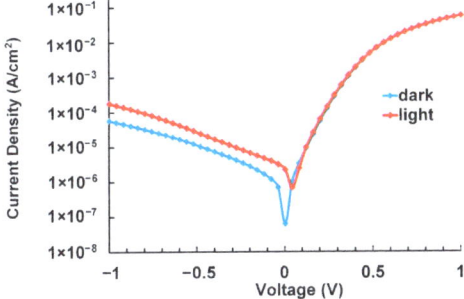

Figure 10. I–V characteristics of the PEDOT–NaFl based device in the dark (blue curve) and under near-IR illumination (brown curve) in the 700 nm–900 nm bandwidth at a power of 15 mW/cm^2.

Table 3. Conductivity of composites under irradiation with the 700 nm–900 nm near-IR bandwidth. The incident power was 15 mW/cm^2 and the reverse bias voltage was 0.04 V.

	PEDOT–NaFl	**PEDOT–KPCF**	**PEDOT–PAMPSA**
Photocurrent current, µA/cm^2	2.5	0.54	0.02
Photosensitivity, µA/W	160	36	1.4

4. Conclusions

A comparative study of EDOT electropolymerization in aqueous solutions containing water-soluble fullerene derivatives with different ionogenic groups without adding supporting electrolytes was carried out for the first time. In this case, fullerenes were incorporated into the PEDOT films and acted as doping anions, compensating for the positive charges on the growing PEDOT chains.

It appeared that one of the composite layers (PEDOT–KPCF) can only be electrodeposited onto a thin PEDOT sublayer. Moreover, such sublayers usually served as hole-transporting and electron-blocking layers in various organic electronic devices.

It was demonstrated for the first time that the fullerene structure influences the rate and characteristics of PEDOT electrodeposition, degree of PEDOT doping and the morphology of the composite layers. The PEDOT–KPCF composite has a shorter conjugation length, a lower doping degree, and its surface is rougher with some hollows and hills. On the contrary, NaFl creates more favorable conditions for the formation of PEDOT chains and a more uniform and dense layer.

Photoconductivity of the composites of PEDOT with fullerene derivatives in the near IR spectral range has been demonstrated. The main result of the action of near-IR radiation on the composite films is the photogeneration of charge carriers due to the transfer of a photoexcited electron from PEDOT to an acceptor fullerene molecule.

Supplementary Materials: The following supporting information can be downloaded at: https://www.mdpi.com/article/10.3390/polym17010014/s1. Figure S1: Results of Zeta-potential measurements in 0.0005 M of NaFl (a) and KPCF (b). Electrolyte added: 10^{-3} M NaCl; Figure S2: Cyclic voltammogram measured during the PEDOT film deposition from an aqueous solution of 0.01 M EDOT and 0.02 M PAMPSA; Figure S3: Electron absorption spectra of the PEDOT film formed on the working electrode during the polymerization of EDOT in the PS mode at the potential of 0.9 V in aqueous solutions of 0.02 M PAMPSA; Figure S4: Eelectron absorption spectra of the PEDOT-PAMPSA film measured at fixed potentials in 0.5 M NaClO4 aqueous solution; Figure S5: Frame-averaged EDX spectra of PEDOT-KPCF (a) and PEDOT-NaFl (b) films electrodeposited on ITO transparent electrodes; Figure S6: Cyclic voltammetry of PEDOT-NaFl (a) and PEDOT-KPCF (b) electrodeposited into Pt-electrodes with PEDOT-PAMPSA sublayer. Electrolyte: 0.2 M Bu4NBF4 in AN. Scan rate: 20 mV/s; Figure S7: UPS spectra: secondary electron cutoff and valence band region of (a) NaFl, (b) KPCF and (c) PEDOT-PAMPSA films. The inset shows an enlarged view of spectrum on the semi-log plot to observe the shift of HOMO w. r. t. EF. Each energy value was determined with an experimental error of ±0.2 eV; Figure S8: HOMO and LUMO energy levels of NaFl, KPCF and PEDOT-PAMPSA obtained basing on UPS data and optical bandgap listed in Table 2; Figure S9: AFM image (a) and SEM (b) image of the surface of PEDOT-PAMPSA film deposited onto FTO-electrodes. Table S1. Frame-averaged content of the representative chemical elements in PEDOT-KPCF and PEDOT-NaFl films electrodeposited on ITO transparent electrodes.

Author Contributions: Conceptualization: O.G. and A.T.; experimental investigation: A.N., V.K. and I.S.; data analysis and interpretation: O.G., A.T. and A.N.; writing—original draft preparation: O.G.; writing—review and editing: A.N. and A.T.; supervision: A.T.; project administration: A.N. and A.T.; funding acquisition: A.N. and A.T. All authors have read and agreed to the published version of the manuscript.

Funding: This study was supported by the Russian Science Foundation (project no. 23-1900884 for the investigation of photophysical properties) and the Ministry of Science and Higher Education of Russia (project no. 122011300052-1 for the Institute of Physical Chemistry and Electrochemistry of the Russian Academy of Sciences for the electrosynthesis and studying electrochemical and spectroscopic properties).

Institutional Review Board Statement: Not applicable.

Informed Consent Statement: Not applicable.

Data Availability Statement: Data are contained within the article and Supplementary Materials.

Acknowledgments: We are grateful to P.A. Troshin and O. Kraevaya (Federal Research Center of Problems of Chemical Physics and Medicinal Chemistry of the Russian Academy of Sciences) for supplying the fullerene derivatives and M.T. Teplonogova (Kurnakov Institute of General and Inorganic Chemistry of the Russian Academy of Sciences) for the SEM measurements. The UV–Vis–NIR spectroscopy and AFM measurements were performed using the equipment of the CKP FMI IPCE RAS. The SEM measurements were performed using the equipment of the JRC PMR IGIC RAS. The UPS was conducted at St. Petersburg State University, Research Center for Physical Methods of Surface Investigations.

Conflicts of Interest: The authors declare no conflicts of interest. The funders had no role in the design of the study; in the collection, analyses, or interpretation of the data; in the writing of the manuscript; or in the decision to publish the results.

References

1. Jolt Oostra, A.; Blom, P.W.M.; Michels, J.J. Prevention of Short Circuits in Solution-Processed OLED Devices. *Org. Electron.* **2014**, *15*, 1166–1172. [CrossRef]
2. Huang, Z.; Zhong, Z.; Peng, F.; Ying, L.; Yu, G.; Huang, F.; Cao, Y. Copper Thiocyanate as an Anode Interfacial Layer for Efficient Near-Infrared Organic Photodetector. *ACS Appl. Mater. Interfaces* **2021**, *13*, 1027–1034. [CrossRef]
3. Fan, B.; Wang, P.; Wang, L.; Shi, G. Polythiophene/Fullerene Bulk Heterojunction Solar Cell Fabricated via Electrochemical Co-Deposition. *Sol. Energy Mater. Sol. Cells* **2006**, *90*, 3547–3556. [CrossRef]
4. Nasybulin, E.; Cox, M.; Kymissis, I.; Levon, K. Electrochemical Codeposition of Poly(Thieno[3,2-b]Thiophene) and Fullerene: An Approach to a Bulk Heterojunction Organic Photovoltaic Device. *Synth. Met.* **2012**, *162*, 10–17. [CrossRef]
5. Reynoso, E.; Durantini, A.M.; Solis, C.A.; Macor, L.P.; Otero, L.A.; Gervaldo, M.A.; Durantini, E.N.; Heredia, D.A. Photoactive Antimicrobial Coating Based on a PEDOT-Fullerene C60polymeric Dyad. *RSC Adv.* **2021**, *11*, 23519–23532. [CrossRef]
6. Suárez, M.B.; Aranda, C.; Macor, L.; Durantini, J.; Heredia, D.A.; Durantini, E.N.; Otero, L.; Guerrero, A.; Gervaldo, M. Perovskite Solar Cells with Versatile Electropolymerized Fullerene as Electron Extraction Layer. *Electrochim. Acta* **2018**, *292*, 697–706. [CrossRef]
7. Dominguez-Alfaro, A.; Jénnifer Gómez, I.; Alegret, N.; Mecerreyes, D.; Prato, M. 2D and 3D Immobilization of Carbon Nanomaterials Into Pedot Via Electropolymerization of a Functional Bis-Edot Monomer. *Polymers* **2021**, *13*, 436. [CrossRef] [PubMed]
8. Alegret, N.; Dominguez-Alfaro, A.; Salsamendi, M.; Gomez, I.J.; Calvo, J.; Mecerreyes, D.; Prato, M. Effect of the Fullerene in the Properties of Thin PEDOT/C60 Films Obtained by Co-Electrodeposition. *Inorganica Chim. Acta* **2017**, *468*, 239–244. [CrossRef]
9. Akiyama, T.; Yoneda, H.; Fukuyama, T.; Sugawa, K.; Yamada, S.; Takechi, K.; Shiga, T.; Motohiro, T.; Nakayama, H.; Kohama, K. Facile Fabrication and Photocurrent Generation Properties of Electrochemically Polymerized Fullerene-Poly(Ethylene Dioxythiophene) Composite Films. *Jpn. J. Appl. Phys.* **2009**, *48*, 04C172. [CrossRef]
10. Dumitriu, C.; Mousavi, Z.; Latonen, R.M.; Bobacka, J.; Demetrescu, I. Electrochemical Synthesis and Characterization of Poly(3,4-Ethylenedioxythiophene) Doped with Sulfonated Calixarenes and Sulfonated Calixarene-Fullerene Complexes. *Electrochim. Acta* **2013**, *107*, 178–186. [CrossRef]
11. Lv, X.; Huang, C.; Tameev, A.; Qian, L.; Zhu, R.; Katin, K.; Maslov, M.; Nekrasov, A.; Zhang, C. Electrochemical Polymerization Process and Excellent Electrochromic Properties of Ferrocene-Functionalized Polytriphenylamine Derivative. *Dye. Pigment.* **2019**, *163*, 433–440. [CrossRef]
12. Gribkova, O.L.L.; Kabanova, V.A.A.; Yagodin, A.V.V.; Averin, A.A.A.; Nekrasov, A.A.A. Water-Soluble Phthalocyanine with Ionogenic Groups as a Molecular Template for Electropolymerization of 3,4-Ethylenedioxythiophene. *Russ. J. Electrochem.* **2022**, *58*, 957–967. [CrossRef]
13. Gribkova, O.L.; Kabanova, V.A.; Kormshchikov, I.D.; Tameev, A.R.; Nekrasov, A.A. Electrodeposition of Photosensitive Layers Based on Conducting Polymers and Zinc Phthalocyaninate, Their Structure and Photoelectrical Properties. *Russ. J. Electrochem.* **2024**, *60*, 448–458. [CrossRef]
14. Kabanova, V.A.; Gribkova, O.L.; Tameev, A.R.; Nekrasov, A.A. Hole Transporting Electrodeposited PEDOT–Polyelectrolyte Layers for Perovskite Solar Cells. *Mendeleev Commun.* **2021**, *31*, 454–455. [CrossRef]
15. Gribkova, O.L.; Sayarov, I.R.; Kabanova, V.A.; Nekrasov, A.A.; Tameev, A.R. Electrodeposited Composite of Poly-3,4-Ethylenedioxythiophene with Fullerenol Photoactive in the Near-IR Range. *Russ. J. Electrochem.* **2024**, *60*, 813–822. [CrossRef]
16. Bobylev, A.G.; Kornev, A.B.; Bobyleva, L.G.; Shpagina, M.D.; Fadeeva, I.S.; Fadeev, R.S.; Deryabin, D.G.; Balzarini, J.; Troshin, P.A.; Podlubnaya, Z.A. Fullerenolates: Metallated Polyhydroxylated Fullerenes with Potent Anti-Amyloid Activity. *Org. Biomol. Chem.* **2011**, *9*, 5714–5719. [CrossRef]
17. Troshin, P.A.; Astakhova, A.S.; Lyubovskaya, R.N. Synthesis of Fullerenols from Halofullerenes. *Fuller. Nanotub. Carbon Nanostructures* **2005**, *13*, 331–343. [CrossRef]
18. Fedorova, N.E.; Klimova, R.R.; Tulenev, Y.A.; Chichev, E.V.; Kornev, A.B.; Troshin, P.A.; Kushch, A.A. Carboxylic Fullerene C60 Derivatives: Efficient Microbicides Against Herpes Simplex Virus And Cytomegalovirus Infections In Vitro. *Mendeleev Commun.* **2012**, *22*, 254–256. [CrossRef]
19. Husebo, L.O.; Sitharaman, B.; Furukawa, K.; Kato, T.; Wilson, L.J. Fullerenols Revisited as Stable Radical Anions. *J. Am. Chem. Soc.* **2004**, *126*, 12055–12064. [CrossRef] [PubMed]
20. Huang, H.-J.; Chetyrkina, M.; Wong, C.-W.; Kraevaya, O.A.; Zhilenkov, A.V.; Voronov, I.I.; Wang, P.-H.; Troshin, P.A.; Hsu, S. Identification of Potential Descriptors of Water-Soluble Fullerene Derivatives Responsible for Antitumor Effects on Lung Cancer Cells via QSAR Analysis. *Comput. Struct. Biotechnol. J.* **2021**, *19*, 812–825. [CrossRef] [PubMed]

21. Gribkova, O.L.; Nekrasov, A.A. Spectroelectrochemistry of Electroactive Polymer Composite Materials. *Polymers* **2022**, *14*, 3201. [CrossRef]
22. Garreau, S.; Duvail, J.L.; Louarn, G. Spectroelectrochemical Studies of Poly(3,4-Ethylenedioxythiophene) in Aqueous Medium. *Synth. Met.* **2001**, *125*, 325–329. [CrossRef]
23. Zozoulenko, I.; Singh, A.; Singh, S.K.; Gueskine, V.; Crispin, X.; Berggren, M. Polarons, Bipolarons, And Absorption Spectroscopy of PEDOT. *ACS Appl. Polym. Mater.* **2019**, *1*, 83–94. [CrossRef]
24. Janssen, R.A.J.; Smilowitz, L.; Sariciftci, N.S.; Moses, D. Triplet-State Photoexcitations of Oligothiophene Films and Solutions. *J. Chem. Phys.* **1994**, *101*, 1787–1798. [CrossRef]
25. Peintler-Kriván, E.; Tóth, P.S.; Visy, C. Combination of in Situ UV–Vis-NIR Spectro-Electrochemical and a.c. Impedance Measurements: A New, Effective Technique for Studying the Redox Transformation of Conducting Electroactive Materials. *Electrochem. Commun.* **2009**, *11*, 1947–1950. [CrossRef]
26. Kabanova, V.; Gribkova, O.; Nekrasov, A. Poly(3,4-Ethylenedioxythiophene) Electrosynthesis in the Presence of Mixtures of Flexible-Chain and Rigid-Chain Polyelectrolytes. *Polymers* **2021**, *13*, 3866. [CrossRef] [PubMed]
27. Yamato, H.; Kai, K.I.; Ohwa, M.; Wernet, W.; Matsumura, M. Mechanical, Electrochemical and Optical Properties of Poly(3,4-Ethylenedioxythiophene)/Sulfated Poly(β-Hydroxyethers) Composite Films. *Electrochim. Acta* **1997**, *42*, 2517–2523. [CrossRef]
28. Randriamahazaka, H.; Noël, V.; Chevrot, C. Nucleation and Growth of Poly(3,4-Ethylenedioxythiophene) in Acetonitrile on Platinum under Potentiostatic Conditions. *J. Electroanal. Chem.* **1999**, *472*, 103–111. [CrossRef]
29. Lyutov, V.; Kabanova, V.; Gribkova, O.; Nekrasov, A.; Tsakova, V. Electrochemically Obtained Polysulfonates Doped Poly(3,4-Ethylenedioxythiophene) Films—Effects of the Dopant's Chain Flexibility and Molecular Weight Studied by Electrochemical, Microgravimetric and XPS Methods. *Polymers* **2021**, *13*, 2438. [CrossRef]
30. Nekrasov, A.A.; Nekrasova, N.V.; Savel'ev, M.A.; Khuzin, A.A.; Barachevsky, V.A.; Tulyabaev, A.R.; Tuktarov, A.R. Electrochemical Investigation of a Photochromic Spiropyran Containing a Pyrrolidinofullerene Moiety. *Mendeleev Commun.* **2023**, *33*, 505–508. [CrossRef]
31. Meskers, S.C.J.; van Duren, J.K.J.; Janssen, R.A.J. Stimulation of Electrical Conductivity in a π-Conjugated Polymeric Conductor with Infrared Light. *J. Appl. Phys.* **2002**, *92*, 7041–7050. [CrossRef]
32. Meskers, S.C.J.; van Duren, J.K.J.; Janssen, R.A.J.; Louwet, F.; Groenendaal, L. Infrared Detectors with Poly(3,4-ethylenedioxy Thiophene)/Poly(Styrene Sulfonic Acid) (PEDOT/PSS) as the Active Material. *Adv. Mater.* **2003**, *15*, 613–616. [CrossRef]
33. Sariciftci, N.S.; Smilowitz, L.; Heeger, A.J.; Wudl, F. Photoinduced Electron Transfer from a Conducting Polymer to Buckminsterfullerene. *Science* **1992**, *258*, 1474–1476. [CrossRef] [PubMed]

Disclaimer/Publisher's Note: The statements, opinions and data contained in all publications are solely those of the individual author(s) and contributor(s) and not of MDPI and/or the editor(s). MDPI and/or the editor(s) disclaim responsibility for any injury to people or property resulting from any ideas, methods, instructions or products referred to in the content.

Review

Emerging Robust Polymer Materials for High-Performance Two-Terminal Resistive Switching Memory

Bixin Li [1,2,3], Shiyang Zhang [1], Lan Xu [1,*], Qiong Su [1] and Bin Du [4,*]

1. School of Physics and Chemistry, Hunan First Normal University, Changsha 410205, China; lbxin86@hotmail.com (B.L.)
2. Shaanxi Institute of Flexible Electronics (SIFE), Northwestern Polytechnical University (NPU), Xi'an 710072, China
3. School of Physics, Central South University, 932 South Lushan Road, Changsha 410083, China
4. School of Materials Science and Engineering, Xi'an Polytechnic University, Xi'an 710048, China
* Correspondence: xulan@hnfnu.edu.cn (L.X.); dubin@xpu.edu.cn (B.D.)

Abstract: Facing the era of information explosion and the advent of artificial intelligence, there is a growing demand for information technologies with huge storage capacity and efficient computer processing. However, traditional silicon-based storage and computing technology will reach their limits and cannot meet the post-Moore information storage requirements of ultrasmall size, ultrahigh density, flexibility, biocompatibility, and recyclability. As a response to these concerns, polymer-based resistive memory materials have emerged as promising candidates for next-generation information storage and neuromorphic computing applications, with the advantages of easy molecular design, volatile and non-volatile storage, flexibility, and facile fabrication. Herein, we first summarize the memory device structures, memory effects, and memory mechanisms of polymers. Then, the recent advances in polymer resistive switching materials, including single-component polymers, polymer mixtures, 2D covalent polymers, and biomacromolecules for resistive memory devices, are highlighted. Finally, the challenges and future prospects of polymer memory materials and devices are discussed. Advances in polymer-based memristors will open new avenues in the design and integration of high-performance switching devices and facilitate their application in future information technology.

Keywords: polymer; biomaterials; resistive switching memory; data storage; memristors

Citation: Li, B.; Zhang, S.; Xu, L.; Su, Q.; Du, B. Emerging Robust Polymer Materials for High-Performance Two-Terminal Resistive Switching Memory. *Polymers* **2023**, *15*, 4374. https://doi.org/10.3390/polym15224374

Academic Editor: Tengling Ye

Received: 9 October 2023
Revised: 7 November 2023
Accepted: 7 November 2023
Published: 10 November 2023

Copyright: © 2023 by the authors. Licensee MDPI, Basel, Switzerland. This article is an open access article distributed under the terms and conditions of the Creative Commons Attribution (CC BY) license (https://creativecommons.org/licenses/by/4.0/).

1. Introduction

In the current big data era, there is a growing demand for exploring data-storage technologies [1,2]. Therefore, various innovative data storage devices have been developed, such as dynamic random-access memory (DRAM), static random-access memory (SRAM), and flash memory. These memory devices are mainly based on conventional Si-based technologies, which suffer from physical miniaturizing limitations due to technical complexity and high cost [3–5]. They cannot fulfill the ever-increasing demands of high data density and fast switching speed. Therefore, striving for next-generation information storage solutions with faster speed, higher density, and lower power is imperative.

Recently, phase change memory (PCM), spin-transfer torque magnetoresistive random-access memory (STT-MRAM), and resistive random-access memory (RRAM) have been developed as next-generation non-volatile memory technologies [6–13]. PCM and STT-MRAM suffer from long write latency and low reliability, respectively. In contrast, RRAM devices with the advantages of low power consumption, high scalability, simple structure, easy fabrication, and low cost are considered promising candidates for future memory technology. The International Semiconductor Technology Roadmap identifies RRAM as one of the new memory technologies with the greatest potential for commercialization. Generally, RRAMs are fabricated as a sandwiched structure with a functional active layer sandwiched between two electrodes. The resistance states can be switched between high-resistance state (HRS, OFF state) and low-resistance state (LRS, ON state) in response to an external electrical stimulus, which is equivalent to "0" to "1" binary conversion. When more than two resistance states show in one material (e.g., "0", "1", "2"), multilevel storage can be expected, and it will increase the storage capacity within one memory cell exponentially [14–16].

To date, various functional materials have been explored for RRAMs, including organic materials, inorganic oxide, and organic–inorganic hybrid materials. The inorganic oxide-based RRAMs exhibit remarkable and stable memory characteristics, while they are limited by non-flexibility, non-recyclability, and environmental unfriendliness in the application of future wearable electronics. Organic polymer materials with the advantage of designable molecular structures, low costs, intrinsic flexibility, solution processability, 3D-stacking capability, and good biocompatibility, have been developed as favorable devices in next-generation memory technology [17]. Numerous efforts have been made to seek high-performance memory devices with a large ON–OFF ratio, low operation voltage, long retention time, as well as high endurance. In particular, the intrinsic flexibility and softness of polymers, especially biomacromolecules, make them ideal for stretchable and wearable electronics in the artificially intelligent lifestyle of the future [18–20]. These devices are mainly based on polymer composites and single-polymer materials. Compared with polymer composites, single-polymer materials can remove the phase separation problem. In order to explain the memory mechanisms, space charge limited current (SCLC), charge transfer, charge trapping/detrapping, filament conduction, and conformational change have been proposed [6,21,22].

In this review, we focus on the recent advances in polymer materials for resistive switching (RS) memory applications and aim to provide comprehensive concepts to develop highly efficient devices for next-generation information technology. First, this review gives a brief introduction on the structure of memory devices and the RS effect. Then, the recent progress of RS devices based on polymer materials, including single-component polymer materials, polymer composites, 2D covalent polymers, and biomacromolecules is summarized. Finally, we outline the challenges and outlook for the further development of polymer-based RS devices.

2. Overview of Polymer-Based RS Memory Devices

2.1. Device Structure

Generally, there are two types of device structures, namely, the vertical metal–insulator–metal (MIM) structure and the lateral field-effect transistor (FET) structure in the polymer-based RS memory devices. These two structures are also called two-terminal and three-terminal devices, respectively (Figure 1). The electrodes are most widely made of Al, Au, Ag, Pt, p- or n-doped Si, indium tin oxide (ITO), or fluorine-doped tin oxide (FTO) [17]. Highly conductive reduced graphene oxide (rGO) has also been developed as an electrode material in addition to traditional metal electrodes. In a typical vertical MIM structure, the RS active layer is sandwiched between the bottom and top electrodes, which can be classified by crossbar and cross-array structures. These structures can provide massive device cell arrays, leading to high-density data storage. In particular, the cross-array geometry possesses more potential to scale down each device cell and realizes a highly integrated RRAM architecture. As a typical structure, Song et al. demonstrated a 3D-stacked crossbar arrayed RS device with a multilayer structure of Al/polyimide(PI):6-phenyl-C61 butyric acid methyl ester (PCBM)/Al/PI:PCBM/Al/PI:PCBM/Al. This structure presented a high storage density [23].

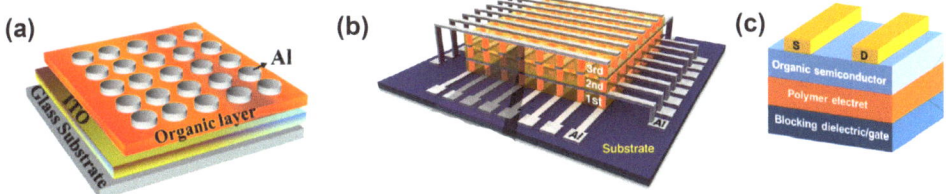

Figure 1. Schematic illustration of three typical RS memory device configurations. (**a**) Crossbar structure. Reproduced with permission [24]. Copyright 2021, Wiley-VCH GmbH. (**b**) Cross-array structure. Reproduced with permission [23]. Copyright 2010, Wiley-VCH GmbH. (**c**) Three-terminal FET RS device structure. Reproduced with permission [25]. Copyright 2019, The Royal Society of Chemistry.

The three-terminal FET RS device is a transverse device structure with two laterally distributed electrodes and a semiconductor channel, as well as a gate electrode. The transverse configuration shows the advantage of being compatible with commercial complementary metal-oxide-semiconductor (CMOS) circuits. There are three types of FET RS devices, e.g., charge trap FET, floating-gate FET, and ferroelectric FET memory [26–31]. However, the operating voltage is relatively high in the FET RS device and this is a major issue that remains to be overcome. In this contribution, we mainly focused on the simple two-terminal MIM RS devices.

2.2. RS Memory Effect

In the RS device, the writing operation is generated by applying a voltage bias or pulse to the device, leading to the conductance switching between the ON and OFF state. Depending on whether external electric power is required to maintain the ON state or not, the RS effect can be classified into volatile and non-volatile memory effects (Figure 2). The representative volatile memory types are the dynamic random-access memory (DRAM) and static random-access memory (SRAM) [32–35]. In DRAM, the ON state could be retained for a short period after turning off the applied voltage, while in SRAM, the device could sustain the ON state for a longer time than that observed in the DRAM device after the removal of the external power supply. Generally, in volatile memory, the ON state can be relaxed to the OFF state without an erasing process. This effect has potential for secure semiconductors and integrated electronic circuits.

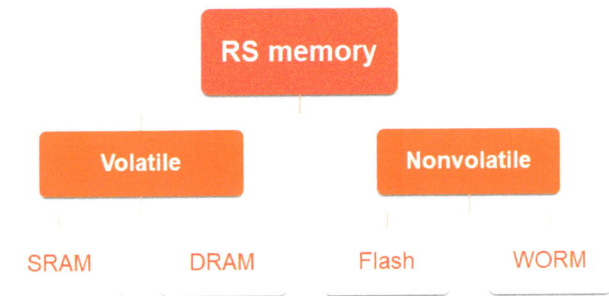

Figure 2. Schematic illustration of the RS memory types.

In contrast to volatile memory, non-volatile memory can hold the stored information for quite a long time after the removal of the electric power. It can be divided into write-once-read-many-times (WORM) memory and rewritable flash memory [36–38]. In the former, the conductance switches from HRS to LRS under a certain voltage, and the LRS cannot be erased even if the external electric field is withdrawn. It is capable of maintaining the ON state permanently and shows extensive applications in rapid archival storage equipment, secure databases, as well as electronic labels. For the latter, the conductance switches between HRS and LRS. Flash memory is a promising candidate in data storage devices such as USB drives, hard disks, and other relevant rewritable digital storage.

2.3. Memory Mechanisms

Tremendous research has been devoted to elucidating the RS phenomena associated with polymer materials. However, the underlying mechanisms of the RS characteristics are still controversial. Based on the theoretical calculations and experimental analysis, some well-established memory mechanisms have been proposed, such as charge transfer, conformational change, charge trapping/de-trapping, filamentary conduction, and redox reaction, as shown in Figure 3.

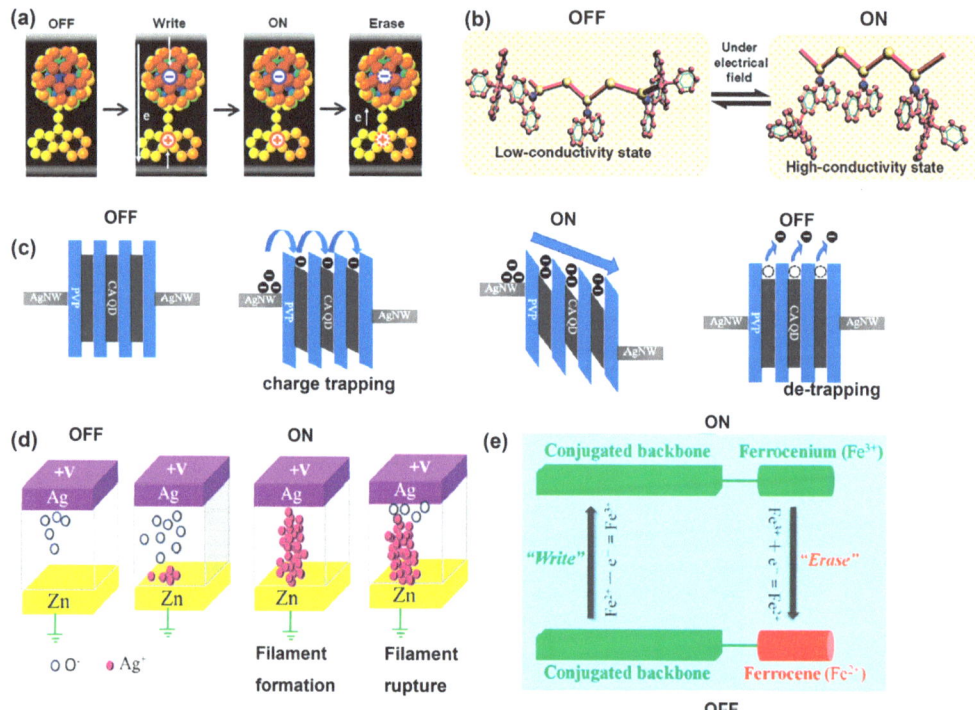

Figure 3. Schematic illustration of various switching mechanisms. (a) Charge transfer. Reproduced with permission [39]. Copyright 2007, American Chemical Society. (b) Conformational change. Reproduced with permission [40]. Copyright 2008, American Chemical Society. (c) Charge trapping/de-trapping. Reproduced with permission [41]. Copyright 2018, The Royal Society of Chemistry. (d) Filamentary conduction. Reproduced with permission [42]. Copyright 2018, American Institute of Physics. (e) Redox reaction. Reproduced with permission [43]. Copyright 2019, American Chemical Society.

2.3.1. Charge Transfer

In a polymer with donor–acceptor (D–A) moieties, the charge transfer process usually occurs under an external electric field. The electronic charge will transfer from the donor to the acceptor moiety, leaving positively charged holes residing on the donors and the molecular orbitals partially filled. Hence, this process leads to the increase of the concentration of free charge carriers with high mobility, resulting in a high-conductive state

(LRS). When an opposite electric field returns charges back to the donor group, the free charge carrier concentration will decrease and the D–A system will return back to HRS, showing the convertible charge transfer interaction. In order to obtain direct evidence of the charge transfer process, some theoretical calculations and experimental investigations have been conducted to show this process, e.g., density functional theory (DFT) calculations, UV–Visible absorption spectra, in situ fluorescence spectra and transmission electron microscope (TEM) images [39,44–46].

Interestingly, by tuning the dipole moment caused in the charge transfer process in D–A systems, volatile and non-volatile memory can be achieved. The strong dipole moment in polymers helps to sustain the high-conductive state, usually leading to non-volatile behavior. In contrast, a weak dipole moment leads to the unstable conductive state, and the volatile memory device will be realized [21].

2.3.2. Conformational Change

The conformational change mechanism is mostly seen in polymers containing carbazole groups in the side chain such as poly(N-vinylcarbazole) (PVK) and its derivatives [40,47]. Initially, the random orientation of carbazole groups hinders the ordered π–π stacking and the charge transport is insufficient, indicating the HRS. With the external electric field, the carbazole groups are able to be rearranged into a nearly face-to-face π–π stacking, the ordered conformation can switch the polymers to LRS. In this process, excellent RS performances can be easily realized by the modification of polymer structures and conformational changes. The HRS to LRS transition is reversible by changing the electric field polarity, possibly due to thermal injecting at the electrode/polymer interface.

There are various characterization techniques to prove the conformational change of polymers, including DFT theoretical calculations, UV–Visible absorption spectra, in situ fluorescence spectra, X-ray diffraction (XRD), cyclic voltammetry (CV), TEM, and Raman spectra.

2.3.3. Charge Trapping/De-Trapping

When metal nanoparticles, quantum dots, fullerenes, or organic semiconducting molecules are doped into polymer matrices, these dopants can act as trapping centers for charge carrier transport [41,48–52]. With increasing external voltage, charge carriers would be gradually injected into the trapping centers and local percolation networks will be formed. A continuous carrier hopping pathway will then switch the device from HRS to LRS.

By continually increasing the external electric field in the same polarity, the trapped charge carriers will exceed the capacity of the conductive channel and induce coulomb repulsion between the trapped charges to rupture the charge transport channel. Then, the device will switch back to HRS, behaving as unipolar RS memory. In contrast, when applying the reverse electric field to the device, the trapped charge carriers release from the trapping centers to rupture the conductive channel, leading to bipolar RS memory.

2.3.4. Filamentary Conduction

Electrochemical metallization memory (ECM) and valence change memory (VCM) are the two representative types of filamentary conduction modes observed in some polymers for RS behavior [42,53–55]. By applying an electric voltage to the top electrode of active metals (e.g., Ag and Cu), the top electrode metal can be oxidated, and metal cations will be released from the top electrode to migrate through the active layer, which hence forms metal filament between top electrode and bottom electrode. The reverse voltage can deteriorate the conductive filament. Generally, VCM forms in donor-type defects such as mobile oxygen vacancies. Under an electric voltage, oxygen vacancies will be gathered at the cathode and diffused into the active layer to create a continuous conductive channel, which leads to a HRS to LRS transition. Under an opposite electric field, the conductive path is ruptured, which is known as a reset process [56].

The experimental methods to verify ECM and VCM mechanisms are relatively mature, e.g., the high-resolution TEM, scanning electron microscopy (SEM), and in situ scanning probe microscopy (SPM).

2.3.5. Redox Reaction

In polymers with transition metal atoms, such as Fe, Co, or Mn in the backbone, the redox reactions of active materials are prone to occur between molecular reduction and oxidation states to alter the conductivity [43,57]. Unpaired or lone pair electrons can be removed, introducing impurity energy levels into the bandgap of active materials. These transition metal atoms usually possess various valences and can be switched between with the external voltage, resulting in binary or multilevel RS behavior. The positive charges can be balanced by reduction of environmental oxygen in the atmosphere or additional counter electrode materials, which may contribute to the stability of the redox system to enhance the endurance characteristics [58]. This electrochemical redox reaction phenomena are often certified by the CV technique to provide experimental evidence.

3. Polymer-Based RS Memory Devices

Polymer materials with the advantages of easy solution processability and intrinsic flexibility, are proving to be attractive for RS memory applications. Various facile low-cost solution methods, such as spin-coating, spray-coating, dip-coating, drop-casting, blade casting, and ink-jet printing are used to deposit polymer films. These materials used in RS memory devices can be classified into single-component polymers and polymer mixtures.

3.1. Single-Component Polymers

3.1.1. Conjugated Polymers

Single-component polymers with donor–acceptor structures have potential charge transfer features, which are effective for realizing RS memories. The donor and acceptor moieties may be incorporated in the polymer backbone, linked as a side functional group, or dangling at the end of the macromolecular chain. By adjusting the strength and loading ratio, as well as spatial arrangement of the donor/acceptor moieties, volatile to non-volatile memory performance can be realized. To date, various conjugated and non-conjugated polymers have been reported to present RS properties.

The majority of RS memory polymers are based on π-conjugated nitrogen atoms, hydrocarbons, and their combinations. Polyazomethine (PAM) materials with imine groups (C=N) in the backbone are a typical type of conjugated polymers in RS memories. Li et al. synthesized two PAM derivatives, PA-1 and PA-2, as shown in Figure 4a [59]. In PA-1, the triphenylamine and oxadiazole moieties act as donor and acceptor, respectively. However, in PA-2, the oxadiazole acts as donor and the acceptor is 3,3′-dinitro-diphenylsulfone. With the device structure of Pt/active layer/Pt, PA-1 shows a rewritable RRAM memory effect with poor endurance, while PA-2 performs a WORM behavior (Figure 4b,c). Interestingly, by changing the bottom electrode to Al, both active layers exhibit the rewritable memory effect. For PA-1, the charge transfer interaction between the triphenylamine and the moderate electron withdrawing ability of oxadiazole is reversible, which results in rewritable memory behavior, whereas the charge transfer interaction between oxadiazole and 3,3′-dinitro-diphenylsulfone is rather strong, causing WORM switching behavior in PA-2. When the Al electrodes are introduced, the Schottky barrier at the Al–polymer interfaces become smaller with the presence of an ultrathin layer of Al_2O_3, which gives a lower reset voltage.

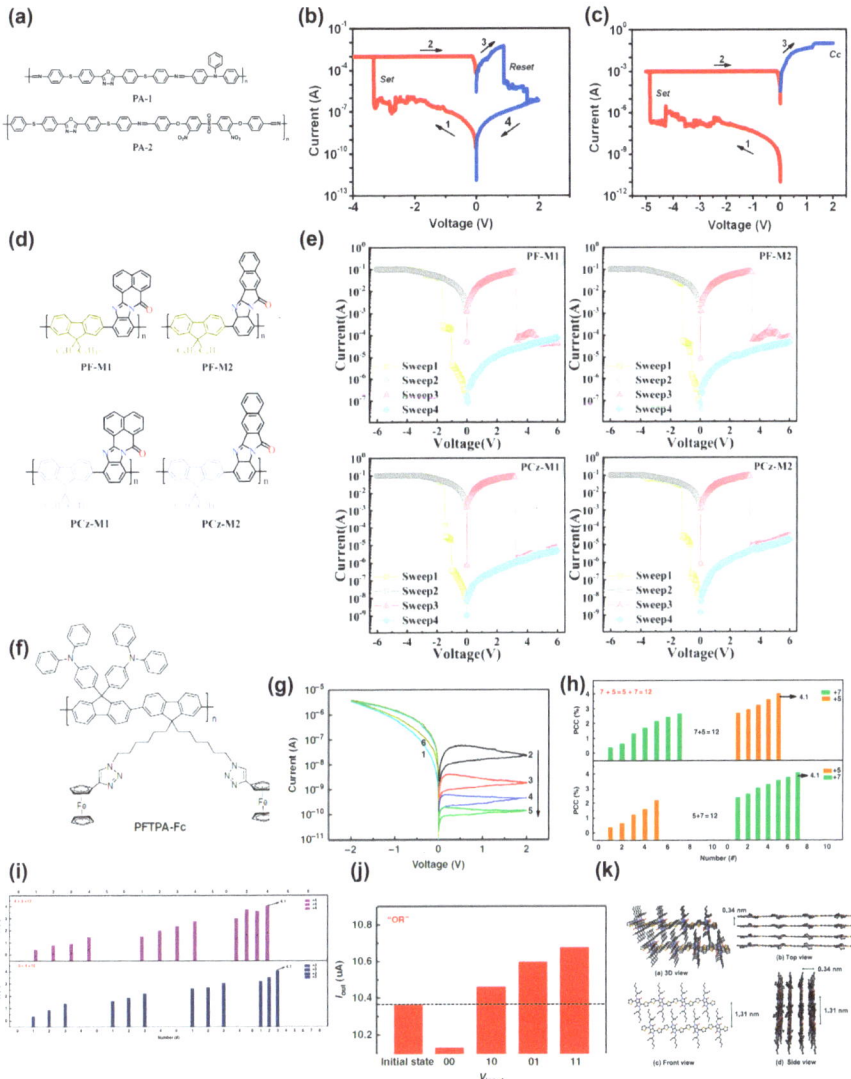

Figure 4. (**a**) Molecule structures of PA-1 and PA-2. *I–V* characteristics of the (**b**) Pt/PA-1/Pt and (**c**) Pt/PA-2/Pt devices. The arrows show the scanning direction of the applied voltage. Reproduced with permission [59]. Copyright 2013, The Royal Society of Chemistry. (**d**) Molecule structures of four donor−acceptor conjugated copolymers. (**e**) *I–V* characteristics of the memory devices based on the conjugated copolymers depicted. Reproduced with permission [60]. Copyright 2021, American Chemical Society. (**f**) Molecule structures of PFTPA–Fc. (**g**) *I–V* curves of the ITO/PFTPA–Fc/Pt device. The number 1-6 mean the six consecutive scanning of the voltage. (**h**) Demonstration of arithmetic commutative addition with the PFTPA–Fc memristor. (**i**) Demonstration of arithmetic multiplication with the PFTPA–Fc memristor. (**j**) Realization of the OR logic gate function with PFTPA–Fc memristor. The dotted line shows the initial device current of 10.36 μA. Reproduced with permission [61]. Copyright 2019, Springer Nature. (**k**) Schematic representations of molecular-chain conformation and packing structure in nanoscale poly(dtDPP) films. Reproduced with permission [62]. Copyright 2021, American Chemical Society.

The majority of polymer-based memory devices present binary storage. In order to improve the capacity, Liu et al. synthesized the first single-polymer-based ternary memory device of iamP6 in 2012, combining the charge transfer and conformational change mechanisms [47]. Afterwards, interest in multilevel storage in a single polymer has been excited. The above research are the pioneering works on single-component polymer-based RS devices.

Recently, Zhang et al. connected naphthalene benzimidazole acceptor units to fluorene/carbazole donor, and four donor−acceptor conjugated copolymers were synthesized by the Suzuki reaction (Figure 4d) [60]. All of these polymers exhibit ternary electronic memory compared with pure fluorene/carbazole counterparts, as depicted in Figure 4e. After introducing monomers, two charge traps occur in the polymers. With increasing voltage to V_{th1}, the injected charge carriers can transfer from the donor to the acceptor to fill the trap. Because the charge depth is associated with the acceptor group, resulting in different charge trap size, these traps cannot be filled simultaneously. The weak electron absorption ability of benzimidazole contributes to a small charge trap, which will be filled up at first, leading to the ON_1 state. In contrast, the naphthalene structure has a stronger electron absorption ability with a large charge trap, which needs more energy to fill all charge traps to reach the ON_2 state.

It is beneficial to incorporate redox active moieties onto the pendants of the polymers, to obtain astounding memory behavior. Zhang et al. introduced triphenylamine (TPA) and ferrocene (Fc) onto the sidechains of fluorene skeletons through the Suzuki reaction and "Click" chemistry, respectively, to achieve the final conjugated polymer PFTPA–Fc (Figure 4f) [61]. The ITO/PFTPA–Fc/Pt device exhibits four consecutive levels of RS characteristics by the electric-field-induced electrochemical reactions through three redox active moieties (Figure 4g). Moreover, four basic decimal arithmetic operations of addition, subtraction, multiplication, and division can be realized in the device (Figure 4h–j). This finding proves the feasibility of integrating multilevel memory and computing capability into a single memristive device by ingenious molecular design. Anchoring Fc in fluorene derivatives with porphyrin- and benzene-based diethynyl ligand, Roy et al. synthesized two metallopolymers (P1 and P2) [63]. Both polymers demonstrate WORM memory characteristics. Fc is effective in memristive polymer molecular design.

To achieve information storage and processing multifunctional memristors, Ren et al. introduced Ir complexes as electron-withdrawing groups on the polyfluorene backbone and synthesized a poly(9,9-dioctyl-9H-fluorene-*alt*-1,3-bis(2-ethylhexyl)-5,7-di(thiophen-2-yl)-4H,8H-benzo[1,2-c:4,5-c′]dithiophene-4,8-dione)-*alt*-(2,4-Pentanedionato)bis(2-(thiophen-2-yl)-pyridine)iridium) (PFTBDD-IrTPy) copolymer [64]. The synergetic electrochemical metallization and charge transfer effect between donors and acceptors are responsible for the memory behavior. The as-fabricated device can behave as an artificial synaptic emulation, simple Boolean logic, and decimal arithmetic calculation, which means it has so-called multibit data storage and processing capabilities in one memristor, like the findings reported in PFTPA–Fc.

Jung et al. designed several dithienyl-diketopyrrolopyrrole-based (dtDPP-based) narrow bandgap polymers, i.e., poly(2,5-bis(2-ethylhexyl)-3,6-di(thiophen-2-yl)-2,5-dihydropyrrolo[3,4-c]pyrrole-1,4-dione)) (poly(dtDPP)), poly(3-(5-(9,9-dioctyl-9H-fluoren-2-yl)thiophen-2-yl)-2,5-bis(2-ethylhexyl)-6-(thiophen-2-yl)-2,5-dihydropyrrolo[3,4-c]pyrrole-1,4-dione) (poly(dtDPP-FL)), and poly(3-(5-(4-(diphenylamino)phenyl)-thiophen-2-yl)-2,5-bis(2-ethylhexyl)-6-(thiophen-2-yl)-2,5-dihydropyrrolo[3,4-c]pyrrole-1,4-dione) (poly-(dtDPP-TPA)) [62]. Due to the self-assembled capability, poly(dtDPP) forms an edge-on layer structure, as can be seen from Figure 4k, whereas the others are preferentially oriented within the film. By adjusting the donor−acceptor powers in the backbone, poly(dtDPP) and poly-(dtDPP-TPA) show non-volatile WORM behavior, in comparison with poly(dtDPP-FL) exhibiting a DRAM behavior. Balancing of electron donor and acceptor powers is, in effect, adjusting memory type. Moreover, a redox active entity with thiophene-DPP donor and anthraquinone acceptor can also be used to obtain electrical bistability in the design of polymer memristive materials [65].

3.1.2. Non-Conjugated Polymers

In addition to conjugated polymers, usually with a rigid backbone, non-conjugated polymers have the merits of environmental stability and flexibility and also enable RS behaviors. Beyond traditional hydrocarbons and nitrogen-based polymers, oxygen-containing electroactive polymers are superior for novel RS memory. In 2020, Ree et al. synthesized a series of poly(ethylene-*alt*-maleate)s derivatives with oxygen constituents and their derivatives as side groups through the postmodification reactions: poly(ethylenealt-di(3-methoxylbenzyl) maleate) (PEM-BzOMe), poly-(ethylene-*alt*-di(3,5-dimethoxylbenzyl) maleate) (PEM-BzOMe$_2$), poly(ethylene-*alt*-dipiperonyl maleate) (PEM-BzO$_2$C), and poly(ethylene-*alt*-di(3,4,5-trimethoxybenzyl) maleate) (PEM-BzOMe$_3$), as depicted in Figure 5a [66]. The oxygen-containing polymers show superior thermal stability up to 180 °C, and exhibit reliable *p*-type unipolar volatile or non-volatile RS characteristics with high ON–OFF ratios ranging from 1.0×10^3 to 1.0×10^8. PEM-BzOMe and PEM-BzOMe$_2$ exhibit excellent unipolar DRAM behavior in a limited thickness range. However, for PEM-BzO$_2$C, WORM memory behavior can be observed in 17.7 nm devices and changes to DRAM memory mode with a film thickness less than <136 nm. A similar phenomenon also shows in PEM-BzOMe$_3$. Figure 5b,c demonstrate the combination of Schottky emission and trap-limited SCLC conductions in the OFF-state and hopping conduction in the ON-state between charge trap sites are responsible for the memory behavior. This contribution proves that the RS behavior is controllable by tailoring the number of oxyphenyl units and/or oxy atomic components in the phenyl unit.

Ryu et al. first demonstrated 2-pyrrolidone and succinimide as electroactive elements in memory application [67]. Four polymers with and without 2-pyrrolidone and succinimide moieties were synthesized: poly(ethylene-*alt*-di(2-pyrrolidone-5-ethyl) maleate) (PEM-EP), poly(ethylene-*alt*-di(acetamidoethyl) maleate) (PEM-EA), poly(ethylene-*alt*-di(succinimido-N-ethyl) maleate) (PEM-ES), and poly(ethylene-alt-di(3-oxo-1-butyl) maleate) (PEM-EB). The chemical structures are presented in Figure 5d. It can be clearly seen from Figure 5e that PEM-EP and PEM-ES show non-volatile WORM memory over the film thickness range of 10–30 nm and 10–80 nm, respectively, whereas volatile digital memory can be seen in these two polymers over a rather narrower range of film thickness (Figure 5f). The succinimide moiety has relatively higher affinity and stabilization power, resulting in better memory performance.

By introducing iridium(III) complex as pendant groups in non-conjugated polymers, the model polymer is shown in Figure 5g. Yang et al. also realized volatile and non-volatile memory [68]. The concentrations of iridium(III) complex in the polymers affect the memory behavior. The polymers without iridium(III) complex shows SRAM behavior. The 4%, 8%, 12%, and 16% concentrations of iridium(III) complex in the polymers exhibit flash memory.

As a kind of polyanionic nano-cluster, polyoxometalate (POM) exhibits several discrete redox states in a narrow potential range, which can be used in multilevel memories. Hu et al. prepared a ternary redox POM-based inorganic–organic hybrid polymer, polymethyl methacrylate (PMMA)-MAPOM, by the copolymerization of MMA and MAPOM with 1,1′-Azobis(cyclohexanecarbonitrile) (ABCN) as the initiator [69]. The ITO/PMMA-MAPOM/Pt device presents rewriteable switching properties among three redox states under different reset voltages, showing multilevel properties and endurance over 50 cycles. The multi-redox states of manganese centers in polyoxoanion altering the effective carrier density in the switching layer is responsible for the switching behavior.

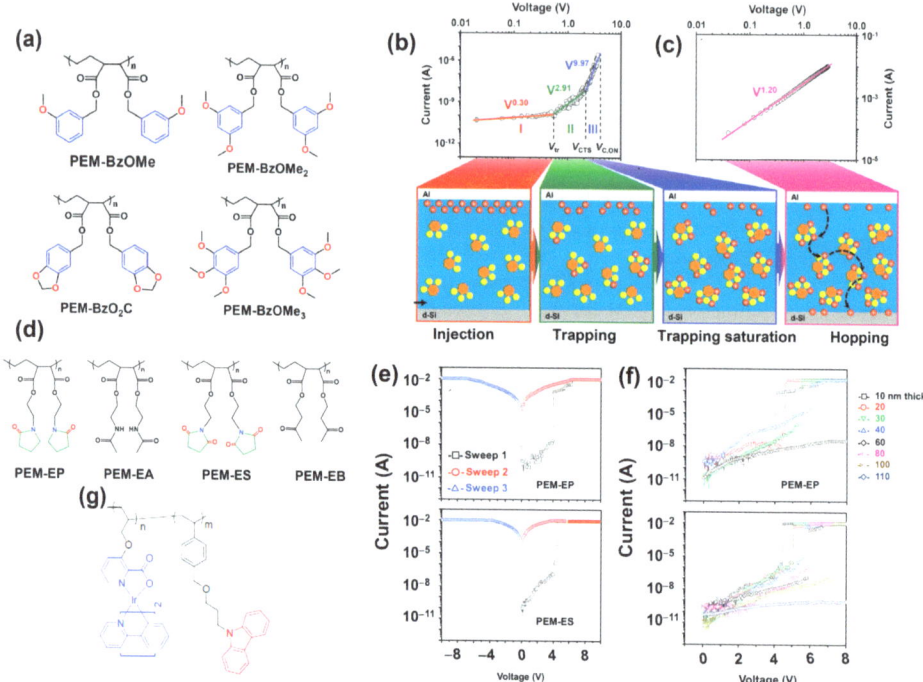

Figure 5. (**a**) Molecule structures of PEM-BzOMe, PEM-BzOMe$_2$, PEM-BzO$_2$C, and PEM-BzOMe$_3$. (**b**) OFF-state of the measured I–V curves where the symbols are the measured data, and the solid lines represent the fit results using Schottky emission. Schematic diagrams of charge injection, trap, and trap-saturation are shown below. (**c**) ON-state of the measured I–V curves where the symbols are the measured data, and the solid lines represent the fit results using a hopping conduction model. Schematic diagrams of the hopping process are shown below. Reproduced with permission [66]. Copyright 2020, American Chemical Society. (**d**) Molecule structures of PEM-EP, PEM-EA, PEM-ES, and PEM-EB. (**e**) I–V characteristics of the memory devices based on PEM-EP and PEM-ES. (**f**) I–V curves of polymers of various thicknesses in sandwiched devices with a d-Si top electrode and an Al top electrode: PEM-EP (10–60 nm thick), and PEM-ES (10–110 nm thick). Reproduced with permission [67]. Copyright 2021, Wiley-VCH GmbH. (**g**) The chemical structures of iridium(III) complex as pendant groups. Reproduced with permission [68]. Copyright 2020, The Royal Society of Chemistry.

3.2. Polymer Mixtures

Polymers mixed with small molecules, nanoparticles, quantum dots, and hybrid perovskites to form polymer mixtures are widely reported in RS memory devices. They can integrate the merits of these materials to obtain high-performance memory devices.

Polymers mixed with small molecular semiconductors can achieve the donor–acceptor structure, which is beneficial for memristive application. PVK and poly(3-hexylthiophene) (P3HT) are typical donors. The composite films of PVK mixed with carbon nanotubes (CNTs), and C$_{60}$ have been reported [39,70]. The donor–acceptor system plays a vital role in binary RS devices. In particular, it is important to note that the donor–acceptor–acceptor system might be promising in ternary RS memristors. Pan et al. incorporated small molecular acceptors of 1,3-bis[2-(4-tert-butylphenyl)-1,3,4-oxadiazo-5-yl]benzene (OXD-7) into PVK to study the memory behavior, as illustrated in Figure 6a [71]. In Figure 6b, under 25 wt% of OXD-7, the composite films show stable memory curves with set and reset voltages of −3.1 and 2.3 V, respectively. Similar memristive behavior can also be seen with 30 wt% of OXD-7. When the concentration of OXD-7 increases to 40 wt%, no RS characteristic appears. They further introduced 2-(4-*tert*-butylphenyl)-5-(4-biphenylyl)-1,3,4-oxadiazole (PBD) to construct the donor–acceptor–acceptor system. The composite film of PVK (24 wt% OXD-7:6 wt% PBD) exhibits remarkable ternary RS behavior with a switching ratio of $1:10:10^4$ in Figure 6c. The ON–OFF ratios in the donor–acceptor–acceptor systems are generally higher than those in donor–acceptor systems. The electric-field-induced charge transfer between PVK donor and oxadiazole moiety-formed OXD-7 and PBD is responsible for the RS effect. Li et al. reported memristive behavior based on P3HT mixed with 2,4,5,6-tetrakis(carbazol-9-yl)-1,3-dicyanobenzene (4CzIPN) or 4,5-bis(carbazol-9-yl)-1,2-dicyanobenzene (2CzPN) composites (Figure 6d) [72]. The two carbazolyl dicyanobenzenes with low intrinsic mobility and high steric hindrance might inhibit the leakage current of the HRS. Dramatically, these composite films show switching ratios higher than 10^5, retention times of more than 5×10^4 s, and endurance cycles of 150 times. The charge trapping and detrapping process leads to the charge transport channels, which are responsible for the memory behavior. The memory mechanism is comprehensively illustrated in Figure 6e. This finding reveals the effect of intermolecular interaction on RS behavior. Sun et al. implanted 2-Amino-5-methyl-1,3,4-thiadiazole into poly(4-vinylphenol) (PVP) to construct a Al/PVP:thiadiazole/Al device [73]. Both non-volatile WORM and flash memory behaviors are present in a single device. From Figure 6f, in the forward voltage sweep, the device first shows WORM behavior. Then, in the reverse voltage sweep, the device shows a second "write" operation and the "erasing" operation appears in the following forward scanning direction. Therefore, the device could realize "0"–"1"–"2"–"1"–"2" tri-stable resistance states and the device-to-device variations and tri-state variations in 109 cycles are presented in Figure 6g. The "0" state is not the electroforming state, resulting in an electroforming-free device. This work provides a new strategy for designing ternary data storage utilities.

Polymers could also act as matrixes for nanoparticles or quantum dots for memory operations. Nanoparticles usually act as a trap center in the switching layer. Kim et al. reported a flexible and stable memristive devices consisting of hexagonal boron nitride nanosheets (h-BN NSs):PMMA nanocomposites [74]. H-BN has a smooth atomic surface, no dangling bonds and excellent thermal stability. The strong electron binding force of h-BN makes it suitable for use as a carrier trapping center in the memory. The device shows a WORM character with an ON–OFF ratio of 10^3. The flexible device on polyethylene naphthalate (PEN) substrate can maintain the memory properties over 2×10^3 bending cycles. The discrete energy level state causes a strong quantum confinement effect in the h-BN NSs, as shown in Figure 7a–c, which is responsible for the WORM effect. Zhou et al. investigated WS_2 nanoparticles, which have a large specific surface area and good conductivity, to be doped in poly[2,7-9-(9-heptadecanyl)-9*H*-carbazole-*co*-benzo[4,5] imidazole[2,1-α] isoindol-11-one] (PIIO) matrix [75]. The incorporation of 6 wt% WS_2 nanoparticles showed the best non-volatile ternary storage features with the switching ratios of $1:1.11 \times 10^1:2.03 \times 10^4$ for three resistance states (Figure 7d). The WS_2 nanoparticles lower the charge injection barrier and induce conductive pathways and conductive filaments.

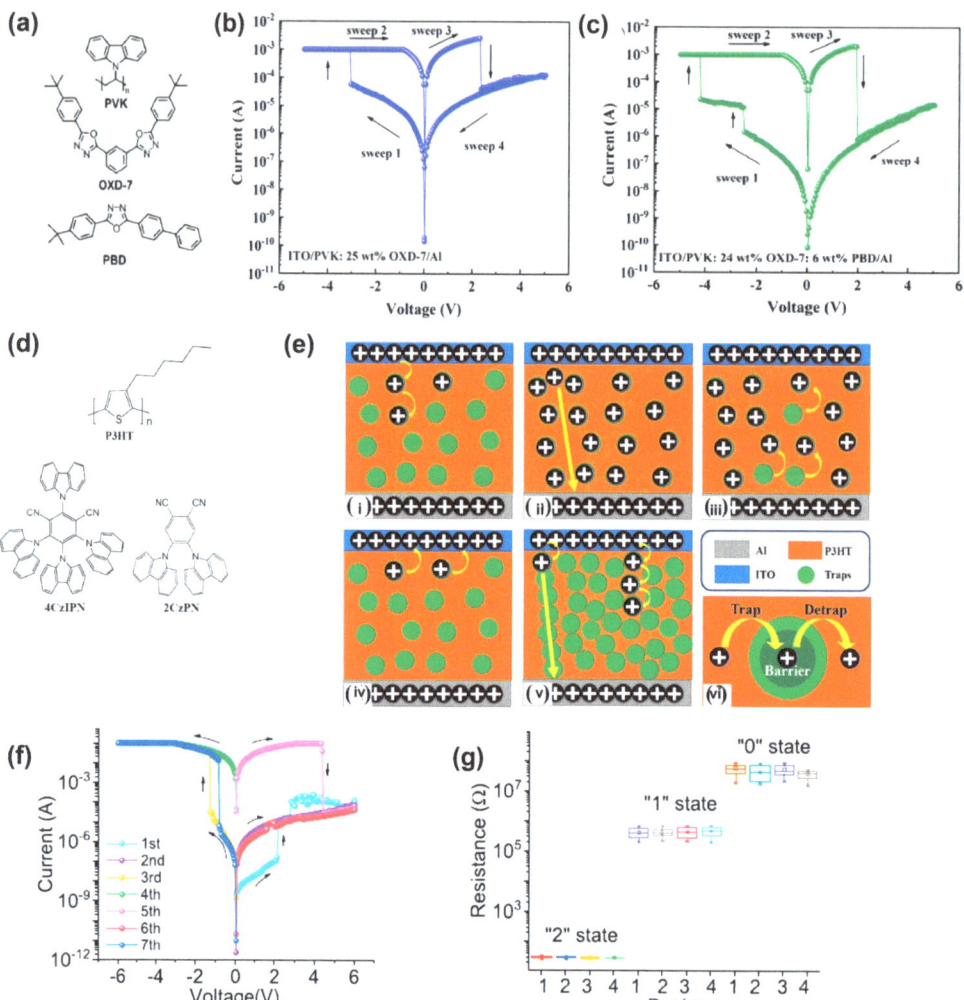

Figure 6. (**a**) Chemical structures of PVK, PBD, and OXD-7. (**b**) *I–V* characteristics of the ITO/PVK: 25 wt% OXD-7/Al devices. (**c**) *I–V* characteristics of the devices incorporated with PVK: 24 wt% OXD-7: 6 wt% PBD. Reproduced with permission [71]. Copyright 2021, The Royal Society of Chemistry. (**d**) Chemical structures of P3HT, 4CzIPN, and 2CzPN. (**e**) Schematic illustration of the switching mechanism. Charge transfer processes of (**i**) trap filling, (**ii**) fully filling trap, (**iii**) trap pumping, (**iv**) vacant trap, and (**v**) current leakage. (**vi**) Schematic illustration of the trap, de-trap, and charge barrier. Reproduced with permission [72]. Copyright 2022, American Chemical Society. (**f**) *I–V* curves of the fabricated multifunctional Al/PVP:thiadiazole/Al device with an initial positive voltage sweep. (**g**) Device-to-device "0", "1", and "2" tri-states variations in 109 cycles of four devices with initially applied positive voltage. Reproduced with permission [73]. Copyright 2019, Elsevier.

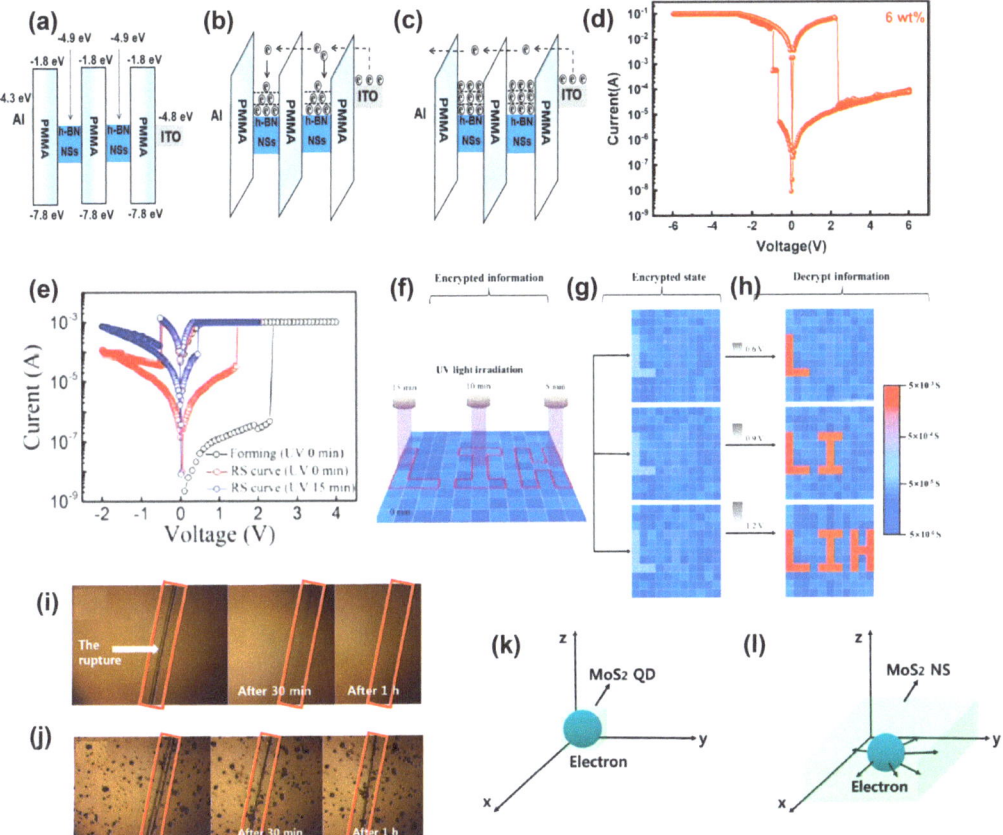

Figure 7. (a) The energy band diagram for the Al/h-BN NSs:PMMA/ITO/PEN devices and the carrier transport processes of (b) HRS, and (c) LRS. Reproduced with permission [74]. Copyright 2021, Elsevier. (d) $I-V$ curves of PIIO:6 wt% WS$_2$ nanoparticles. Reproduced with permission [75]. Copyright 2022, American Chemical Society. (e) The effects of UV irradiation on RS behaviors of an ITO/PVP-NCQDs/Al memory device. (f) The UV irradiation represented the process of information encryption, in which three regions (the image "L", "I", and "H") underwent 15, 10, and 5 min UV irradiation, respectively. (g) Read pulses with different amplitudes (0.6, 0.9, 1.2 V) were applied to the encrypted state. (h) Diverse images can be decrypted including image "L", image "LI", and image "LIH", respectively. Reproduced with permission [76]. Copyright 2020, The Royal Society of Chemistry. (i) PVA-IMGQD film heated at 50 °C for 1 h showed almost complete healing. (j) PVA-pure GQD film heated at 50 °C for 1 h showed almost no healing. Reproduced with permission [77]. Copyright 2021, The Royal Society of Chemistry. Wiley-VCH GmbH. Schematic diagrams of electrons confined in (k) MoS$_2$ quantum dots and (l) MoS$_2$ NSs. Reproduced with permission [78]. Copyright 2021, Wiley-VCH GmbH.

Carbon quantum dots with small size, high electron transfer efficiency, and attractive optical properties are promising in optoelectronic applications. Lin et al. investigated a memory device based on PVP and N-doped carbon quantum dot nanocomposites to observe photo-tunable memory behavior [76]. The set voltages and the switching ratios decrease with the increasing time of UV light irradiation (Figure 7e). UV light induces local conductive amorphous carbon region, which can enhance the internal electrical field and shorten the charge tunneling channel. As shown in Figure 7f–h, with a 9 × 9 RRAM array, this device can realize encrypted image storage. The inputted images of letters "L", "I", and "H" were irradiated by UV light for 15, 10, and 5 min, respectively. Herein, they can be readout through electric voltage pulses with different amplitudes on the encrypted RRAM array. Jiang et al. reported a carbon dot:PVP nanocomposite-based RS device with silver nanowires as top and bottom electrodes on the flexible gelatin film substrate [79]. The memory behavior shows negligible fluctuations over 100 bending cycles. The trap-related SCLC is attributed to the RS mechanism. The all-biocompatible materials exhibit excellent degradability. The device can dissolve completely within 90 s after being submerged in deionized water at 55 °C and degrade naturally in soil within 6 days. This work paves the way for carbon dots in flexible and wearable green electronics.

Wearable electronic devices may be damaged under repetitious mechanical stress. Kim's group presented a self-healable RS device based on a composite layer composed of a PVA matrix and imidazole-modified graphene quantum dots (IMGQD) [77]. The device exhibits WORM behavior with a set voltage of 1.7 V. Figure 7i,j depict that the PVA–IMGQD films can be completely self-healed after 1 h at 50 °C. Meanwhile, the PVA film doped with pure GQDs does not show self-healing. The imidazole groups in the IMGQDs are the key factor to obtain self-healing. At the crack interface of the films, the PVA–IMGQD chains on both sides are gradually recombined due to hydrogen bonding, leading to self-healing. When the device is completely cut off, including the ITO bottom electrode, the current will gradually recover with the progress of self-healing. At the same time, the retention and durability of the device are nearly unchanged. This excellent result is of great importance for the development of portable electronic systems.

To enhance the thermal stability of the memristive device, molybdenumdisulfide (MoS_2) quantum dots with high temperature resistance and strong quantum confinement effect were incorporated into the polyimide (PI) matrix [78]. The ITO/PEDOT:PSS/PI-MoS_2 quantum dot/Al structure exhibits WORM characteristics in the voltage range from -6 to 3 V. However, the PI–MoS_2 nanosheet (NS)-based devices show bipolar flash memory. The WORM behavior is attributed to the strong binding force of the quantum confinement effect in the doped MoS_2 quantum dots. The electrons are trapped in three dimensional directions in the quantum dots, while for the nanosheets, the electrons are only limited in the z direction and they can be released under a certain reverse voltage (Figure 7k,l). No significant degradation of the memory characteristics can be observed under high annealing temperatures of 50, 100, and 200 °C, showing the high thermal stability of MoS_2 quantum dot.

Chalcogenide-based colloidal CdSe quantum dots were embedded in PVP matrix to enhance the switching ratio, which was reported by Sahu et al [80]. They found a nanoscale heterostructure formed with colloidal monodispersed CdSe quantum dots and PVP, which helps to achieve the high ON–OFF ratio of 10^5. The memory mechanism is analyzed from fitting the I–V curves. The charge conduction in the HRS state is due to hot-charge-injection and space-charge-injection conduction. This switches to Ohmic conduction in the LRS state.

Nowadays, organic–inorganic hybrid perovskites (OIHPs) have attracted increasing attention in optoelectronic devices due to their high linear absorption coefficient, tunable bandgap, long exciton diffusion length, high electron mobility, high crystallinity, and solution processability [81–83]. Great efforts have been devoted to OIHP-based RS memory devices [84–86]. Pristine perovskite or perovskite quantum dots can also be utilized in a polymer matrix to improve the memristive behavior. In 2016, Chen et al. pioneered the use of $CH_3NH_3PbI_3$:PVK for blending active layers to construct a bulk heterojunction (BHJ) concept as popularly used in solar cells [87]. This BHJ-based device exhibits a non-volatile WORM memory with a large ON–OFF ratio of more than 10^3. PVK and $CH_3NH_3PbI_3$ act as a donor and acceptor, respectively. The intermolecular charge transfer between the donor and the acceptor induced by the electric field has been attributed to the WORM properties. Polymer matrix may influence the polymer–perovskite composite-based memory behaviors. Zhang et al. compared the PMMA and polyethylene oxide (PEO) matrixes with respect to the RS performance of $Cs_2AgBiBr_6$ double-perovskite nanofillers [88]. The PEO-based device does not depict significant changes in RS performance upon $Cs_2AgBiBr_6$ doping. However, an obvious impact of 2 wt% $Cs_2AgBiBr_6$ in PMMA-based devices could be achieved with low set and reset voltages, and a high ON–OFF ratio of 10^4 (Figure 8a,b). The pristine PEO has higher ionic conductivity than that of PMMA, which is nearly an insulator. Herein, a higher HRS current caused by ionic conductivity could be observed in a PEO-based device. The conductivity of the polymer matrix has a crucial effect on RS performance.

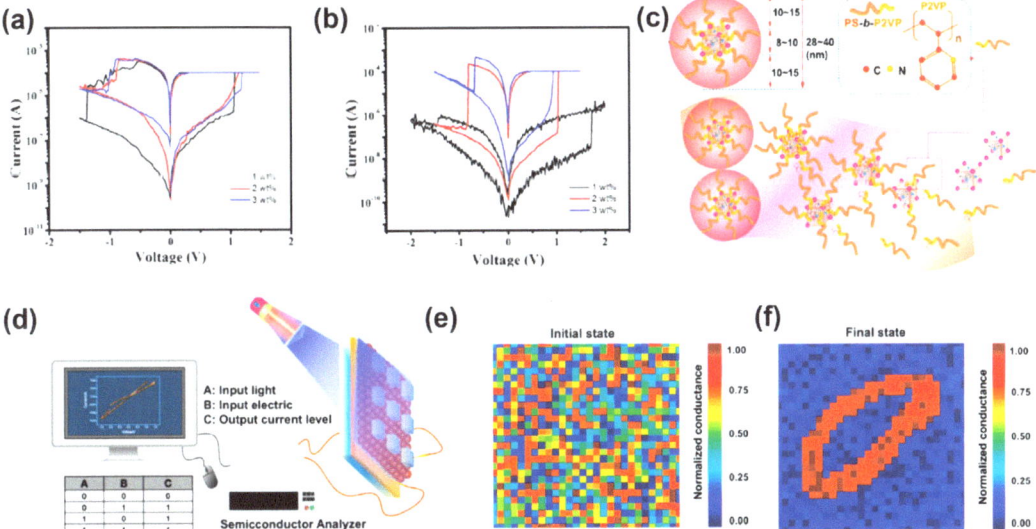

Figure 8. I–V curves of the composite devices in various concentrations for (a) an ITO/$Cs_2AgBiBr_6$@PEO/Au device, and (b) an ITO/$Cs_2AgBiBr_6$@PMMA/Au device. Reproduced with permission [88]. Copyright 2023, Elsevier. (c) Schematic illustration of S2VP:$CsPbBr_3$ quantum dots core-shell nanosphere composite. (d) Schematic illustration of the S2VP–$CsPbBr_3$ quantum-dot-based logic OR device. Mapping a representative input digit of 784 synaptic weights connected to the output digit "0" shown at the (e) initial and (f) final states of training. Reproduced with permission [89]. Copyright 2023, American Chemical Society.

$CsPbBr_3$ quantum dots are easy to decompose with environmental water and oxygen. Very recently, Jiang et al. used an amphiphilic diblock copolymer polystyrene-poly2-vinyl pyridine (PS-*b*-P2VP, S2VP) to protect the $CsPbBr_3$ quantum dots and realized a core–shell nanosphere composite (Figure 8c), leading to a robust and light-tunable memristor [89]. The device demonstrates ultra-stable RS behavior over 5000 cycles and over 5 million seconds, rendering it favorable for light tunability. Light and external electric fields can effectively change the resistance state of the device to realize the logical operation (Figure 8d). This light-tunable behavior can be used in biologically visually inspired neuromorphic computing. Simple machine learning was illustrated by simulating optoelectronic neural network learning. They used a single-layer perceptron model to categorize 28 × 28 pixels of handwritten digital images from a National Institute of Standards and Technology (MNIST) dataset using a backpropagation algorithm to perform supervised learning, as shown in Figure 8e,f. After 900 training epochs, the device demonstrates a maximum recognition accuracy of 97%, higher than the previous reports.

4. 2D Covalent Polymer-Based RS Memory Devices

Different from traditional organic polymers, 2D covalent polymers are synthesized through covalent bonding of pre-designed molecular building blocks to form covalently linked networks of monomers with periodic structures in two orthogonal directions [90]. They have been successfully prepared by the Langmuir–Blodgett (LB) method, chemical vapor deposition (CVD), and surface confined synthesis [91–93]. Covalent bonds are very stable, rendering these polymers with high stability in many solvents and harsh environments. Additionally, 2D polymers (2DPs) are stable in monolayers with sub-nanometer thickness, which is similar to graphene. Hence, the monolayer 2DPs are expected to reduce the filament length and minimize the energy to form and rupture the conductive filaments. These unique properties of 2DPs make them promising candidates for the advanced RS memories [94].

In 2019, Liu et al. first investigated an innovative 2DP-based non-volatile memristive device [95]. A wafer-scale ultrathin 2D imine polymer film was synthesized through a Schiff base polycondensation reaction from benzene-1,3,5-tricarbaldehyde (BTA) and p-phenylenediamine (PDA) building blocks (Figure 9a). Notably, the construction of a memory device based on $2DP_{BTA+PDA}$ films presents superior RS behaviors with an ON–OFF ratio from 10^2 to 10^5, an impressive retention time of 8×10^4 s and a stable endurance of 200 cycles (Figure 9b). The superior thermal stability allows the $2DP_{BTA+PDA}$-based device to show an increasing ON–OFF ratio with increasing annealing temperature at 300 °C, and this non-volatile memory behavior is also stable in polar and non-polar solvents. A flexible device with graphene/$2DP_{BTA+PDA}$/Ag structure on a polyimide (PI) substrate exhibits excellent memory behavior and mechanical durability over 500 bending cycles. Good flexibility and thermal stability give $2DP_{BTA+PDA}$ great potential in wearable electronics. The Ag conductive filament resistances are responsible for the switching mechanism of this device, which is certified by annular dark-field scanning transmission electron microscopy (ADF STEM) and electron energy loss spectrum (EELS), as shown in Figure 9c. These findings indicate the promising potential of 2DP thin films in next-generation memory devices and lead to the development of reliable memory devices.

Later, to demonstrate the effect of chemical structures of microporous polymers (MP) on the performance of memristor, they continued to synthesize two microporous covalent polymers ($MP_{TPA+TAPB}$, $MP_{OTPA+TAPB}$) with tris(4-aminophenyl)-benzene (TAPB) to react with terephthalaldehyde (TPA) and 2,5-dioctyloxyterephthalaldehyde (OTPA), as schemed in Figure 9d [96]. The incorporation of octyloxy groups within the dialdehyde monomer reduces the band gap and changes the pore environment. Thereafter, the HRS resistance is reduced, resulting in an increase in the ON–OFF ratio by an order of magnitude. The switching mechanism is attributed to the electrochemical metallized Ag conductive filaments that connect the top and bottom electrodes. Their findings provide a design criterion between molecular structures and memory properties.

Hu's group further explored the ternary electronic memory of 2DP with 2,5-bis(3-(9H-carbazol-9-yl)propoxy)terephthalaldehyde (TPAZ) and TAPB as monomers in 2021 (Figure 9e) [97]. The intrinsic subnanometer channel of pillar [5] arene and nanometer channel of 2DP construct multilevel channels for ternary memory devices based on $2DP_{TPAZ+TAPB}$ (Figure 9f). The device exhibits a high ON–OFF ratio of $1:10:10^3$, and a high ternary yield of 75%, as shown in Figure 9g. The $2DP_{TPAZ+TAPB}$ also presents stable flexible ternary memory after bending for 500 cycles, as well as thermal stability to a high temperature of 300 °C.

The memory mechanisms of the above research are mainly based on conductive filaments. Hu's group integrated the conformational change mechanism in a 2DP reaction between 2,5-bis(3-(9H-carbazol-9-yl)propoxy)terephthalaldehyde (TPAK) and TAPB [98]. By controlling the compliance current (I_{CC}), three RS memory behaviors, including non-volatile WORM, non-volatile FLASH memory, and volatile dynamic DRAM behavior, were achieved with the configuration of ITO/$2DP_{TPAK+TAPB}$/Au. Specifically, the devices display volatile DRAM at $I_{CC} = 10^{-4}$ A, while they show non-volatile FLASH and WORM memory behaviors at $I_{CC} = 10^{-3}$ and $I_{CC} = 10^{-1}$ A, respectively. Figure 9h shows the conformation change in the carbazole groups through their rotation, leading to more regular π–π stacking, which is confirmed by UV–Visible spectra.

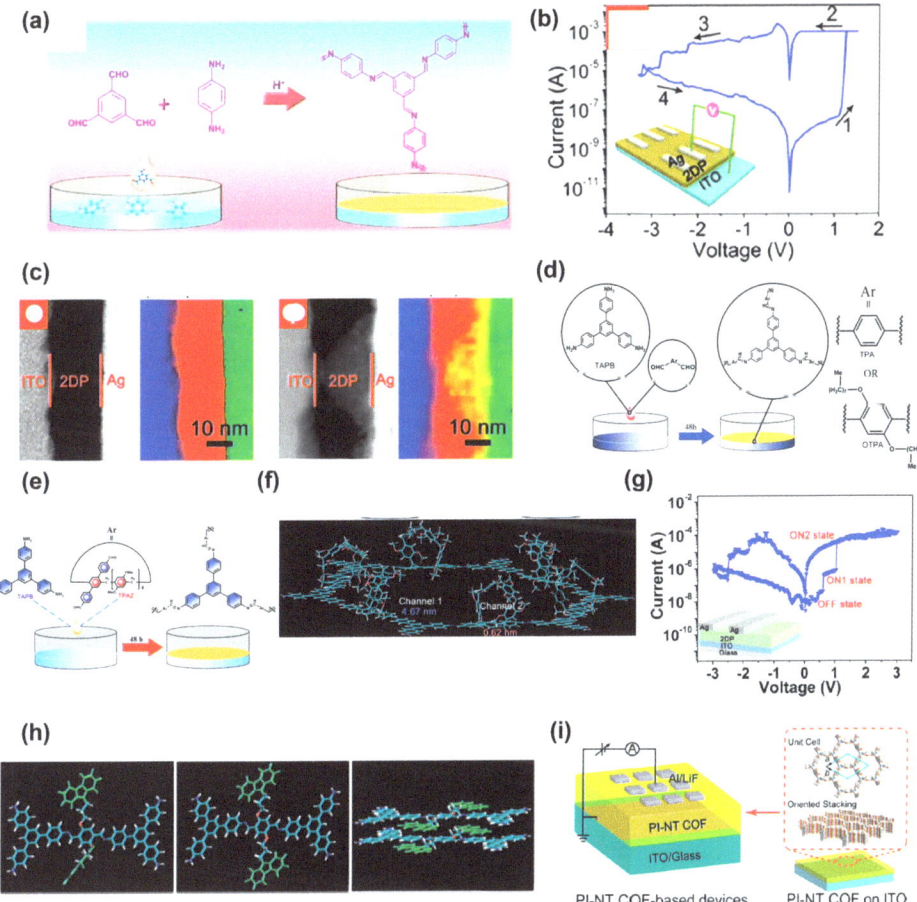

Figure 9. (**a**) A schematic illustration of the synthesis of 2DP$_{BTA+PDA}$ films through the Schiff-base reaction of the monomers. (**b**) *I–V* curves for the ITO/2DP$_{BTA+PDA}$/Ag configuration. (**c**) From left to right: ADF STEM image and chemical maps of ITO/2DP$_{BTA+PDA}$/Ag device in the initial and on state. Reproduced with permission [95]. Copyright 2019, WILEY-VCH. (**d**) The preparation diagram of MP$_{TPA+TAPB}$, MP$_{OTPA+TAPB}$ on the solution/air interface. Reproduced with permission [96]. Copyright 2020, The Royal Society of Chemistry. (**e**) The preparation diagram of 2DP$_{TPAZ+TAPB}$ at the solution/air interface. (**f**) Simulated structure of the 2DP$_{TPAZ+TAPB}$ material part unit. (**g**) *I–V* characteristics of the ITO/2DP$_{TPAZ+TAPB}$/Ag device. Reproduced with permission [97]. Copyright 2021, WILEY-VCH. (**h**) The conformation change memory mechanism. Reproduced with permission [98]. Copyright 2022, The Royal Society of Chemistry. (**i**) Schematic diagram of the PI–NT COF film stacking and the ITO/PI-NT COF film/LiF/Al configuration. Reproduced with permission [99]. Copyright 2020, American Chemical Society.

In 2023, Hu's group demonstrated highly crystalline single-layer 2D polymers (SL-2DPs) via the condensation of TAPB and terephthalaldehyde decorated with different lengths of alkoxy chains (TPOC$_x$, x = 0, 2, 4, 8, 12, 16, 22) using a LB method [100]. The long alkoxy chains were incorporated to enhance the crystalline structure of 2DP. The devices based on SL-2DPs show an ultralow working voltage (0.6 V), good endurance (324 cycles), and long retention times (10^5 s). Additionally, the device exhibits excellent mechanical flexibility and electrical reliability. The memory behavior is still maintained under the strains of 2.6%. The research by Hu's group suggests the potential for 2DP in emerging applications such as dense data storage and ultra-thin, highly stable, flexible electronics.

The 2DPs mentioned above typically exhibit a π-conjugated structure. However, incorporating a donor–acceptor moiety into 2DPs has been shown to result in more effective and durable RS memory behaviors. In 2020, Sun et al. demonstrated the first donor–acceptor two-dimensional polyimide covalent organic framework (2D PI–NT COF) films with 4,4′,4″-triaminotriphenylamine (TAPA) and naphthalene-1,4,5,8-tetracarboxylic dianhydride (NTCDA) as donor and acceptor, respectively (Figure 9i) [99]. The high-quality film exhibits high crystallinity, good orientation, and tunable thickness. The memory device shows superior WORM performance, with an ON–OFF ratio exceeding 10^6, a retention time of 10^4 s, and high stability. Due to the donor–acceptor moieties, an electric-field-induced intramolecular charge transfer is attributed to the memory behavior.

Li and coworkers further introduced an electron acceptor, [2,2′-bithiophene]-5,5′-dicarbaldehyde (BTDD) or (E)-5,5′-(ethene-1,2-diyl)bis(thiophene-2-carbaldehyde) (TVTDD), and an electron donor, 4,4′,4″-(1,3,5-triazine-2,4,6-triyl)trianiline (TAPT), into 2D imine frameworks to achieve COF–TT–BT and COF–TT–TVT [101]. The 100 nm thick COF–TT–BT and COFTT–TVT films show rewritable memories with high ON–OFF ratios (10^5 and 10^4) and low driving voltages (1.30 and 1.60 V), which are different from the WORM of 2D PI–NT COF. The energy level alignment between the COF and the electrodes may impact the mechanism of RS. Chen's group developed COF-based redox activity memory materials, namely COF–Azu, which is made by combining TAPB, azulene 1,3-dicarbaldehyde (Azu), and TFPB–PDAN by the co-condensation of 1,3,5-tris(4-formylphenyl)-benzene (TFPB) and p-phenylenediacetonitrile (PDAN) [102,103]. The electric-field-induced charge transfer in the donor–acceptor structure is responsible for the memory. These results further fulfill the application of 2D COF materials in high-performance RS memory devices.

The 2D structural homogeneity may promote effective charge transport and improve the performance of the polymer-based memory device. To overcome the low production yield and reliability of the device, Zhang et al. designed a 2D conjugated polymer, PBDTT–BQTPA, composed of redox active triphenylamine moieties anchored on the coplanar bis(thiophene)-4,8-dihydrobenzo[1,2-b:4,5-b]dithiophene (BDTT) donor and quinoxaline acceptor [104]. The coplanar structure and ordered π–π stacking of the macromolecule backbone render the polymer with enhanced crystalline uniformity; therefore, the production yield rises to 90%, with switching parameter variation decreasing to 3.16–8.29% and potential scalability changing into a 100 nm scale.

5. Biomacromolecules-Based RS Memory Devices

Biomacromolecules can be easily obtained by extraction from living organisms, without needing a complex chemical synthesis route. With the advantages of biocompatibility, environmental friendliness, and flexibility, the electronic devices based on biomaterials are generally cost-effective, ecofriendly, and sustainable. Biomaterials show interesting applications in implantable biomedical devices. In addition, their self-decomposition behaviors make them valuable for applications related to security circumstances. Various studies have been devoted to biomaterial-based RS memory devices, including chitosan, starch, lignin, protein, glucose, enzyme, and DNA [105,106].

As a common biomaterial in our daily life, egg albumen exhibits promising memristive behavior in transparent and flexible memristor devices. A flexible polyethylene terephthalate (PET)/ITO/egg albumen/tungsten device (Figure 10a), shows outstanding memristive operation under mechanical bending without significant performance degradation and possesses a transparency of more than 90% with visible light in a wavelength range of 230–850 nm (Figure 10b,c) [107]. This device can mimic certain neural bionic behaviors to present typical synapse performance, i.e., short-term plasticity (STP), long-term plasticity (LTP), and transitions between STP and LTP. Additionally, it can be dissolved in deionized water within 1 day. These results show that albumen-based devices are attractive as biocompatible and biodegradable electronics. As shown in Figure 10d,e, Zhou et al. fabricated albumen-based large-area paper substrates exhibiting both robust physical flexibility and excellent electric properties [108]. The crossbar arrays show an ON–OFF ratio of 10^4, and a retention time of 10^4 s even after bending 10^4 times. This device can realize complete memory logic blocks containing NOT, OR, and AND gates (Figure 10f). Wang et al. further explored egg albumen's application in a logic circuit [109]. With the configuration of Al/PMMA/egg albumen:Au nanoparticles/PMMA/Al, the ON–OFF ratio is enhanced dramatically, to 2.86×10^5, compared with the device without PMMA as insulating layer. The oxygen-vacancy-dominated conductive filaments are responsible for the RS mechanism. The device also presents full basic logic functions, including AND, OR, NOT, NAND, and NOR, based on auxiliary logic. By introducing multiwalled carbon nanotubes (MWCNTs) into egg albumen, the switching ON–OFF ratio of ITO/egg albumen:MWCNTs/Al device increases as the concentration of MWCNTs decreases [110]. By controlling the compliance current, the multilevel RS memory realizes 2-bit data storage to increase the storage density. These egg-albumen-based functional materials meet most of the required standards of electrical characteristics for an RRAM including a high ON–OFF ratio, high electrical endurance, long retention time, fast switching speed, and low power consumption. Therefore, it plays a pivotal role in the non-volatile memory device.

Silk fibroin is derived from natural silk. In order to improve the stability and power consumption of silk-fibroin-based memristors, Zhang et al. doped the silk fibroin with Ag and an ethanol-based post-treatment to form microcrystal regions in the bulk of the silk fibroin, as can be seen in Figure 10g [111]. The microcrystal regions make the charge carriers transport through the fixed and short paths, resulting in a high stability and low power consumption (0.7 µW) memristor. The switching mechanism is attributed to the SCLC mechanism. The non-linear transmission function of synapses shows the great potential of silk fibroin in artificial synapses. Wang et al. synthesized a novel protein-based polymer by the polymerization of silk fibroin and 2-isocyanatoethyl methacrylate [112]. The polymer acquired an analogue-type computing behavior characterized by more than 32 programmable conductance states. The analogue property does not show any depression in ambient air for 4 months (Figure 10h–j). A physical model consisting of the trapping and de-trapping of the injected electrons may be responsible for the analogue-type behavior. The electrodes play a vital role in memory behavior. With a GO/silk fibroin/GO structure, Liu et al. reported multilevel storage with binary and ternary switching behaviors in a single device [113]. For $I_{cc} \leq 0.01$ A, the device exhibits binary switching because of the SCLC mechanism, while I_{cc} exceeds 0.01 A, the device changes to ternary switching behavior, due to the SCLC and Poole–Frenkel emission mechanisms. By analyzing handwritten numbers obtained from the Modified National Institute of Standards and Technology database, this memristor-based ternary weight network exhibits a recognition accuracy of 92.3%, which is better than that based on a binary neural network. This study shows the application of silk fibroin in neural morphological computing.

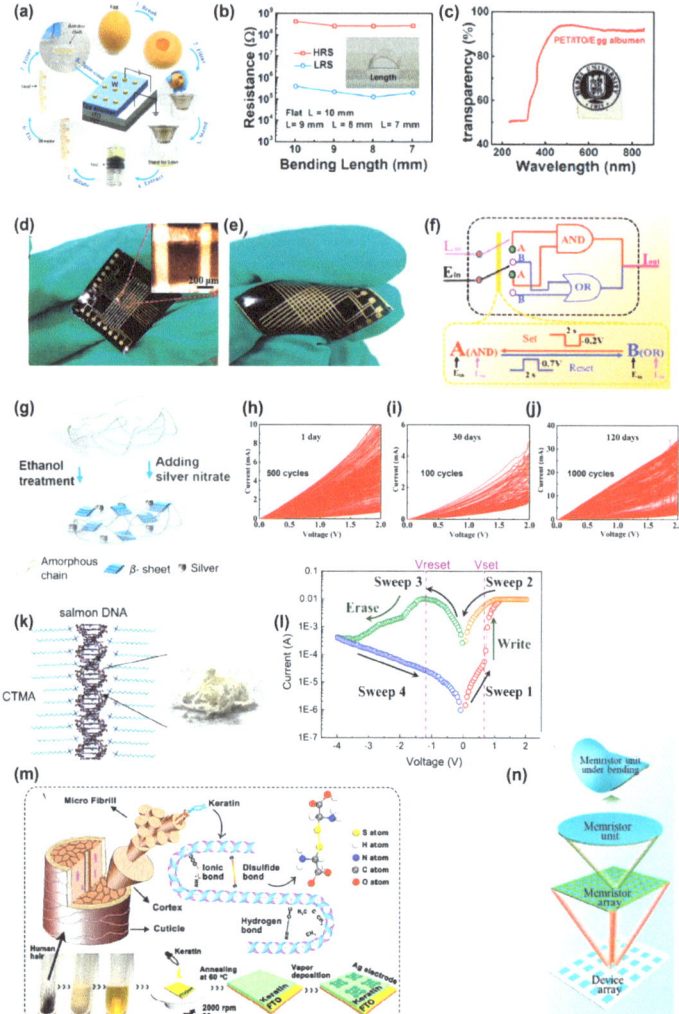

Figure 10. (a) Proposed fabrication process and schematic of a PET/ITO/egg albumen/W memristor. (b) The device's HRS and LRS versus bending length. (c) Transparency demonstration of the device. Reproduced with permission [107]. Copyright 2019, American Chemical Society. Photographs of the Au/albumen/Au crossbar arrays in (d) flat, and (e) moderate-angle bending states. (f) A schematic diagram of the memory logic gate switching between "OR" and "AND" states. Reproduced with permission [108]. Copyright 2019, The Royal Society of Chemistry. (g) Schematic of the conformational transition of silk fibroin with Ag doping and ethanol treatment. Reproduced with permission [111]. Copyright 2021, American Chemical Society. (h–j) The endurance and retention test for the analogue RS memristor after: 1 day; 30 days; and 120 days. Reproduced with permission [112]. Copyright 2021, The Royal Society of Chemistry. (k) The chemical structure of DNA–CTMA. (l) I–V curve of the fabricated device based on DNA–CTMA. Reproduced with permission [114]. Copyright 2018, Elsevier. (m) Schematic diagrams of keratin from human hair, the chemical bonds and structures in keratin, and the fabrication process of FTO/keratin/Ag memory devices. Reproduced with permission [115]. Copyright 2019, The Royal Society of Chemistry. (n) Schematic diagram of a device with a crossbar array and a cell being bent. Reproduced with permission [116]. Copyright 2023, Wiley-VCH.

DNA molecules are complete in nature and can be easily extracted from biological species. Since Hung et al. pioneered the investigation into the DNA-based transparent WORM device, the DNA in RS memory have been comprehensively explored [117]. Jeng et al. fabricated a DNA-based RRAM device with DNA–cetyltrimethylammonium (CTMA) macromolecules (Figure 10k) [114]. The ITO/DNA–CTMA/Ag device shows low set and reset voltages of 0.65 V and −1.25 V, respectively. Figure 10l shows that it has an ON–OFF ratio of 10^2, an electrical endurance of 200 cycles, and a long retention time of 10^4 s. Notably, the device exhibits a pronounced negative differential resistance (NDR) region. Multilevel resistances can be observed by adjusting the magnitude of negative biases. The conduction mechanism was due to the formation of a conductive Ag filament, which was ascertained from completely studying the electrical properties at different temperatures, active layer thicknesses, and electrode materials. This finding paves the way toward future integration of natural DNA-based biomaterials. Abbass et al. fabricated a transparent memory device by sandwiching a Cu^{2+}-doped salmon DNA molecule between FTO and Pt [118]. The device shows the set and reset voltages of −3.5 and 2.5 V, respectively. Unlike the majority of RRAM devices, the Cu^{2+}-doped device shows LRS initially with an ON–OFF ratio of 10^3, a retention time of 10^4 s, and an endurance of 10^5 pulses. The migration of Cu^{2+} under externally electric power induces the RS memory. In addition, this device has good thermal stability in the temperature range of 25 to 85 °C.

The human hair keratin with the configuration of FTO/keratin/Ag exhibits good RS performance, high transmittance, as well as physically transient properties (Figure 10m) [115]. The non-volatile memory performance exhibits a resistance window larger than 10^3, switching ratio, and retention time of more than 10^4 s. The keratin-based RRAM device can be dissolved within 30 min in deionized water. Cellulose can act as the functional layer material in the RS layer and it can be also used as a substrate to construct a flexible self-supporting memristor, which was studied by Xia et al. [119]. The cellulose membrane/Cr/Au/cellulose/Ag/Au device presents volatile and non-volatile RS behaviors by controlling the compliance current in the set process. Under a compliance current of 1.1×10^{-7} A, the device shows volatile threshold switching behavior with a set voltage in the range 0.2–0.6 V. The non-volatile bipolar switching behavior can be seen with the compliance current of 1.0×10^{-4} A, causing it to exhibit an ON–OFF ratio of 10^6 and retention time of more than 1000 s. The device can work stably in the bending mode and in the temperature range from 20 to 100 °C. This study provides a facile strategy for constructing a natural polymer-based RS device.

Recently, Sun et al. studied the corn-starch-based biomaterial flexible RS device [116]. The corn starch was extracted directly from corn plants and mixed with PVDF to prepare a flexible ITO/corn-starch:PVDF/Ag sandwich structure. The device shows a capacitive effect at different values of V_{max}. Notably, at the memory window of −12 V, the largest switching ratio is 3.5×10^2, and almost symmetrical capacitive-coupled memristive I–V characteristics are achieved. By applying the appropriate pulse amplitude, width, and frequency, the device also presents synaptic behavior due to the presence of a large number of conducting ions in the active layer. The memristive device array may be used to map shape changes caused by subtle pressure changes through measuring the current changes, as schemed in Figure 10n. A word, "WATERLOO", can be sensed by writing it with a stylus pen to record the current changes of the device cells. This work elucidates the role of corn starch in realizing eco-friendly green wearable electronics and its potential application in artificial intelligence.

6. Summary and Outlook

In conclusion, this review demonstrates a summary of recent progresses of polymer-based RS memory devices, including single-component polymers, polymer mixtures, 2D covalent polymers, and biomacromolecules. The advantages and disadvantages of these devices are listed in Table 1. The device structures, switching types, and mechanisms are thoroughly introduced. Based on these enlightening investigations, we have faith that the polymer-based RS materials and devices will open novel opportunities in the information storage and information processing fields to realize multifunctional memristors in future RS technology.

Table 1. Summary of the advantages and disadvantages of different polymer-based RS memory devices.

	Advantages	Disadvantages
Single-component polymers	1. Easy molecule design. 2. Avoids the probability of phase separation.	1. Switching performance is restricted. 2. Complex synthesis route.
Polymer mixtures	1. Electronic properties, processability, and mechanical flexibility can be easily adjusted. 2. Facile thin film preparation.	1. Phase separation.
2D covalent polymers	1. High thermal and chemical stabilities. 2. Excellent scalability.	1. The quality of the thin film is limited. 2. Slower write speeds.
Biomacromolecules	1. Self-decomposition, biodegradability. 2. Environmentally friendly. 3. Flexible.	1. A short life span. 2. Complex purification of biofilms. 3. Poor thermal stability.

Generally, designing redox-active entities with donors and acceptors in polymer molecules is a popular strategy for achieving switching behavior. In order to obtain high –storage-capacity devices, a multilevel or ternary memory device is urgently needed. In addition to the traditional compliance current control in the device testing, designing a donor–acceptor–acceptor structure provides an innovative insight to realize ternary memory. Moreover, the unique properties of recently developed 2D covalent polymers with high chemical and thermal stabilities, large surface areas, and tunable electronic properties, renders it easy to obtain high ON–OFF ratios, low power consumption, and high stability memristors. Biomacromolecule-based RRAMs show interesting applications in mimicking the human brain, implantable devices, memory storage, and wearable electronics. They open a new window for the development of wearable and implantable memory devices.

Although tremendous advances have been witnessed in the investigation of switching polymer materials, there are still some fundamental issues that need to be conquered for further development of polymer switching. Firstly, to date, although the switching parameters (e.g., ON–OFF ratio, operational voltage, cycling endurance, and retention time) have been greatly improved by various methods, they are not very competitive with inorganic counterparts. The large cycle-to-cycle and device-to-device variations in the switching parameters result in low production yields of devices on crossbar arrays. This is far from valuable integration into large-scale integrated circuits. The fabrication technology of polymer memory devices is not mature enough to compete with the existing silicon-based memory devices for high reproducibility and low-cost production. Especially in 2D covalent polymers, achieving high-quality thin films with good reproducibility and crystallinity remains a challenge. Secondly, the stability of polymer memory devices is readily influenced by ambient moisture and temperature. For example, 2D covalent polymers and biomaterials are sensitive to moisture, light, and heat. There is still a long way for polymer materials to go with regards to commercial application, which requires more research efforts to optimize. Thirdly, some switching mechanisms have been proposed to account for the polymer materials. However, a definite mechanism with both theoretical and experimental evidence is not yet completely accepted. A deeper understanding of the different types of mechanisms is essential for molecular structure design and modification. For example, the charge transfer is a well-known switching mechanism in donor–acceptor structure. However, direct physical evidence is lacking to confirm the existence of a charge transfer state. Its long lifetime in a memory device is in conflict with the transient spectroscopy measurements performed on solar cells. Only with a clear understanding of the memory mechanism to guide the rational molecular structure–property relationship can the polymer memory device make rapid progress in the future.

Emerging RS memory devices based on polymer materials have advanced rapidly over the past years, through addressing the aforementioned scientific and technical issues. This research needs physicists', chemists', material scientists', and electrical engineers' infinite efforts in the field of polymer materials and devices in the upcoming era of artificial intelligence.

Author Contributions: Funding acquisition, writing—original draft preparation, B.T., L.X. and B.D.; writing—review and editing, S.Z., L.X. and Q.S.; validation and supervision, L.X.; formal analysis and investigation, B.D. All authors have read and agreed to the published version of the manuscript.

Funding: This work was financially supported by the China Postdoctoral Science Foundation (2021M690127), the Natural Science Foundation of Hunan Province (Grant No. 2021JJ40141), Shaanxi Natural Science Foundation of China (2023-JC-QN-0680), the Scientific Research Fund of Hunan Provincial Education Department (21B0815), and the Training Project of Changsha City for Distinguished Young Scholars (kq2107023).

Data Availability Statement: Not applicable.

Conflicts of Interest: The authors declare no conflict of interest.

References

1. Philip Chen, C.L.; Zhang, C.-Y. Data-intensive applications, challenges, techniques and technologies: A survey on Big Data. *Inform. Sci.* **2014**, *275*, 314–347. [CrossRef]
2. Service, R.F. Chipmakers look past Moore's law, and silicon. *Science* **2018**, *361*, 321. [CrossRef] [PubMed]
3. Lundstrom, M. Moore's Law Forever? *Science* **2003**, *299*, 210–211. [CrossRef] [PubMed]
4. Kolar, J.; Macak, J.M.; Terabe, K.; Wagner, T. Down-scaling of resistive switching to nanoscale using porous anodic alumina membranes. *J. Mater. Chem. C* **2014**, *2*, 349–355. [CrossRef]
5. Zahoor, F.; Azni Zulkifli, T.Z.; Khanday, F.A. Resistive Random Access Memory (RRAM): An Overview of Materials, Switching Mechanism, Performance, Multilevel Cell (mlc) Storage, Modeling, and Applications. *Nanoscale Res. Lett.* **2020**, *15*, 90. [CrossRef]
6. Zhang, Z.; Wang, Z.; Shi, T.; Bi, C.; Rao, F.; Cai, Y.; Liu, Q.; Wu, H.; Zhou, P. Memory materials and devices: From concept to application. *InfoMat* **2020**, *2*, 261–290. [CrossRef]
7. Salinga, M.; Kersting, B.; Ronneberger, I.; Jonnalagadda, V.P.; Vu, X.T.; Le Gallo, M.; Giannopoulos, I.; Cojocaru-Mirédin, O.; Mazzarello, R.; Sebastian, A. Monatomic phase change memory. *Nat. Mater.* **2018**, *17*, 681–685. [CrossRef]
8. Ciocchini, N.; Laudato, M.; Boniardi, M.; Varesi, E.; Fantini, P.; Lacaita, A.L.; Ielmini, D. Bipolar switching in chalcogenide phase change memory. *Sci. Rep.* **2016**, *6*, 29162. [CrossRef]
9. Zhu, J.G.; Shadman, A. Resonant Spin-Transfer Torque Magnetoresistive Memory. *IEEE Trans. Magn.* **2019**, *55*, 3400407. [CrossRef]
10. Khang, N.H.D.; Ueda, Y.; Hai, P.N. A conductive topological insulator with large spin Hall effect for ultralow power spin–orbit torque switching. *Nat. Mater.* **2018**, *17*, 808–813. [CrossRef]
11. Ando, K.; Fujita, S.; Ito, J.; Yuasa, S.; Suzuki, Y.; Nakatani, Y.; Miyazaki, T.; Yoda, H. Spin-transfer torque magnetoresistive random-access memory technologies for normally off computing. *J. Appl. Phys.* **2014**, *115*, 172607. [CrossRef]
12. Lastras-Montaño, M.A.; Cheng, K.-T. Resistive random-access memory based on ratioed memristors. *Nat. Electron.* **2018**, *1*, 466–472. [CrossRef]
13. Wan, W.; Kubendran, R.; Schaefer, C.; Eryilmaz, S.B.; Zhang, W.; Wu, D.; Deiss, S.; Raina, P.; Qian, H.; Gao, B.; et al. A compute-in-memory chip based on resistive random-access memory. *Nature* **2022**, *608*, 504–512. [CrossRef] [PubMed]
14. Rao, K.D.M.; Sagade, A.A.; John, R.; Pradeep, T.; Kulkarni, G.U. Defining Switching Efficiency of Multilevel Resistive Memory with PdO as an Example. *Adv. Electron. Mater.* **2016**, *2*, 1500286. [CrossRef]
15. Li, H.; Xu, Q.; Li, N.; Sun, R.; Ge, J.; Lu, J.; Gu, H.; Yan, F. A Small-Molecule-Based Ternary Data-Storage Device. *J. Am. Chem. Soc.* **2010**, *132*, 5542–5543. [CrossRef] [PubMed]

16. Gu, P.-Y.; Gao, J.; Lu, C.-J.; Chen, W.; Wang, C.; Li, G.; Zhou, F.; Xu, Q.-F.; Lu, J.-M.; Zhang, Q. Synthesis of tetranitro-oxacalix[4]arene with oligoheteroacene groups and its nonvolatile ternary memory performance. *Mater. Horiz.* **2014**, *1*, 446–451. [CrossRef]
17. Lin, W.-P.; Liu, S.-J.; Gong, T.; Zhao, Q.; Huang, W. Polymer-Based Resistive Memory Materials and Devices. *Adv. Mater.* **2014**, *26*, 570–606. [CrossRef]
18. Han, S.-T.; Zhou, Y.; Roy, V.A.L. Towards the Development of Flexible Non-Volatile Memories. *Adv. Mater.* **2013**, *25*, 5425–5449. [CrossRef]
19. Sun, G.; Liu, J.; Zheng, L.; Huang, W.; Zhang, H. Preparation of Weavable, All-Carbon Fibers for Non-Volatile Memory Devices. *Angew. Chem. Int. Ed.* **2013**, *52*, 13351–13355. [CrossRef]
20. Lee, J.H.; Park, S.P.; Park, K.; Kim, H.J. Flexible and Waterproof Resistive Random-Access Memory Based on Nitrocellulose for Skin-Attachable Wearable Devices. *Adv. Funct. Mater.* **2020**, *30*, 1907437. [CrossRef]
21. Li, Y.; Qian, Q.; Zhu, X.; Li, Y.; Zhang, M.; Li, J.; Ma, C.; Li, H.; Lu, J.; Zhang, Q. Recent advances in organic-based materials for resistive memory applications. *InfoMat* **2020**, *2*, 995–1033. [CrossRef]
22. Gao, S.; Yi, X.; Shang, J.; Liu, G.; Li, R.W. Organic and hybrid resistive switching materials and devices. *Chem. Soc. Rev.* **2018**, *48*, 1531–1565. [CrossRef] [PubMed]
23. Song, S.; Cho, B.; Kim, T.-W.; Ji, Y.; Jo, M.; Wang, G.; Choe, M.; Kahng, Y.H.; Hwang, H.; Lee, T. Three-Dimensional Integration of Organic Resistive Memory Devices. *Adv. Mater.* **2010**, *22*, 5048–5052. [CrossRef] [PubMed]
24. Han, J.; Lian, H.; Cheng, X.; Dong, Q.; Qu, Y.; Wong, W.-Y. Study of Electronic and Steric Effects of Different Substituents in Donor–Acceptor Molecules on Multilevel Organic Memory Data Storage Performance. *Adv. Electron. Mater.* **2021**, *7*, 2001097. [CrossRef]
25. Zhang, G.; Lee, Y.-J.; Gautam, P.; Lin, C.-C.; Liu, C.-L.; Chan, J.M.W. Pentafluorosulfanylated polymers as electrets in nonvolatile organic field-effect transistor memory devices. *J. Mater. Chem. C* **2019**, *7*, 7865–7871. [CrossRef]
26. Liu, H.; Wang, C.; Han, G.; Li, J.; Peng, Y.; Liu, Y.; Wang, X.; Zhong, N.; Duan, C.; Wang, X.; et al. ZrO2 Ferroelectric FET for Non-volatile Memory Application. *IEEE Electr. Device Lett.* **2019**, *40*, 1419–1422. [CrossRef]
27. Zhou, L.; Mao, J.; Ren, Y.; Han, S.-T.; Roy, V.A.L.; Zhou, Y. Recent Advances of Flexible Data Storage Devices Based on Organic Nanoscaled Materials. *Small* **2018**, *14*, 1703126. [CrossRef]
28. Xu, T.; Guo, S.; Qi, W.; Li, S.; Xu, M.; Xie, W.; Wang, W. High-performance flexible organic thin-film transistor nonvolatile memory based on molecular floating-gate and pn-heterojunction channel layer. *Appl. Phys. Lett.* **2020**, *116*, 023301. [CrossRef]
29. Higashinakaya, M.; Nagase, T.; Abe, H.; Hattori, R.; Tazuhara, S.; Kobayashi, T.; Naito, H. Electrically programmable multilevel nonvolatile memories based on solution-processed organic floating-gate transistors. *Appl. Phys. Lett.* **2021**, *118*, 103301. [CrossRef]
30. Wang, H.; Wen, Y.; Zeng, H.; Xiong, Z.; Tu, Y.; Zhu, H.; Cheng, R.; Yin, L.; Jiang, J.; Zhai, B.; et al. Two-Dimensional Ferroic Materials for Non-volatile Memory Applications. *Adv. Mater.* **2023**, 2305044. [CrossRef]
31. Hong, Z.; Zhao, J.; Huang, K.; Cheng, B.; Xiao, Y.; Lei, S. Controllable switching properties in an individual CH3NH3PbI3 micro/nanowire-based transistor for gate voltage and illumination dual-driving non-volatile memory. *J. Mater. Chem. C* **2019**, *7*, 4259–4266. [CrossRef]
32. Ling, Q.-D.; Song, Y.; Lim, S.-L.; Teo, E.Y.-H.; Tan, Y.-P.; Zhu, C.; Chan, D.S.H.; Kwong, D.-L.; Kang, E.-T.; Neoh, K.-G. A Dynamic Random Access Memory Based on a Conjugated Copolymer Containing Electron-Donor and -Acceptor Moieties. *Angew. Chem. Int. Ed.* **2006**, *45*, 2947–2951. [CrossRef] [PubMed]
33. Zhu, M.-G.; Zhang, Z.; Peng, L.-M. High-Performance and Radiation-Hard Carbon Nanotube Complementary Static Random-Access Memory. *Adv. Electron. Mater.* **2019**, *5*, 1900313. [CrossRef]
34. Avila-Niño, J.A.; Patchett, E.R.; Taylor, D.M.; Assender, H.E.; Yeates, S.G.; Ding, Z.; Morrison, J.J. Stable organic static random access memory from a roll-to-roll compatible vacuum evaporation process. *Org. Electron.* **2016**, *31*, 77–81. [CrossRef]
35. Spessot, A.; Oh, H. 1T-1C Dynamic Random Access Memory Status, Challenges, and Prospects. *IEEE Trans. Electron. Dev.* **2020**, *67*, 1382–1393. [CrossRef]
36. Saha, M.; Dey, S.; Nawaz, S.M.; Mallik, A. Environment-friendly resistive memory based on natural casein: Role of electrode and bio-material concentration. *Org. Electron.* **2023**, *121*, 106869. [CrossRef]
37. Hsieh, H.-C.; Wu, N.; Chuang, T.-H.; Lee, W.-Y.; Chen, J.-Y.; Chen, W.-C. Eco-Friendly Polyfluorene/Poly(butylene succinate) Blends and Their Electronic Device Application on Biodegradable Substrates. *ACS Appl. Polym. Mater.* **2020**, *2*, 2469–2476. [CrossRef]
38. Kumar, D.; Aluguri, R.; Chand, U.; Tseng, T.Y. Metal oxide resistive switching memory: Materials, properties and switching mechanisms. *Ceram. Int.* **2017**, *43*, S547–S556. [CrossRef]
39. Ling, Q.-D.; Lim, S.-L.; Song, Y.; Zhu, C.-X.; Chan, D.S.-H.; Kang, E.-T.; Neoh, K.-G. Nonvolatile Polymer Memory Device Based on Bistable Electrical Switching in a Thin Film of Poly(N-vinylcarbazole) with Covalently Bonded C60. *Langmuir* **2007**, *23*, 312–319. [CrossRef]
40. Xie, L.-H.; Ling, Q.-D.; Hou, X.-Y.; Huang, W. An Effective Friedel−Crafts Postfunctionalization of Poly(N-vinylcarbazole) to Tune Carrier Transportation of Supramolecular Organic Semiconductors Based on π-Stacked Polymers for Nonvolatile Flash Memory Cell. *J. Am. Chem. Soc.* **2008**, *130*, 2120–2121. [CrossRef]
41. Zhou, Z.; Mao, H.; Wang, X.; Sun, T.; Chang, Q.; Chen, Y.; Xiu, F.; Liu, Z.; Liu, J.; Huang, W. Transient and flexible polymer memristors utilizing full-solution processed polymer nanocomposites. *Nanoscale* **2018**, *10*, 14824–14829. [CrossRef] [PubMed]

42. Kadhim, M.S.; Yang, F.; Sun, B.; Wang, Y.; Guo, T.; Jia, Y.; Yuan, L.; Yu, Y.; Zhao, Y. A resistive switching memory device with a negative differential resistance at room temperature. *Appl. Phys. Lett.* **2018**, *113*, 053502. [CrossRef]
43. Li, Y.; Zhu, X.; Li, Y.; Zhang, M.; Ma, C.; Li, H.; Lu, J.; Zhang, Q. Highly Robust Organometallic Small-Molecule-Based Nonvolatile Resistive Memory Controlled by a Redox-Gated Switching Mechanism. *ACS Appl. Mater. Interfaces* **2019**, *11*, 40332–40338. [CrossRef] [PubMed]
44. Li, Y.-Q.; Fang, R.-C.; Zheng, A.-M.; Chu, Y.-Y.; Tao, X.; Xu, H.-H.; Ding, S.-J.; Shen, Y.-Z. Nonvolatile memory devices based on polyimides bearing noncoplanar twisted biphenyl units containing carbazole and triphenylamine side-chain groups. *J Mater. Chem.* **2011**, *21*, 15643–15654. [CrossRef]
45. Ling, Q.-D.; Chang, F.-C.; Song, Y.; Zhu, C.-X.; Liaw, D.-J.; Chan, D.S.-H.; Kang, E.-T.; Neoh, K.-G. Synthesis and Dynamic Random Access Memory Behavior of a Functional Polyimide. *J. Am. Chem. Soc.* **2006**, *128*, 8732–8733. [CrossRef]
46. Zhang, B.; Liu, Y.-L.; Chen, Y.; Neoh, K.-G.; Li, Y.-X.; Zhu, C.-X.; Tok, E.-S.; Kang, E.-T. Nonvolatile Rewritable Memory Effects in Graphene Oxide Functionalized by Conjugated Polymer Containing Fluorene and Carbazole Units. *Chem. Eur. J.* **2011**, *17*, 10304–10311. [CrossRef]
47. Liu, S.-J.; Wang, P.; Zhao, Q.; Yang, H.-Y.; Wong, J.; Sun, H.-B.; Dong, X.-C.; Lin, W.-P.; Huang, W. Single Polymer-Based Ternary Electronic Memory Material and Device. *Adv. Mater.* **2012**, *24*, 2901–2905. [CrossRef]
48. Bozano, L.D.; Kean, B.W.; Beinhoff, M.; Carter, K.R.; Rice, P.M.; Scott, J.C. Organic Materials and Thin-Film Structures for Cross-Point Memory Cells Based on Trapping in Metallic Nanoparticles. *Adv. Funct. Mater.* **2005**, *15*, 1933–1939. [CrossRef]
49. Zhang, P.; Xu, B.; Gao, C.; Chen, G.; Gao, M. Facile Synthesis of Co9Se8 Quantum Dots as Charge Traps for Flexible Organic Resistive Switching Memory Device. *ACS Appl. Mater. Interfaces* **2016**, *8*, 30336–30343. [CrossRef]
50. Ouyang, J.; Chu, C.-W.; Szmanda, C.R.; Ma, L.; Yang, Y. Programmable polymer thin film and non-volatile memory device. *Nat. Mater.* **2004**, *3*, 918–922. [CrossRef]
51. Qi, S.; Iida, H.; Liu, L.; Irle, S.; Hu, W.; Yashima, E. Electrical Switching Behavior of a [60]Fullerene-Based Molecular Wire Encapsulated in a Syndiotactic Poly(methyl methacrylate) Helical Cavity. *Angew. Chem. Int. Ed.* **2013**, *52*, 1049–1053. [CrossRef] [PubMed]
52. Sun, Y.; Li, L.; Wen, D.; Bai, X. Bistable electrical switching and nonvolatile memory effect in mixed composite of oxadiazole acceptor and carbazole donor. *Org. Electron.* **2015**, *25*, 283–288. [CrossRef]
53. Cho, B.; Yun, J.-M.; Song, S.; Ji, Y.; Kim, D.-Y.; Lee, T. Direct Observation of Ag Filamentary Paths in Organic Resistive Memory Devices. *Adv. Funct. Mater.* **2011**, *21*, 3976–3981. [CrossRef]
54. Gao, S.; Song, C.; Chen, C.; Zeng, F.; Pan, F. Formation process of conducting filament in planar organic resistive memory. *Appl. Phys. Lett.* **2013**, *102*, 141606. [CrossRef]
55. Chen, X.; Huang, P.; Zhu, X.; Zhuang, S.; Zhu, H.; Fu, J.; Nissimagoudar, A.S.; Li, W.; Zhang, X.; Zhou, L.; et al. Keggin-type polyoxometalate cluster as an active component for redox-based nonvolatile memory. *Nanoscale Horiz.* **2019**, *4*, 697–704. [CrossRef]
56. Shan, Y.; Lyu, Z.; Guan, X.; Younis, A.; Yuan, G.; Wang, J.; Li, S.; Wu, T. Solution-processed resistive switching memory devices based on hybrid organic-inorganic materials and composites. *Phys. Chem. Chem. Phys.* **2018**, *20*, 23837–23846. [CrossRef]
57. Choi, T.-L.; Lee, K.-H.; Joo, W.-J.; Lee, S.; Lee, T.-W.; Chae, M.Y. Synthesis and Nonvolatile Memory Behavior of Redox-Active Conjugated Polymer-Containing Ferrocene. *J. Am. Chem. Soc.* **2007**, *129*, 9842–9843. [CrossRef]
58. Liu, G.; Wang, C.; Zhang, W.; Pan, L.; Zhang, C.; Yang, X.; Fan, F.; Chen, Y.; Li, R.-W. Organic Biomimicking Memristor for Information Storage and Processing Applications. *Adv. Electron. Mater.* **2016**, *2*, 1500298. [CrossRef]
59. Pan, L.; Hu, B.; Zhu, X.; Chen, X.; Shang, J.; Tan, H.; Xue, W.; Zhu, Y.; Liu, G.; Li, R.-W. Role of oxadiazole moiety in different D–A polyazothines and related resistive switching properties. *J. Mater. Chem. C* **2013**, *1*, 4556–4564. [CrossRef]
60. Zhang, H.; Zhou, Y.; Wang, C.; Wang, S. Realizing the Conversion of Resistive Switching Behavior from Binary to Ternary by Adjusting the Charge Traps in the Polymers. *ACS Appl. Electron. Mater.* **2021**, *3*, 2807–2817. [CrossRef]
61. Zhang, B.; Fan, F.; Xue, W.; Liu, G.; Fu, Y.; Zhuang, X.; Xu, X.-H.; Gu, J.; Li, R.-W.; Chen, Y. Redox gated polymer memristive processing memory unit. *Nat. Commun.* **2019**, *10*, 736. [CrossRef] [PubMed]
62. Jung, Y.; Li, W.; Kim, J.; Michinobu, T.; Ree, M. n-Type Digital Memory Characteristics of Diketopyrrolopyrrole-Based Narrow Bandgap Polymers. *J. Phys. Chem. C* **2021**, *125*, 27479–27488. [CrossRef]
63. Cheng, X.; Md, A.; Lian, H.; Zhong, Z.; Guo, H.; Dong, Q.; Roy, V.A.L. Study on synthesis, characterization, and nonvolatile memory behavior of ferrocene-containing metallopolymers. *J. Organomet. Chem.* **2019**, *892*, 34–40. [CrossRef]
64. Ren, Y.; Lin, W.-C.; Ting, L.-Y.; Ding, G.; Yang, B.; Yang, J.-Q.; Chou, H.-H.; Han, S.-T.; Zhou, Y. Iridium-based polymer for memristive devices with integrated logic and arithmetic applications. *J. Mater. Chem. C* **2020**, *8*, 16845–16857. [CrossRef]
65. Barman, B.K.; Ghosh, N.G.; Giri, I.; Kumar, C.; Zade, S.S.; Vijayaraghavan, R.K. Incorporating a redox active entity to attain electrical bistability in a polymer semiconductor. *Nanoscale* **2021**, *13*, 6759–6763. [CrossRef]
66. Ree, B.J.; Isono, T.; Satoh, T. Chemically Controlled Volatile and Nonvolatile Resistive Memory Characteristics of Novel Oxygen-Based Polymers. *ACS Appl. Mater. Interfaces* **2020**, *12*, 28435–28445. [CrossRef]
67. Ryu, W.; Xiang, L.; Jin, K.S.; Kim, H.-J.; Kim, H.-C.; Ree, M. Newly Found Digital Memory Characteristics of Pyrrolidone- and Succinimide-Based Polymers. *Macromol. Rapid Comm.* **2021**, *42*, 2100186. [CrossRef]
68. Yang, B.; Tao, P.; Ma, C.; Tang, R.; Gong, T.; Liu, S.; Zhao, Q. Iridium(iii) complex-containing non-conjugated polymers for non-volatile memory induced by switchable through-space charge transfer. *J. Mater. Chem. C* **2020**, *8*, 5449–5455. [CrossRef]

69. Hu, B.; Wang, C.; Wang, J.; Gao, J.; Wang, K.; Wu, J.; Zhang, G.; Cheng, W.; Venkateswarlu, B.; Wang, M.; et al. Inorganic–organic hybrid polymer with multiple redox for high-density data storage. *Chem. Sci.* **2014**, *5*, 3404–3408. [CrossRef]
70. Liu, G.; Ling, Q.-D.; Teo, E.Y.H.; Zhu, C.-X.; Chan, D.S.-H.; Neoh, K.-G.; Kang, E.-T. Electrical Conductance Tuning and Bistable Switching in Poly(N-vinylcarbazole)−Carbon Nanotube Composite Films. *ACS Nano* **2009**, *3*, 1929–1937. [CrossRef]
71. Pan, S.; Zhu, Z.; Yu, H.; Lan, W.; Wei, B.; Guo, K. Switching the resistive memory behavior from binary to ternary logic via subtle polymer donor and molecular acceptor design. *J. Mater. Chem. C* **2021**, *9*, 5643–5651. [CrossRef]
72. Li, W.; Zhu, H.; Sun, T.; Qu, W.; Fan, X.; Gao, Z.; Shi, W.; Wei, B. High On/Off Ratio Organic Resistive Switching Memory Based on Carbazolyl Dicyanobenzene and a Polymer Composite. *J. Phys. Chem. C* **2022**, *126*, 12897–12905. [CrossRef]
73. Sun, Y.; Wen, D. Nonvolatile WORM and rewritable multifunctional resistive switching memory devices from poly(4-vinyl phenol) and 2-amino-5-methyl-1,3,4-thiadiazole composite. *J. Alloys Compd.* **2019**, *806*, 215–226. [CrossRef]
74. Li, M.; An, H.; Kim, T.W. Highly flexible and stable memristive devices based on hexagonal boron-nitride nanosheets: Polymethyl methacrylate nanocomposites. *Org. Electron.* **2021**, *99*, 106322. [CrossRef]
75. Zhou, Y.; Zhao, X.; Chen, J.; Gao, M.; He, Z.; Wang, S.; Wang, C. Ternary Flash Memory with a Carbazole-Based Conjugated Copolymer: WS2 Composites as Active Layers. *Langmuir* **2022**, *38*, 3113–3121. [CrossRef]
76. Lin, Y.; Zhang, X.; Shan, X.; Zeng, T.; Zhao, X.; Wang, Z.; Kang, Z.; Xu, H.; Liu, Y. Photo-tunable organic resistive random access memory based on PVP/N-doped carbon dot nanocomposites for encrypted image storage. *J. Mater. Chem. C* **2020**, *8*, 14789–14795. [CrossRef]
77. An, H.; Kim, Y.; Li, M.; Kim, T.W. Highly Self-Healable Write-Once-Read-Many-Times Devices Based on Polyvinylalcohol-Imidazole Modified Graphene Nanocomposites. *Small* **2021**, *17*, 2102772. [CrossRef]
78. An, H.; Ge, Y.; Li, M.; Kim, T.W. Storage Mechanisms of Polyimide-Molybdenum Disulfide Quantum Dot Based, Highly Stable, Write-Once-Read-Many-Times Memristive Devices. *Adv. Electron. Mater.* **2021**, *7*, 2000593. [CrossRef]
79. Jiang, T.; Meng, X.; Zhou, Z.; Wu, Y.; Tian, Z.; Liu, Z.; Lu, G.; Eginlidil, M.; Yu, H.-D.; Liu, J.; et al. Highly flexible and degradable memory electronics comprised of all-biocompatible materials. *Nanoscale* **2021**, *13*, 724–729. [CrossRef]
80. Pradhan, R.R.; Bera, J.; Betal, A.; Dagar, P.; Sahu, S. Hot Injection-Based Synthesized Colloidal CdSe Quantum Dots Embedded in Poly(4-vinylpyridine) (PVP) Matrix Form a Nanoscale Heterostructure for a High On–Off Ratio Memory-Switching Device. *ACS Appl. Mater. Interfaces* **2021**, *13*, 25064–25071. [CrossRef]
81. Jeon, N.J.; Noh, J.H.; Yang, W.S.; Kim, Y.C.; Ryu, S.; Seo, J.; Seok, S.I. Compositional engineering of perovskite materials for high-performance solar cells. *Nature* **2015**, *517*, 476. [CrossRef] [PubMed]
82. Liu, M.; Johnston, M.B.; Snaith, H.J. Efficient planar heterojunction perovskite solar cells by vapour deposition. *Nature* **2013**, *501*, 395–398. [CrossRef]
83. Tsai, H.; Nie, W.; Blancon, J.C.; Stoumpos, C.C.; Asadpour, R.; Harutyunyan, B.; Neukirch, A.J.; Verduzco, R.; Crochet, J.J.; Tretiak, S.; et al. High-efficiency two-dimensional Ruddlesden-Popper perovskite solar cells. *Nature* **2016**, *536*, 312–316. [CrossRef] [PubMed]
84. Yoo, E.J.; Lyu, M.; Yun, J.H.; Kang, C.J.; Choi, Y.J.; Wang, L. Resistive Switching Behavior in Organic-Inorganic Hybrid CH3 NH3 PbI3-x Clx Perovskite for Resistive Random Access Memory Devices. *Adv. Mater.* **2015**, *27*, 6170–6175. [CrossRef]
85. Guan, X.; Lei, Z.; Yu, X.; Lin, C.-H.; Huang, J.-K.; Huang, C.-Y.; Hu, L.; Li, F.; Vinu, A.; Yi, J.; et al. Low-Dimensional Metal-Halide Perovskites as High-Performance Materials for Memory Applications. *Small* **2022**, *18*, 2203311. [CrossRef]
86. Liu, Q.; Gao, S.; Xu, L.; Yue, W.J.; Zhang, C.W.; Kan, H.; Li, Y.; Shen, G.Z. Nanostructured perovskites for nonvolatile memory devices. *Chem. Soc. Rev.* **2022**, *51*, 3341–3379. [CrossRef] [PubMed]
87. Wang, C.; Chen, Y.; Zhang, B.; Liu, S.; Chen, Q.; Cao, Y.; Sun, S. High-efficiency bulk heterojunction memory devices fabricated using organometallic halide perovskite: Poly(N-vinylcarbazole) blend active layers. *Dalton Trans.* **2016**, *45*, 484–488. [CrossRef]
88. Zhang, D.; Zhu, S.; Zeng, J.; Ma, H.; Gao, J.; Yao, R.; He, Z. Significance of polymer matrix on the resistive switching performance of lead-free double perovskite nanocomposite based flexible memory device. *Ceram. Int.* **2023**, *49*, 25105–25112. [CrossRef]
89. Jiang, Q.; Ren, Y.; Cui, Z.; Li, Z.; Hu, L.; Guo, R.; Duan, S.; Xie, T.; Zhou, G.; Xiong, S. CsPbBr3 Perovskite Quantum Dots Embedded in Polystyrene-poly2-vinyl Pyridine Copolymer for Robust and Light-Tunable Memristors. *ACS Appl. Nano Mater.* **2023**, *6*, 8655–8667. [CrossRef]
90. Zhuang, X.; Mai, Y.; Wu, D.; Zhang, F.; Feng, X. Two-Dimensional Soft Nanomaterials: A Fascinating World of Materials. *Adv. Mater.* **2015**, *27*, 403–427. [CrossRef]
91. Murray, D.J.; Patterson, D.D.; Payamyar, P.; Bhola, R.; Song, W.; Lackinger, M.; Schlüter, A.D.; King, B.T. Large Area Synthesis of a Nanoporous Two-Dimensional Polymer at the Air/Water Interface. *J. Am. Chem. Soc.* **2015**, *137*, 3450–3453. [CrossRef]
92. Muñoz, R.; Gómez-Aleixandre, C. Review of CVD Synthesis of Graphene. *Chem. Vap. Depos.* **2013**, *19*, 297–322. [CrossRef]
93. Liu, C.; Park, E.; Jin, Y.; Liu, J.; Yu, Y.; Zhang, W.; Lei, S.; Hu, W. Separation of Arylenevinylene Macrocycles with a Surface-Confined Two-Dimensional Covalent Organic Framework. *Angew. Chem. Int. Ed.* **2018**, *57*, 8984–8988. [CrossRef] [PubMed]
94. Zhou, P.-K.; Yu, H.; Li, Y.; Yu, H.; Chen, Q.; Chen, X. Recent advances in covalent organic polymers-based thin films as memory devices. *J Polym. Sci.* **2023**, 1–18. [CrossRef]
95. Liu, J.; Yang, F.; Cao, L.; Li, B.; Yuan, K.; Lei, S.; Hu, W. A Robust Nonvolatile Resistive Memory Device Based on a Freestanding Ultrathin 2D Imine Polymer Film. *Adv. Mater.* **2019**, *31*, 1902264. [CrossRef]
96. Song, Y.; Liu, J.; Li, W.; Liu, L.; Yang, L.; Lei, S.; Hu, W. Effect of functional groups on microporous polymer based resistance switching memory devices. *Chem. Commun.* **2020**, *56*, 6356–6359. [CrossRef]

97. Song, Y.; Feng, G.; Sun, C.; Liang, Q.; Wu, L.; Yu, X.; Lei, S.; Hu, W. Ternary Conductance Switching Realized by a Pillar[5]arene-Functionalized Two-Dimensional Imine Polymer Film. *Chem. Eur. J.* **2021**, *27*, 13605–13612. [CrossRef]
98. Song, Y.; Feng, G.; Wu, L.; Zhang, E.; Sun, C.; Fa, D.; Liang, Q.; Lei, S.; Yu, X.; Hu, W. A two-dimensional polymer memristor based on conformational changes with tunable resistive switching behaviours. *J. Mater. Chem. C* **2022**, *10*, 2631–2638. [CrossRef]
99. Sun, B.; Li, X.; Feng, T.; Cai, S.; Chen, T.; Zhu, C.; Zhang, J.; Wang, D.; Liu, Y. Resistive Switching Memory Performance of Two-Dimensional Polyimide Covalent Organic Framework Films. *ACS Appl. Mater. Interfaces* **2020**, *12*, 51837–51845. [CrossRef]
100. Liu, L.; Geng, B.; Ji, W.; Wu, L.; Lei, S.; Hu, W. A Highly Crystalline Single Layer 2D Polymer for Low Variability and Excellent Scalability Molecular Memristors. *Adv. Mater.* **2023**, *35*, 2208377. [CrossRef]
101. Li, C.; Li, D.; Zhang, W.; Li, H.; Yu, G. Towards High-Performance Resistive Switching Behavior through Embedding a D-A System into 2D Imine-Linked Covalent Organic Frameworks. *Angew. Chem. Int. Ed.* **2021**, *60*, 27135–27143. [CrossRef] [PubMed]
102. Wu, D.; Che, Q.; He, H.; El-Khouly, M.E.; Huang, S.; Zhuang, X.; Zhang, B.; Chen, Y. Room-Temperature Interfacial Synthesis of Vinylene-Bridged Two-Dimensional Covalent Organic Framework Thin Film for Nonvolatile Memory. *ACS Mater. Lett.* **2023**, *5*, 874–883. [CrossRef]
103. Zhao, Z.; El-Khouly, M.E.; Che, Q.; Sun, F.; Zhang, B.; He, H.; Chen, Y. Redox-Active Azulene-based 2D Conjugated Covalent Organic Framework for Organic Memristors. *Angew. Chem. Int. Ed.* **2023**, *62*, e202217249. [CrossRef] [PubMed]
104. Zhang, B.; Chen, W.; Zeng, J.; Fan, F.; Gu, J.; Chen, X.; Yan, L.; Xie, G.; Liu, S.; Yan, Q.; et al. 90% yield production of polymer nano-memristor for in-memory computing. *Nat. Commun.* **2021**, *12*, 1984. [CrossRef] [PubMed]
105. Rehman, M.M.; Ur Rehman, H.M.M.; Kim, W.Y.; Sherazi, S.S.H.; Rao, M.W.; Khan, M.; Muhammad, Z. Biomaterial-Based Nonvolatile Resistive Memory Devices toward Ecofriendliness and Biocompatibility. *ACS Appl. Electron. Mater.* **2021**, *3*, 2832–2861. [CrossRef]
106. Zhang, Y.; Fan, S.; Zhang, Y. Bio-memristors based on silk fibroin. *Mater. Horiz.* **2021**, *8*, 3281–3294. [CrossRef]
107. Yan, X.; Li, X.; Zhou, Z.; Zhao, J.; Wang, H.; Wang, J.; Zhang, L.; Ren, D.; Zhang, X.; Chen, J.; et al. Flexible Transparent Organic Artificial Synapse Based on the Tungsten/Egg Albumen/Indium Tin Oxide/Polyethylene Terephthalate Memristor. *ACS Appl. Mater. Interfaces* **2019**, *11*, 18654–18661. [CrossRef]
108. Zhou, G.; Ren, Z.; Wang, L.; Sun, B.; Duan, S.; Song, Q. Artificial and wearable albumen protein memristor arrays with integrated memory logic gate functionality. *Mater. Horiz.* **2019**, *6*, 1877–1882. [CrossRef]
109. Wang, L.; Zhu, H.; Zuo, Z.; Wen, D. Full-function logic circuit based on egg albumen resistive memory. *Appl. Phys. Lett.* **2022**, *121*, 243505. [CrossRef]
110. Wang, L.; Yang, T.; Wen, D. Tunable Multilevel Data Storage Bioresistive Random Access Memory Device Based on Egg Albumen and Carbon Nanotubes. *Nanomaterials* **2021**, *11*, 2085. [CrossRef]
111. Zhang, Y.; Han, F.; Fan, S.; Zhang, Y. Low-Power and Tunable-Performance Biomemristor Based on Silk Fibroin. *ACS Biomater. Sci. Eng.* **2021**, *7*, 3459–3468. [CrossRef] [PubMed]
112. Wang, W.; Zhou, G.; Wang, Y.; Sun, B.; Zhou, M.; Fang, C.; Xu, C.; Dong, J.; Wang, F.; Duan, S.; et al. An analogue memristor made of silk fibroin polymer. *J. Mater. Chem. C* **2021**, *9*, 14583–14588. [CrossRef]
113. Liu, S.; Cheng, Y.; Han, F.; Fan, S.; Zhang, Y. Multilevel resistive switching memristor based on silk fibroin/graphene oxide with image reconstruction functionality. *Chem. Eng. J.* **2023**, *471*, 144678. [CrossRef]
114. Jeng, H.-Y.; Yang, T.-C.; Yang, L.; Grote, J.G.; Chen, H.-L.; Hung, Y.-C. Non-volatile resistive memory devices based on solution-processed natural DNA biomaterial. *Org. Electron.* **2018**, *54*, 216–221. [CrossRef]
115. Lin, Q.; Hao, S.; Hu, W.; Wang, M.; Zang, Z.; Zhu, L.; Du, J.; Tang, X. Human hair keratin for physically transient resistive switching memory devices. *J. Mater. Chem. C* **2019**, *7*, 3315–3321. [CrossRef]
116. Sun, B.; Chen, Y.; Zhou, G.; Zhou, Y.; Guo, T.; Zhu, S.; Mao, S.; Zhao, Y.; Shao, J.; Li, Y. A Flexible Corn Starch-Based Biomaterial Device Integrated with Capacitive-Coupled Memristive Memory, Mechanical Stress Sensing, Synapse, and Logic Operation Functions. *Adv. Electron. Mater.* **2023**, *9*, 2201017. [CrossRef]
117. Hung, Y.-C.; Hsu, W.-T.; Lin, T.-Y.; Fruk, L. Photoinduced write-once read-many-times memory device based on DNA biopolymer nanocomposite. *Appl. Phys. Lett.* **2011**, *99*, 253301. [CrossRef]
118. Abbas, Y.; Dugasani, S.R.; Raza, M.T.; Jeon, Y.-R.; Park, S.H.; Choi, C. The observation of resistive switching characteristics using transparent and biocompatible Cu2+-doped salmon DNA composite thin film. *Nanotechnology* **2019**, *30*, 335203. [CrossRef]
119. Xia, J.; Zhang, Z.; He, H.; Xu, Y.; Dong, D.; Yang, R.; Miao, X. Environment-friendly regenerated cellulose based flexible memristive device. *Appl. Phys. Lett.* **2021**, *119*, 201904. [CrossRef]

Disclaimer/Publisher's Note: The statements, opinions and data contained in all publications are solely those of the individual author(s) and contributor(s) and not of MDPI and/or the editor(s). MDPI and/or the editor(s) disclaim responsibility for any injury to people or property resulting from any ideas, methods, instructions or products referred to in the content.

MDPI AG
Grosspeteranlage 5
4052 Basel
Switzerland
Tel.: +41 61 683 77 34

Polymers Editorial Office
E-mail: polymers@mdpi.com
www.mdpi.com/journal/polymers

Disclaimer/Publisher's Note: The title and front matter of this reprint are at the discretion of the Guest Editors. The publisher is not responsible for their content or any associated concerns. The statements, opinions and data contained in all individual articles are solely those of the individual Editors and contributors and not of MDPI. MDPI disclaims responsibility for any injury to people or property resulting from any ideas, methods, instructions or products referred to in the content.

www.ingramcontent.com/pod-product-compliance
Lightning Source LLC
LaVergne TN
LVHW072330090526
838202LV00019B/2389